Metapopulation Ecology

Oxford Series in Ecology and Evolution

Edited by Robert M. May and Paul H. Harvey

Metapopulation Ecology

Ilkka Hanski

Research Professor,
Academy of Finland

OXFORD

UNIVERSITY PRESS

OXFORD

UNIVERSITY PRESS

Great Clarendon Street, Oxford OX2 6DP

Oxford University Press is a department of the University of Oxford.
It furthers the University's objective of excellence in research, scholarship,
and education by publishing worldwide in

Oxford New York

Auckland Bangkok Buenos Aires Cape Town Chennai
Dar es Salaam Delhi Hong Kong Istanbul Karachi Kolkata
Kuala Lumpur Madrid Melbourne Mexico City Mumbai
Nairobi São Paulo Shanghai Taipei Tokyo Toronto

Oxford is a registered trade mark of Oxford University Press
in the UK and in certain other countries

Published in the United States
by Oxford University Press Inc., New York

First published 1999
Reprinted 2001 (twice), 2002, 2003

A catalogue record for this book is available from the British Library

Library of Congress Cataloging in Publication Data
Hanski, Ilkka.
Metapopulation ecology/Ilkka Hanski.
p. cm.—(Oxford series in ecology and evolution)
Includes bibliographical references (p. 266) and index.
1. Population biology. 2. Ecology. I. Title. II. Series.
QH352.H358 1999 577.8'8—dc21 98–31723
ISBN 0 19 854065 5 (Pbk)

Printed in Great Britain
on acid-free paper by
Bookcraft (Bath) Ltd., Midsomer Norton, Avon

Preface

The two subjects in population ecology that have had the greatest appeal to me include some very general questions about the distribution and abundance of species and the more specific subject of dynamics of spatially structured populations. No doubt the years I spent collecting insects in my youth left a lasting impression. As an ecologist, I started with the causes and effects of small-scale patchiness in insect populations, but ever since I discovered metapopulations 20 years ago much of my research has been guided by this concept. This volume is my synthesis of metapopulation ecology, inspired especially by insects living in highly fragmented landscapes. Other ecologists would have arrived at different syntheses, for which there is both need and plenty of scope. Two alternative syntheses that are only cursorily touched in this volume are landscape ecology of organisms living in complex mosaic landscapes and a more theoretical approach unifying metapopulation ecology as presented in this book with other areas of population biology, such as epidemiological theory, which share similar conceptual and theoretical underpinnings.

Interest in spatial issues in population ecology, and in population biology more generally, has increased greatly in recent years. Several volumes have appeared that thoroughly cover many new and exciting theoretical advances in spatial ecology. In my own research, and in this book, I have attempted to maintain a close dialogue between models and ecological field studies. The theory and models that are described here do not represent the state of the art in mathematical ecology, and this book has not been written primarily for theoreticians. Similarly, my treatment of field studies might not always satisfy empirical ecologists. It is the combination of the two that should be considered; this book has been written for those ecologists who realize that it is the combination of theory and field studies that really matters.

Many colleagues have greatly contributed to the contents of this volume. Although I have never met Richard Levins, his writings have had a lasting influence on me. Conversations with Michael Gilpin in the autumn 1988 were important in deepening my interest in metapopulations. Collaboration with Mats Gyllenberg and Atte Moilanen during the past six years has been instrumental for the development of many ideas and models. Discussions and joint work with Chris Thomas have been equally valuable. The Glanville fritillary butterfly project has functioned as a critical test bench of concepts and models since 1991, and I have greatly enjoyed and benefited from the innumerable interactions with the students, post-doctoral research workers and colleagues involved in this project; special thanks are due to Mark Camara, Wille Fortelius, Tad Kawecki, Mikko Kuussaari, Guangchun Lei, Atte Moilanen, Marko Nieminen, Timo Pakkala, Juha Pöyry, Ilik Saccheri, Mike Singer and Niklas Wahlberg. My metapopulation studies have been generously funded by the Academy of Finland (the Finnish National Research Council).

I am indebted to the following ecologists and population biologists for reading, and commenting on, parts of the manuscript: Nick Barton, David Boughton, Cajo ter Braak, Mark Camara, Mats Gyllenberg, Mike Hassell, Pelle Ingvarsson, Mikko Kuussaari, Volker Loeschcke, Pekka Pamilo, Juha Pöyry, Chris Ray, Ilik Saccheri and Mike Singer. I am especially grateful to Sean Nee for reading and commenting on the entire manuscript. Mikko Kuussaari, Atte Moilanen, Juha Pöyry and Niklas Wahlberg provided analyses and material for this book. I am grateful to Pia Vikman and Christian Kotkavuori for all their help during the preparation of the manuscript.

Much of this book was written at the Imperial College Centre for Population Biology, Ascot, UK, and at the National Center for Ecological Analysis and Synthesis in Santa Barbara. I thank the directors John Lawton and Bill Murdoch for their unfailing support. I was accompanied by Eeva, Katri, Matti and Eveliina, who have not only put up with my business for so long but have also given the support without which this work could not have been completed. Judith and Bob May initially made me start on this project. They, Paul Harvey and Catherine Kennedy have been exceptionally good in making the life of this author as easy as possible.

Helsinki
March 1998   I. H.

Contents

1

Prologue

1.1 The metapopulation approach

The traditional approach to population ecology assumes that the individuals in which we happen to be interested share the same environment—they belong to the same population. We estimate the sizes of populations, and we study the factors affecting population growth rates. Models are constructed to predict how population size changes under different environmental conditions, including the effect of population size itself. The population concept has been absolutely indispensable for the progress that ecologists and other population biologists have made since the early part of this century (Kingland 1985; McIntosh 1985). We cannot even express ourselves without using the term population.

In its simplest interpretation, canonized in the traditional models of population dynamics, the population concept assumes that all individuals in a population interact equally with all other individuals. This assumption of 'mass action' is obviously a simplification, and it has been relaxed in various ways in both models and field studies. For instance, one might wish to account for the consequences of age or size structure in populations (Metz and Diekmann 1986; Ebenman and Persson 1988), or for the spatial aggregation of individuals (de Jong 1979; Hanski 1981; Hassell and May 1985; Anderson and May 1991). An even more fundamental assumption of the traditional approach to population dynamics is that the population is closed to migration. If movements of individuals to and from a population make an important difference, something more inclusive than the traditional population concept is needed. For instance, in a study of the acorn woodpecker, *Melanerpes formicivorus*, known for its habits of hoarding vast numbers of acorns and communal breeding (Stacey 1979), the long-term persistence of a local population in New Mexico, USA, was found to be critically dependent on immigration (Stacey and Taper 1992; Middleton and Nisbet 1997). The acorn woodpecker study is exceptional because of the wealth of detailed information collected on individually marked birds, allowing Stacey and Taper (1992) to draw their conclusions, but there are no reasons to assume that the dynamics of this woodpecker are exceptional. Migration might also affect abundance fluctuations in persisting populations. The renowned vole oscillations in Fennoscandia (Hansson and Henttonen 1985; Hanski *et al.* 1993a) exhibit large-scale spatial synchrony, which is generally attributed to movements of nomadic predators among vole populations (Ims and Steen 1990; Heikkilä *et al.* 1994).

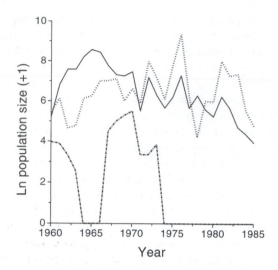

Fig. 1.1 Changes in the sizes of three nearby local populations of the Edith's checkerspot butterfly, *Euphydryas editha*, in Jasper Ridge, California. One population became extinct in 1964 but was re-established in 1967, to go extinct again in 1974 (Ehrlich *et al.* personal communication).

When dealing with small populations with a high risk of extinction, the inadequacy of the traditional approach to population dynamics becomes even more apparent. Paul Ehrlich and his colleagues have studied populations of the Edith's checkerspot butterfly, *Euphydryas editha*, in California (Ehrlich *et al.* 1975; Ehrlich 1984). Like many other insect populations, these butterfly populations show wild oscillations in numbers. Ehrlich's 35-year study documented the extinction of one of the three well-studied populations in 1964 (Fig. 1.1). A new population was established at the same site in 1967, even if only for a few more years, no doubt by immigrants from the remaining populations. After 1985, the decline of the two remaining populations accelerated, and they were both declared extinct by 1996 (McGarrahan 1997). This verdict was premature, as a few butterflies were seen in 1997 (C. Boggs, personal communication), but while you read this text the classical Jasper Ridge checkerspot populations have most likely been irreversibly lost.

In these examples, the focal populations did not occur in isolation—they interacted, via migration, with other conspecific populations within some larger region. In 1970, Richard Levins coined the term *metapopulation* to describe a 'population' consisting of many *local populations*, in the same sense in which a local population is a population consisting of individuals. According to the classical metapopulation concept, which Levins established, all local populations have a substantial probability of extinction, like Ehrlich's butterfly populations, and hence the long-term persistence of the species can only occur at the regional or metapopulation level. The same might be true of populations with such high migration rates, as apparently is the case in Stacey's woodpecker populations, that actual extinctions are not commonly observed. Alternatively, population extinction might not be an issue at all, but migration among local populations might nonetheless affect their dynamics, as in the case of voles and their predators. The metapopulation concept has even been applied, especially in theoretical studies, to large continuous 'populations', which are more or less arbitrarily divided into smaller spatial units, reflecting the restricted movement range of individuals. These latter studies have produced several important insights into spatial population dynamics, but the metapopulation concept is nonetheless of the greatest value when applied to species living in physically patchy environments with sufficiently large patches to support local breeding populations. The metapopulation concept appeals to ecologists because the world is patchy, has always been so, and is sadly becoming, for many species, ever more patchy (Fig. 1.2).

With the increasing interest in metapopulation biology (Hanski and Simberloff 1997), one is often asked to define the term 'metapopulation' in practical terms. However, given the bewildering kaleidoscope of patchy environments and hence of patchy population structures, our task is not so much to classify species into one or another category (hardly any species would exactly fit any particular category anyway), but to find ways of comprehending their biology and of predicting their dynamics. The real question is whether a *metapopulation approach* is useful or not. That is, are the assumptions valid that space is discrete; that ecological processes take place at two scales, local and metapopulation; and that the discrete spatial units of habitat are large and permanent enough to enable the persistence of local breeding populations for at least a few generations. Because metapopulation biology encompasses all that is included in population biology at the metapopulation level, it is clear that metapopulation structure has significant consequences not only for population ecology but also for the behaviour of individuals and for the genetic structure and evolution of (meta)populations (Hanski and Gilpin 1997). This book is primarily about metapopulation ecology—population ecology at the metapopulation level.

1.2 Some related approaches

The metapopulation approach described in this book resides in the middle of two other approaches to large-scale spatial ecology—landscape ecology and spatial dynamics in continuous space. In comparison with landscape ecology, metapopulation studies typically assume an environment consisting of discrete patches of suitable habitat surrounded by uniformly unsuitable habitat (illustrations like Fig. 1.2 reinforce this view). Real landscapes tend to have a much more complex structure, with habitat suitability being a continuous rather than a binary variable. Landscape ecology is focused on describing the structure of real landscapes, but landscape ecologists also study the ecology of species living in these landscapes. Forman and Godron (1986), Forman (1995) and Wiens (1997) list a number of research themes that characterize current landscape ecology, including spatial and

Fig. 1.2 Forest fragmentation in southern Finland (Espoo) during the past 200 years. (Based on Wuorenrinne 1978.)

temporal variation in the quality of landscape elements, boundary effects, connectivity effects, and the impact of habitat patch context on the dynamics of species within patches. Metapopulation ecology as portrayed in this book shares many of the same concerns, but a simpler description of the landscape is assumed to facilitate the building of theory and to guide field studies. In many cases this simplification is an entirely acceptable approximation, in some other cases something else might be needed.

Spatial dynamics in continuous space has been studied with a range of modelling approaches, some of which are briefly discussed in this book and which belong to metapopulation ecology in a broad sense: cellular automata and coupled-map lattice models (Sections 5.1 and 7.3). The shared interest with metapopulation models for discrete space is the role that spatial dynamics play in long-term persistence. Another distinctive challenge for models of spatial dynamics in continuous space, unlike for patch-based metapopulation models, is to demonstrate the emergence of complex dynamics and spatial patterns in abundance due to population dynamic processes alone (Rhodes *et al.* 1996; Tilman and Kareiva 1997; Bascompte and Solé 1998). The formidable empirical challenge of applying these models to natural populations has only just begun. An important general challenge for the future is to advance a more comprehensive synthesis of spatial ecology, incorporating key elements from landscape ecology, metapopulation ecology, epidemiology and a range of theoretical approaches to spatial ecology.

A characteristic feature of the spatial distribution of most species on most spatial scales is patchiness (Taylor 1961; Taylor and Taylor 1979; Wilson 1980; Hanski 1994b; Levin 1994; Brown 1995; Wiens 1997). Spatial patchiness is occasionally considered as synonymous with metapopulation structure, but this is not helpful, because it draws attention to patterns rather than processes. Seemingly similar spatial patterns may be generated by very different processes, hence analyses based on patterns only are liable to be misleading. A case in point is the adaptive deme formation hypothesis, according to which herbivores are locally adapted at the spatial scale of individual host plants (Edmunds and Alstad 1978). Fine-scale local adaptation might indeed occur (Boecklen and Mopper 1998), but insect populations can be demographically and genetically spatially structured for many other reasons also. In the demarcation of the metapopulation approach, I have emphasized discrete habitat patches that are large enough to support local breeding populations. Spatial patchiness occurs also within local populations, in non-random distribution of individuals. One important line of population ecological research has focused on such within-population patchiness. In the case of plants and other sessile organisms, ecological interactions are necessarily restricted to a relatively small number of neighbours even in reproductively panmictic populations, with major consequences for local dynamics. On larger spatial scales, species might show 'patchy' geographical distributions in response to continuous spatial variation in the physical environment, because of their history, or for other reasons. Below, I briefly touch these latter approaches to spatial ecology, which complement rather than compete with the metapopulation approach. For general reviews see Kareiva (1990), Kolasa and Pickett (1990), Shorrocks and Swingland (1990), Levin (1992), Levin *et al.* (1993), Hansson *et al.* (1995), Rhodes *et al.* (1996) and Tilman and Kareiva (1997).

Within-population spatial patchiness

Aggregation of individuals in response to the patchiness of their resources is a common cause of small-scale spatial population structure. In some cases the spatial aggregates are very ephemeral, for instance when foraging bumblebees distribute themselves among clumps of flowers. Behavioural ecologists study the causes of individual movements among resource patches and the ensuing spatial distribution of foragers (Krebs and Davies 1984; Stephens and Krebs 1986). In other cases patchy distributions of individuals are less ephemeral though the aggregates still reflect the movement behaviour of individuals. For instance, vast numbers of insects breed in microhabitats such as mushrooms, dung, carrion and decaying wood (Atkinson and Shorrocks 1984; Hanski 1987; Hanski and Cambefort 1991), but typically only one generation develops in an individual resource patch, and the emerging adults mix freely within a larger area and thereby belong to the same local breeding population. Small-scale spatial aggregation has fundamental consequences for single-species dynamics (de Jong 1979; Hassell and May 1985), interspecific competition (Atkinson and Shorrocks 1981; Hanski 1981, 1987; Ives and May 1985; Shorrocks and Rosewell 1987; Ives 1988, 1991), host–parasitoid interactions (Hassell 1978; Hassell and Pacala 1990), and evolutionary dynamics (Wilson 1980). For a useful general theoretical treatment see Chesson (1998). A major theme in this literature is how small-scale aggregation of individuals has a stabilizing effect on competitive and predator–prey interactions, essentially because aggregation tends to create temporary refuges for inferior competitors and prey. There is no sharp distinction between these 'patchy population' scenarios (Harrison 1991, 1994) and the classical metapopulation scenario. For instance, decaying tree trunks may last for so long and provide such a large quantity of resources that many generations of insects may develop in a single trunk, and the individuals thereby form what should be called a local population (Section 8.5).

Interactions restricted to neighbouring individuals

Competition among plants for light, nutrients and water is typically restricted to relatively few neighbouring individuals. The small-scale spatial structures that evolve as a result of these interactions may significantly affect the dynamics of populations and communities. For instance, Pacala and Deutschman (1995) found that eliminating the horizontal spatial structure in a model of forest dynamics eventually reduced the standing crop to half and accelerated the loss of successional diversity. If the scale of migration of the offspring is also restricted, complex spatial patterns may evolve (Pacala and Levin 1997; Section 5.1). In the case of interspecific competition, the initially surprising result emerges that the more equal two species are in their competitive abilities, the less they will actually compete. The explanation is in the spatial segregation that is likely to evolve when the scales of competitive interactions and migration are restricted—the two species form clusters of their own, whereby most individuals compete only or most frequently with conspecific individuals (briefly touched upon in Section 7.1). Pacala (1997) reviews the available field studies related to this spatial segregation hypothesis. This model and the aggregation model referred to above both involve intraspecific aggregation (and hence interspecific segregation), but the segregation model additionally describes in mechanistic terms how such patterns might evolve in sessile organisms competing for space.

Neighbourhood models of competition are closely related to cellular automata and stochastic lattice models with arbitrary division of continuous space into discrete units. A key feature of these models, which are briefly discussed in Section 5.1, is that colonization is restricted to nearby sites. Like the neighbourhood models for competing species, the lattice models may generate large-scale spatial patchiness in the absence of any environmental heterogeneity (Section 7.3).

Diffusion and invasions

Neighbourhood models describe interactions among more or less sessile individuals. Even in species with mobile individuals, suitable habitat might be so extensive that not all individuals in a large population are equally likely to interact with each other; they are simply too far apart. The dynamics of such populations have been modelled using diffusion models, in which space is continuous rather than discrete (Skellam 1951; Levin 1979; Okubo 1980; Kareiva 1990). These models describe local dynamics for each point in space using continuous-time population models. Migration is included by adding a diffusion term to the model, most simply assuming random movement of individuals. When applied to a single unstructured population, diffusion models predict what an ecologist would intuitively expect—migration tends to obliterate density differences in space. But something different may occur in more complex models involving, for instance, age-structure (Hastings 1992) or interactions among two or more species (Kareiva 1990). Interactions among individuals and migration may now generate spatial variance in density in the absence of any environmental heterogeneity. Such 'diffusive instability' is best known from predator–prey models (Kareiva 1990; Deutschmann *et al.* 1993; Holmes *et al.* 1994). Once again, these results imply that patterns of large-scale aggregation of individuals are not sufficient to imply environmental heterogeneity, or, for that matter, to justify the metapopulation approach as used in this book.

Historically, diffusion models are best known for the contributions that they have made to the study of invading species (Fisher 1937; Skellam 1951; Murray 1988; Okubo *et al.* 1989; Andow *et al.* 1990). In a simple model with logistic local dynamics combined with constant diffusion, the speed of invasion, V (km year^{-1}), is given by:

$$V = 2\sqrt{rD} \qquad\qquad (1.1)$$

where r is the intrinsic growth rate (year^{-1}) and D is the diffusion coefficient (km^2 year^{-1}). Lubina and Levin (1988) and Andow *et al.* (1990) describe studies of mammals and insects which estimated r and D independently and tested predictions based on eqn 1.1 with empirical data. The results were encouraging in both studies. On the other hand, the spread of Holocene trees at the end of the Pleistocene was far too rapid, exceeding 0.1 km year^{-1} for many tree genera, to be compatible with classical diffusion models (Clark *et al.* 1998). The reason for this conflict seems to be highly leptokurtic (fat-tailed) seed dispersal, implying that a small fraction of seeds disperses much longer distances than the average seed. Models with a fat-tailed dispersal kernel do not predict a constant rate of spread, like eqn 1.1, but an ever accelerating rate of spread (Kot *et al.* 1996; Clark *et al.* 1998).

Classical diffusion models ignore environmental heterogeneity, which can only be seen as an implicit element in the model parameters (D in eqn 1.1). An example in which habitat patchiness seems to have had some effect on invasion rate is reported by Nash *et al.* (1995). If the environment is strongly patchy, it is natural to consider using metapopulation models to predict the invasion of empty patch networks, a situation that is of interest to conservation biologists. Hanski and Thomas (1994) used a metapopulation model to predict the invasion of butterflies into patch networks; one example is described in some detail in Section 5.5. Spreading of species into patchy environments is an area where diffusion models (Shigesada *et al.* 1979, 1987) and metapopulation models touch common ground and where interesting work remains to be done. Unfortunately, habitat patchiness is likely to amplify stochasticity in long-distance migration, with the consequence that accurate prediction of the rate of spread becomes practically impossible (Lewis 1997).

1.3 The past 50 years

Metapopulation ecology has no prominent past. A comprehensive history of ecology (McIntosh 1985) does not even mention metapopulations, nor is there any reference to research that we could belatedly call metapopulation ecology. This is not entirely surprising, because 'the great tradition of balance of nature, going back to antiquity, imputed to nature homogeneity, constancy, or equilibrium and abhorred thoughts of extinction and randomness' (McIntosh 1991).

Sewall Wright and Andrewartha and Birch

Contributions by two individuals, Sewall Wright and Richard Levins, and one text book, 'The Distribution and Abundance of Animals' by Andrewartha and Birch (1954), stand out as truly significant landmarks in the development of metapopulation biology. Sewall Wright (1940) and Andrewartha and Birch (1954) emphasized the 'breeding structure' of populations, often involving sets of 'local populations' connected by migration, the 'metapopulation' in Levins's (1970) terminology.

Curiously, it is not obvious why they did so, or why did other concurrent ecology texts, such as 'Principles of Animal Ecology' (Allee *et al.* 1949) and 'The Natural Regulation of Animal Numbers' (Lack 1954), fail to pay any attention to spatially structured populations. Apparently, as a theoretician, Wright (1931, 1940) was content with the seemingly plausible assumption that species often occur as more or less isolated demes (local populations), and he had the insight that evolution might proceed rapidly in spatially structured populations, especially if there are local extinctions and recolonizations (Section 6.4).

Andrewartha and Birch, in contrast, were experienced insect ecologists, especially interested in the effects of environmental factors on population dynamics. They vehemently opposed 'the dogma of density-dependent factors', which focused on population regulation and the persistence of local populations. Lack of constant density dependence in the dynamics might lead to extinctions, and Andrewartha and Birch (1954) advocated the view that extinctions of local populations are a common phenomenon: 'spots that are occupied today may become vacant tomorrow and reoccupied next week or next year' (Andrewartha and Birch 1954, p. 87). They cited various examples supporting their case: the grasshopper *Austroicetes cruciata* in Australia, which they were personally familiar with; the interaction between the prickly-pear cactus *Opuntia* and its specialist herbivore *Cactoblastis cactorum* in Australia (Nicholson 1947); the interaction between the mink and the muskrat in North America (Errington 1943); and various butterflies, including the checkerspot *Euphydryas aurinia* in England (Ford 1945). One reason why other ecologists might have remained unimpressed by the idea of frequent local extinctions is simply the failure to recognize local populations, or relatively small-scale spatial structure in natural populations. Thus when Allee *et al.* (1949) admit that extinctions have occurred, their examples are from the species level: the Arizona elk, the great auk, the Labrador duck, the passenger pigeon, the Carolina parakeet, the Eskimo curlew and the heath hen.

Andrewartha and Birch (1954) illustrated frequent local extinctions and recolonizations with figures like that reproduced in Fig. 1.3. They reasoned that the fraction of localities occupied by a species at any one time increases with migration rate and the density of suitable localities, thereby anticipating some of the key predictions of the yet non-existing metapopulation models. It is unfortunate that Andrewartha and Birch (1954, pp. 406–407) strongly opposed the use of simple mathematical models in ecology. Had they not done so, their insights might have been clearer and they might have had a far greater impact in ecology.

Gadgil and den Boer

Andrewartha and Birch's work emphasized the risk of extinction of local populations, and they implicitly recognized the possibility of metapopulation persistence in a stochastic balance between local extinctions and recolonizations. This idea was formally developed by Levins (1969, 1970), as will be described below and in more detail in Section 4.1. Another early line of metapopulation

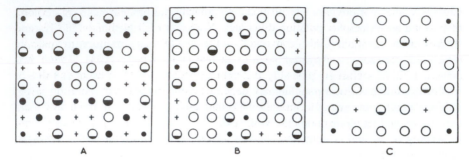

Fig. 1.3 An illustration from Andrewartha and Birch (1954) demonstrating how the fraction of occupied habitat patches depends on the density of the patches (low in C) and on migration rate (high in A). The size of the circle indicates the size of the habitat patch, the extent of shading gives the relative population size.

thinking focused on migration. Gadgil (1971) posed many important questions about metapopulation-level consequences of migration and about the evolution of migration rate (Section 6.1), a theme bridging metapopulation biology and life-history evolution (Olivieri and Gouyon 1997), and a theme that has received much recent attention following a key theoretical paper by Hamilton and May (1977). The much-cited paper by den Boer (1968) on 'spreading of risk' pioneered ideas about metapopulation regulation of species with unstable ('unregulated') local populations and high rate of migration. The current metapopulation concept has emerged as a synthesis of the line of thinking exemplified by Andrewartha and Birch, emphasizing the significance of asynchronous local dynamics and extinctions, and the ideas of Gadgil and den Boer, emphasizing the significance of migration in the 'spreading of risk'.

Nicholson and Huffaker

Though most early studies on metapopulation ecology concerned single species, there are two important exceptions. Nicholson, an ardent opponent of the ideas of his fellow Australians Andrewartha and Birch, stressed the critical role of competition in providing the density-dependent feedback that is required for population regulation. Nicholson is remembered by many because of his experimental studies of blowflies (1954, 1957), but his most influential contributions include also theoretical studies of host–parasitoid dynamics (Nicholson 1933; Nicholson and Bailey 1935). The simplest host–parasitoid model did not predict a stable equilibrium (Hassell 1978), instead it predicted oscillation of numbers with increasing amplitude. In Nicholson's (1933, p. 177) own words, 'the population is broken into widely separated small groups of individuals'. Such 'groups of host increase geometrically for a few generations, but are sooner or later found by parasites and ultimately exterminated'; 'in the meantime there has been a migration of hosts, some of which have established new groups' (p. 163). Here is a clear vision of host–parasitoid metapopulation dynamics. Note that both Nicholson (1933) and Andrewartha and Birch (1954) converged on the idea of metapopulation persistence of unstable but more or less independently fluctuating local populations, though they arrived at this conclusion from very different directions—local fluctuations driven by density-independent versus density-dependent processes (Harrison and Taylor 1997). Twenty years later, Huffaker (1958) conducted very suggestive if not entirely conclusive laboratory experiments with a herbivorous mite and a predatory mite fluctuating in the manner predicted by Nicholson (1933) in a mini-landscape made of oranges (Fig. 1.4).

Levins

Levins's (1969, 1970) well-known contribution was the construction of a simple model which encapsulates the essence of metapopulation-level persistence of an assemblage of extinction-prone local populations, and which has served as a starting point for much subsequent theoretical work (the model is described in Section 4.1). In a series of original and insightful papers Levins applied his approach to questions ranging from single-species dynamics in temporally varying environments (Levins 1969) to optimal pest control policies (Levins 1969), to group selection (Levins 1970), and to interspecific competition (Levins and Culver 1971).

Levins was not alone in the late 1960s in contemplating population extinctions and colonizations. MacArthur and Wilson published in 1963, and in book-length form in 1967, their dynamic theory of island biogeography, which is strikingly similar to Levins's metapopulation theory. There are however two differences— Levins was concerned with the dynamics of a single species living in a network of habitat patches whereas MacArthur and Wilson modelled the dynamics of species richness on true and habitat islands receiving immigrants from a permanent source population (mainland). Given the frequent and close contacts between Levins and MacArthur (Levins 1968; Fretwell 1975), it seems likely that the conceptions of the two theories were tangibly connected, though neither Levins nor MacArthur ever explicitly observed that a special case of Levins's model is just a single-species version of the MacArthur–Wilson model (Section 4.1). A recent dramatic shift from the use of the island biogeography theory to metapopulation theories in conservation biology (Section 10.1) has largely ignored the conceptual link in these ideas. Here is a striking example of Kuhn's (1970) 'paradigm shift' in population biology (Hanski and Simberloff 1997).

1.4 An empiricist's guide to metapopulation models

Mathematical models of metapopulations are constructed in the hope that they will clarify our thinking, reveal unexpected and significant consequences of particular assumptions, and lead to interesting new predictions that could be tested with

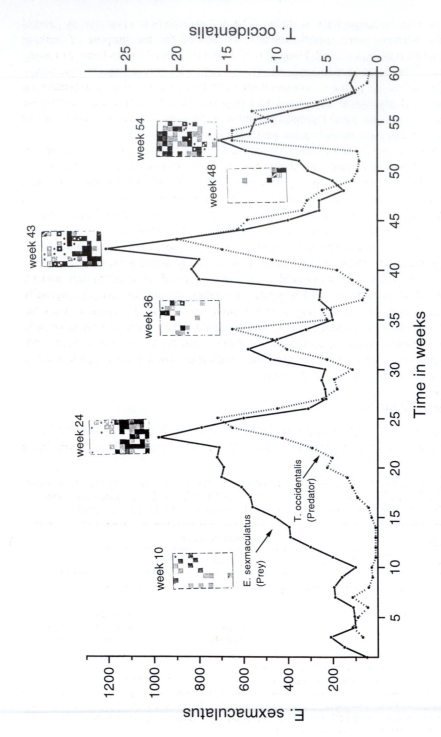

Fig. 1.4 Changing patch-occupancy patterns of the herbivorous mite *Eotetranychus sexmaculatus* and the predatory mite *Typhlodromus occidentalis* in an experimental patchy environment consisting of oranges on a tray. The lines give changes in metapopulation sizes, the small panels show examples of instantaneous spatial patterns. (Based on Huffaker 1958.)

observational and experimental studies. The first objectives are best met by general models, whereas more specific models are needed for the purpose of making quantitative predictions. Both kinds of model are represented in this book. For many ecologists, the most fundamental objective of population modelling is to produce a set of equations, or perhaps a computer program, that would enable us to predict the dynamics of real populations and metapopulations. In conservation, predictive models are needed for the purpose of influencing the way we humans treat our environment and the other species with which we share it.

In this section I introduce the main approaches that ecologists have used to model metapopulation dynamics. Two attributes in particular characterize a metapopulation model, the number of local populations and the way spatial locations of these populations are treated in the model (Table 1.1). Like members of the Australian Arunta tribe, population ecologists tend to count 'one, two, and (infinitely) many' (May 1976). The extreme cases of only two local populations, and very many local populations, turn out to be easier to deal with than the more realistic case of tens of local populations ('several' in Table 1.1). Spatially implicit models make the assumption, called the mean-field assumption, that all populations are equally connected to each other, which greatly simplifies the model analysis. Spatially explicit models assume that migration is distance-dependent, for instance restricted to the neighbouring habitat patches, and the models keep track of who is actually moving where. Finally, spatially realistic models assume the kind of habitat patch geometries represented by real fragmented landscapes. These models can be used to make testable quantitative predictions.

Three basic models

Of the types of models included in Table 1.1, three models are fundamental in the sense that they can be generalized to yield the other models. These three basic

Table 1.1 Classification of metapopulation models in terms of how many local populations are modelled and how their spatial locations are treated. Three basic models are shown by bold face, the other three models can be considered as variants of the basic models

Number of populations	Treatment of space		
	implicit	explicit	realistic
Two	**two-population models**		
Several		**lattice models**	*n*-population simulation models; incidence function models
Many	**Levins model**; structured models		

'Several' means from a few populations to any finite number. 'Many' refers strictly speaking to infinitely many, but in practise the models can be considered as good approximations for metapopulations with of the order of 100 local populations or more.

Table 1.2 Comparison of the attributes of the three basic models

Model attribute	Two-population models	Levins model	Lattice models
Number of patches	two (many[1])	infinite[5]	many
Variation in patch size and quality	yes	no (yes[2])	no (yes[3])
Explicit spatial locations for patches	no (yes[1])	no	yes
Time scales for dynamics	one	two (one[2])	one
Local dynamics modelled explicitly	yes	no (yes[2])	no (yes[4])
Stochastic extinctions and recolonizations	no (yes[1])	yes	yes
Analytical results available	yes (no[1])	yes	some
Testable quantitative predictions	yes	limited	no (yes[3])

1 In n-population simulation models (Section 5.5)
2 In structured metapopulation models (Section 4.3)
3 In incidence function models (Section 5.3)
4 In coupled-map lattice models (Section 5.1)
5 In practice the Levins model (and similar deterministic patch models) can be considered as good
   approximations for metapopulations consisting of the order of 100 local populations or more

models are the two-population model, the Levins model and the lattice model. Each of the three types of model captures an essential element of metapopulation dynamics and generates some key predictions. Like any models, however, these three models give only partial and qualified answers, and none comes close to satisfying all the criteria that one could set for an ideal metapopulation model (Table 1.2).

   The two-population model is an extension of traditional single-population models to two local populations connected by migration. These models are deterministic at the population level, and the main issue is the influence of migration on local dynamics. A two-population model can, of course, be extended to many (n) populations connected by migration, though with the cost of largely lost opportunities for analytical results. Most metapopulation models used by conservationists are n-population simulation models. The danger with these models is that they quickly become far too complex to be properly parameterized and tested.

   A more tractable model is obtained by going to the other extreme, by assuming (infinitely) many habitat patches, by entirely ignoring local dynamics, and by modelling just the fraction of patches occupied. The Levins model is an archetypal example of such patch models. The Levins model is an abstract model with several simplifying assumptions (Table 1.2), including the assumption that all patches are equally connected. Patch models are nonetheless useful for the most general analysis of classical metapopulation dynamics. An important extension of the Levins model is structured metapopulation models (Table 1.1), which include local dynamics and describe changes in the distribution of local population sizes rather than in just the number of extant populations. These models can be used to study the role of migration in metapopulation dynamics, an objective which is shared with the two-population models and *n*-population simulation models. The analytical structured models nonetheless retain the Levins model assumption of (infinitely) many habitat patches and equal mixing of migrants, and hence there is no explicit consideration of the spatial locations of habitat patches and populations in these models.

The third basic approach is based on lattice models, by which I refer to interacting particle system models, cellular automata models and coupled-map lattice models. The set of habitat patches occupied by local populations is represented by a regular lattice. The lattice cells ('patches') may have just two possible states, occupied or empty, like in patch models, or they may have a continuous state (coupled-map lattice models). Cells change state according to simple rules; in cellular automata models colonization of vacant cells is typically a function of how many of the neighbouring cells are currently occupied. Incidence function models blend attributes of deterministic patch models and stochastic lattice models, while assuming a spatially realistic configuration of discrete habitat patches like in *n*-population simulation models. Incidence function models can be parameterized for real metapopulations using widely available census data, which opens up a range of practical applications.

Principal messages

Each of the three basic models has made important contributions to metapopulation theory as will be discussed in Part I of this book. Here I isolate the most fundamental messages (Table 1.3). The two-population models demonstrate how migration can significantly affect local dynamics, and often in surprising ways. In extreme cases, a local population might go extinct only because the emigration rate is too high in comparison with the intrinsic rate of population increase and the rate of immigration. On the other hand, the mere presence of a population is not a sufficient cause to judge that the habitat (patch) is suitable for the species–the population might be present because local growth rate is augmented by immigration (source–sink dynamics). More surprising still, a low rate of migration among local populations can dramatically change the type of dynamics, for instance from chaotic dynamics to limit cycles.

The Levins model has contributed the general notion of metapopulation-level persistence of assemblages of extinction-prone local populations (Table 1.3). The Levins model treats local populations as the basic entities comparable to individuals in traditional population models. The necessary condition for metapopulation survival is that a single local population in a network of empty patches causes the colonization of at least one new patch during its lifetime. The Levins model thereby

Table 1.3 Principal messages stemming from the three basic metapopulation models

Model	Principal message
Two-population models	Emigration and immigration affect local population densities, hence the spatial location of a population in relation to other populations is potentially critical for its dynamics and persistence
Levins model	A metapopulation consisting of extinction-prone local populations may survive in a stochastic balance between extinctions and colonizations, hence any landscape structures that affect extinction and colonization rates are important for regional persistence. A threshold condition exists for metapopulation persistence in terms of extinction and colonization rates, which can be interpreted in terms of patch density and average patch size
Lattice models	Spatially restricted migration and interactions may generate complex spatial patterns in density in the absence of any environmental heterogeneity

predicts a threshold condition for metapopulation survival, which is given in terms of the colonization and extinction parameters but can also be interpreted in terms of the structure of the landscape (average patch size and density). An important corollary for conservationists is that, in landscapes experiencing habitat loss and fragmentation, metapopulation extinction is predicted to occur, deterministically, before all suitable habitat has been destroyed. This prediction clearly distinguishes the metapopulation approach from traditional population dynamic approaches focused on local dynamics. Epidemiologists have for a long time been concerned with a similar threshold condition for the persistence of parasites in host populations (Anderson and May 1991). Indeed, patch models and basic epidemiological models share the same model structure (Section 4.1).

The principal novel message emerging from the various lattice-based models is that complex spatial patterns may emerge from simple local dynamics when combined with spatially restricted interactions and movements. This is especially apparent when modelling the interactions of two or more species, which are often predicted to exhibit strongly patchy distributions in the absence of any environmental heterogeneity. Once again, the models imply that the presence or absence of a species in some particular spot is not necessarily a reliable indicator of the quality of the local habitat.

1.5 A theoretician's guide to field studies

Those who have not spent much time in the field and have no first-hand knowledge of the biology of even a single species might not always appreciate how difficult it generally is to gather useful data about population dynamics. Unlike many other areas of population biology, for instance behavioural ecology, where observations on individuals can be used to test model predictions (Stephens and Krebs 1986), population ecology is often troubled by the large scale of the study phenomena and by the environmental noise constantly hammering most populations. The magnitude of these problems is manifested in the 60-year controversy about population regulation (Nicholson 1933; Andrewartha and Birch 1954; Milne 1957; den Boer 1968, 1987; Strong 1986; Hanski 1990; Godfray and Hassell 1992; Turchin 1995). Often the limited amount of data that is available can be blamed for unsatisfactory progress (Hassell *et al.* 1989; Godfray and Hassell 1992; Woiwod and Hanski 1992), but this is not always so. A case in point is the dilemma in the analysis of time-series data on population dynamics (Royama 1992; Turchin and Taylor 1992; Ellner and Turchin 1995)—one would wish to have long data series to have adequate statistical power, but long data series are likely to exhibit non-stationary dynamics, most likely because of environmental changes (Pimm 1992), making the interpretation of the results difficult (Perry *et al.* 1993; Turchin 1995).

Table 1.3 listed three general messages from metapopulation models. It might be surprising that there are no scores of field studies which have specifically tested these and more refined model predictions. One reason is simply insufficient knowledge of theory by field ecologists, which has spawned empirical studies of limited general significance, but another reason is 'impractical' models developed by theoreticians, utterly out of touch with what is happening in the real world. Below, I catalogue 12 types of evidence which field ecologists have reported in support of the general notion that metapopulation-level processes matter in population dynamics. Field studies are discussed in greater detail in Part II of this book.

- Population size or density is significantly affected by migration. All students of ecology are taught that population size is affected by natality, mortality, emigration and immigration (Ricklefs 1990; Krebs 1994; Begon *et al.* 1996), but ecologists know vastly more about natality and mortality than about migration influencing population size. Two of the better known phenomena specifically related to migration are the Krebs effect (Krebs *et al.* 1969), by which is meant elevated density when migration is obstructed, and source–sink dynamics (Pulliam 1988, 1996), which refers to highly asymmetric migration of individuals among populations.
- Population density is affected by patch area and isolation. The source–sink concept assumes that population density is significantly affected by migration. If so, one could expect density to decline with increasing isolation especially in sink populations, because isolation generally reduces immigration but should affect emigration less. Patch area can affect both emigration and immigration. Patch area and isolation effects on population density have been commonly observed in field studies (Section 9.2).
- Asynchronous local dynamics. A necessary condition for metapopulation persistence in the face of unstable local dynamics is local dynamics sufficiently asynchronous to make simultaneous extinction of all local populations unlikely.

A common observation is that local dynamics are to some extent, but not completely, correlated. One generally important factor increasing asynchrony is interaction between environmental (weather) perturbations and local habitat quality. Weiss *et al.* (1988) describe one good example.

- Population turnover, local extinctions and establishment of new populations, is the hallmark of classical metapopulation dynamics, and turnover has been the subject of dozens of recent empirical studies (Section 8.2). To cite a fascinating early study, Boycott (1930) studied fresh-water mollusca inhabiting small ponds in Hertfordshire, England, using both long-term observations and experiments. In a set of 84 ponds he recorded 64 extinctions and 93 colonizations of 18 species over 10 years.

- Presence of empty habitat. For many biologists, the concept of empty but suitable habitat within the regular migration range of the species is difficult to accept. Some empty habitat must exist, however, if there are population extinctions and recolonization is not immediate. In the Glanville fritillary butterfly metapopulation in Finland, 70% of approximately 1600 habitat patches and roughly 60% of the pooled area of the suitable habitat (2.0 km^2) have been empty at any one time (Hanski *et al.* 1995a). The Glanville fritillary is the focus of Part III of this book.

- Metapopulations persist despite population turnover. If all local populations have a substantial risk of local extinction, long-term survival is possible only at the metapopulation level. It has been argued that there are not many convincing empirical examples (Harrison 1991; Harrison and Taylor 1997), but one could as well emphasize that not many large-scale studies have yet been conducted. Menges's (1990) study on the Furbish lousewort, *Pedicularis furbishae*, an endangered American plant, is one of the better documented case studies (Section 10.5). The Glanville fritillary study provides the perhaps most comprehensive example (Chapter 11).

- Extinction risk depends on patch area. Assuming realistically that the expected population size is positively correlated with patch area, we might expect that the risk of population extinction decreases with increasing patch area, because extinction risk practically always decreases with expected population size (Section 2.2). Indeed, there is overwhelming empirical evidence supporting these predictions.

- Colonization rate depends on patch isolation. A few species have an apparently unlimited capacity for migration, including both very small (spores in the aerial plankton) and very large organisms (some whales). For most species individual movement ranges are limited, and not surprisingly the colonization rate of empty patches has been typically found to decrease with increasing isolation.

- Patch occupancy depends on patch area and isolation. Given that any population turnover occurs, and taking for granted the above relationships between patch area and extinction rate, and patch isolation and colonization rate, we would expect that small and isolated patches are those most likely to be empty. Numerous empirical studies have confirmed the effects of patch area and isolation on occupancy; these findings comprise the key empirical basis on which the incidence function model is built (Section 5.3).

- Spatially realistic metapopulation models can be used to make predictions about metapopulation dynamics in particular fragmented landscapes. Unfortunately, although the development of spatially realistic metapopulation models is presently a small industry (Menges 1990; Thomas *et al.* 1990; Pulliam *et al.* 1992; Beier 1993; Hanski and Thomas 1994; Lindenmayer and Possingham 1994; Akçakaya *et al.* 1995; for a review see Lindenmayer *et al.* 1995), there are not many instances in which model predictions have been properly tested. The Glanville fritillary project described in Part III is one example.

- Metapopulation coexistence of competitors. Two or more competitors which cannot coexist locally might be able to coexist as metapopulations. The most striking example is a completely inferior competitor persisting in a patch network owing to its high colonization rate (fugitive coexistence). There are two convincing empirical examples and several more suggestive ones (Section 7.1).

- Metapopulation coexistence of prey and its predator. Like two competitors, a prey and its predator might be able to persist regionally even if they would oscillate to extinction locally. Predator–prey metapopulation coexistence has been extensively studied in models but empirical evidence remains limited (Harrison and Taylor 1997). One of the best examples is the interaction between the two-spotted spider mite, *Tetranychus urticae*, and the predatory mite *Phytoseiulus persimilis* in greenhouses (Nachman 1988, 1991).

Phenomena for which there is no good evidence

Complex spatial dynamics are difficult to study in the field because of logistical problems owing to large scales and because of environmental heterogeneity. One approach is to use 'experimental model systems' (Ims *et al.* 1993), such as the mite predator–prey system studied in a greenhouse by Nachman (1991), but there are great challenges in scaling up the results of most small-scale experiments to real landscapes and metapopulations. Presently there is very limited evidence for many theoretical predictions about spatial dynamics, either from real or experimental metapopulations. Diffusive instability and the predictions of reaction–diffusion theory more generally have gone largely untested (Kareiva 1990; but see an example in Section 7.3), excepting the predicted rates of population invasion (Okubo *et al.* 1989; Andow *et al.* 1990; Nash *et al.* 1995). Spatial chaos and other complex spatial dynamics predicted by predator–prey models (Hassell *et al.* 1991) have not been convincingly recorded in nature. Alternative stable states in metapopulation dynamics (Gyllenberg and Hanski 1992) is supported by a single example (Hanski *et al.* 1995b). Perhaps most striking is the virtual lack of unambiguous evidence for the threshold condition for metapopulation persistence (Table 1.3). Given that there is extensive population turnover in a metapopulation and that only a fraction of suitable habitat is occupied at any one time, there should be little doubt that metapopulation persistence hinges on the density and the sizes of the habitat patches as predicted. The threshold condition is well supported for parasites in host populations (Anderson and May 1991), but it would be very useful to have many critical studies of free-living animals and plants in fragmented landscapes. A general difficulty in detecting these phenomena for which evidence is presently lacking or is scarce is the influence of heterogeneous space on population dynamics (Steinberg and Kareiva 1997). The phenomena might well exist but they are difficult to demonstrate.

1.6 The structure of this book

In my own research on population ecology, I have attempted to achieve a happy marriage between models (theory) and field studies. The cost is the occasional feeling of being incompetent in both areas. But there is plenty of evidence to suggest that theory without real (as opposed to hypothetical) empirical implications is likely to become forgotten, as is empirical research that does not even have a chance to affect the development of the conceptual and theoretical basis of ecology. Merging of theory and field studies is so critical that it is always worth a serious try. My inclinations are reflected in the structure of this book, which is divided into three parts: theory, field studies and a case study on the Glanville fritillary butterfly. The boundaries between the three parts are not sharp. Thus it was convenient to discuss empirical studies in Chapter 7 on multispecies models and throughout the book I have used examples from the Glanville fritillary study.

Following a brief and eclectic introduction to population dynamics in general (Chapter 2), Chapters 3 to 5 focus in turn on the three modelling approaches stemming from the three basic models described in Section 1.4. These chapters cover the extreme cases of metapopulations consisting of just two populations (Chapter 3), metapopulations consisting of many populations (Chapter 4), and the intermediate cases which are important for practical applications (Chapter 5). In Chapter 6 I venture to areas in which I cannot boast any profound expertise, metapopulation genetics and evolution. This chapter was included because it would be arbitrary to draw boundaries between different areas of population biology, and also to give me an opportunity to make some comments about metapopulation genetics and evolution from the ecological perspective. In Chapter 7 we move from single-species to multispecies models. To avoid unnecessary repetition and jumping back and forth between the same themes, in Chapter 7 I cover empirical and theoretical studies on interacting metapopulations and metacommunities.

Part II surveys field studies of metapopulation ecology and discusses some general conceptual issues. It is appropriate to start by describing how population ecologists view the spatial structure of natural populations, and how the observed structures are thought to be related to the assumptions of the metapopulation models. I review evidence on population turnover, extinctions and colonizations, with particular focus on the causes of local extinction. Chapter 9 is devoted to the consequences of population turnover, including the key notion that not all suitable habitat is likely to be occupied all the time. These questions are close to the concerns of conservationists. Recognizing the current interest in metapopulation biology by conservation biologists, Chapter 10 on the factors influencing metapopulation persistence, or extinction, has been written from the conservation perspective.

Part III tells the story of one species of butterfly, the Glanville fritillary (*Melitaea cinxia*), in Åland, south-western Finland. The purpose of the two chapters in Part III is to relate much of the material in Parts I and II to this particular species, in the spirit of assessing the value of the metapopulation approach with this case study of a species living in a naturally fragmented landscape. Chapter 12 serves another function also; here I illustrate the application of the incidence function model to the study of real metapopulations. Most of us have pet ideas that tend to surface in various disguises in unexpected places. My thinking about metapopulations has been greatly influenced and aided by incidence functions, which I have found a helpful and practical tool both in modelling and in empirical studies.

Part I

THEORY

The following six chapters describe the type of theory of metapopulation ecology that I have been interested in, ranging from simple analytical models to empirically-based models of metapopulations living in real fragmented landscapes. I could have written more than I did on several interesting topics, including source–sink metapopulations, the population dynamic consequences of migration among independently fluctuating persisting populations, lattice-based spatially explicit (but not spatially realistic) metapopulation models, and structured metapopulation models. More could also have been made of the similarities between metapopulation theory for free-living organisms, which this book is about, and the related theory for parasites, usually labelled as epidemiological theory (Grenfell and Harwood 1997). These topics are discussed relatively briefly, and I leave the more thorough coverage to others. The focus of this work is on theory which has had substantial influence on empirical metapopulation research and which has substantially facilitated the interpretation of empirical results.

I start with a fleeting overview of local population dynamics, population extinction and population establishment, the building blocks of classical metapopulation dynamics. I attempt to dispel the widely held misconception that metapopulation dynamics would allow long-term persistence of species without the operation of local density-dependent processes. This is not so, though metapopulation ecology does bring a new perspective to the enduring arguments about density dependence and population regulation. I then advocate a recent general model of population extinction, due to Lande, Foley, Middleton *et al.* and others, which neatly summarizes the influence of key demographic parameters in setting the expected time to population extinction. This model is of particular importance in this book because it facilitates the interpretation of the parameters of my favourite metapopulation model, the incidence function model (Section 5.3). Migration of individuals is a vast topic in population biology, which I can only touch very briefly here, primarily because migration is needed for colonization.

The next three chapters elaborate on the three basic model types introduced in Chapter 1: models of metapopulations consisting of just two (or a few) local populations (Chapter 3), models assuming an infinite number of populations (Chapter 4), and spatially explicit and spatially realistic models assuming a finite number of local populations (Chapter 5). The different types of model are most useful for investigating different questions about metapopulation ecology, and hence are not real alternatives to each other. This is often misunderstood, and a model is rejected for wrong reasons. This is true about the mother of all classical metapopulation models, the Levins model, which has been faulted on the grounds that it does not faithfully describe the structure of real landscapes. It was never meant to. But it is revealing that the very simplest models, like the Levins model, have played a far more useful part in the development of empirical metapopulation ecology than the most complex simulation models.

Chapter 6 on genetics and evolution was included in this book with some hesitation, as this is not my own field of research and I can hence provide only a very slanted summary of what is evolving into a truly active area of population biology. The alternative of saying nothing about genetics and evolution was even less attractive. As in many other chapters in Parts I and II, I use here examples from our studies on the Glanville fritillary butterfly; these examples also serve as a prologue to Part III, which is largely focused on this species. Our studies have demonstrated how the extinction and colonization rates in this butterfly metapopulation are influenced by the genetic composition of local populations, thus making it impossible to restrict our attention to ecology only, even if one would be ultimately interested in ecological questions, such as persistence of metapopulations in fragmented landscapes.

Chapter 7 is another digression from the main theme of this book, single-species metapopulation ecology. In Chapter 7 I discuss competition and predation in metapopulations, as well as some issues in metacommunity ecology. Competition and predation enter into classical metapopulation ecology very naturally in situations in which these interactions influence the basic processes of extinction and colonization—which they often do. Like with single-species theory, a key question here is persistence: assuming that two or more interacting species cannot coexist locally, can they coexist as metapopulations? The idea of fugitive coexistence is well-established in ecology, with the prey or the inferior competitor escaping global though not local extinction by being especially good in locating currently unoccupied habitat patches. Metapopulation ecology provides the appropriate framework for rigorous analysis of these questions. Another generally interesting theme emerging from the multispecies models is spatial pattern formation as a result of interspecific interactions in uniform environments. To many ecologists, this is an equally counter-intuitive idea than the presence of empty but unused suitable habitat. The theory about spatial pattern formation is relatively advanced, but the empirical evidence is very limited. The final metacommunity section in Chapter 7 is focused on the distribution of species. Here my aim is to demonstrate that metapopulation ecology neatly unites two general patterns in the distribution of species—the species-area curve and the distribution–abundance relationship. These patterns have previously been discussed in complete isolation.

2

Preliminaries

Three processes are in the hearth of metapopulation ecology: migration and how it affects local dynamics, population extinction and the establishment of new local populations. A vast literature exists on the dynamics of local populations, including population extinction; a substantially smaller but not insignificant literature exists on population establishment. The purpose of this chapter is to assemble some elements from this knowledge that are particularly useful for the study of metapopulation ecology.

2.1 Local population dynamics

Population ecologists have traditionally been concerned with the factors and processes that contribute to population regulation and hence make it likely that, during some reasonably long period of time, the population does not become excessively large nor go extinct (Varley *et al.* 1973; Sinclair 1989; Hanski 1990; Royama 1992; Turchin 1995; Begon *et al.* 1996). The manner in which the theoretical literature has treated the Nicholson–Bailey host–parasitoid model (Nicholson 1933; Nicholson and Bailey 1935) is a telling example. The original model has an unstable point equilibrium from which population sizes of the host and the parasitoid diverge with increasing amplitude following a perturbation. Much useful theory has been developed by adding factors into the model that tend to stabilize the dynamics (Hassell 1978; Hassell and Pacala 1990), but for a long time the focus remained in local dynamics, even if Nicholson (1933) himself had suggested that long-term persistence might actually hinge on large-scale (metapopulation) dynamics. Now this has all changed, and there is a swelling literature on host–parasitoid metapopulation models (Section 7.2 and 7.3).

The meanings of the terms 'density dependence' and 'population regulation' deserve some clarification. By density dependence, ecologists mean any dependence of per capita growth rate on present and/or past population densities (Murdoch and Walde 1989; Hanski 1990; Turchin 1995). Population regulation is commonly used in the sense of a return tendency, due to density-dependent processes, towards an equilibrium density (Varley *et al.* 1973; Dempster 1983; Sinclair 1989). The term equilibrium density has been a source of confusion, as most populations are characterized by more or less fluctuating population size rather than a constant size, even if there is no long-term increasing or decreasing trend. For this reason, Turchin (1995) and others have identified population regulation as the process whereby a population reaches a stationary distribution of population densities. Though this solves the problem of fluctuating population sizes, whether due to environmental perturbations or endogenous processes (below), this definition is awkward in excluding the possibility of population regulation in a changing environment. The spirit of the population regulation concept is perhaps best captured by identifying regulation with a tendency, due to density-dependent processes, to approach a stationary distribution of population densities, though the population might not have time to reach the stationary distribution before the environment has changed or the population has gone extinct. In practice, population regulation is usually studied by studying the strength and the type of density dependence in population growth rate, though clearly density dependence is only a necessary, not a sufficient condition of population regulation. For instance, delayed density dependence might lead to high-amplitude oscillations and increase the risk of population extinction.

In the context of metapopulation dynamics, the emphasis is somewhat different from that in the study of local population dynamics, as we do not assume that population extinctions are necessarily infrequent, nor do we need unfailing local population regulation to explain long-term persistence. This does not mean that one would expect population regulation and hence density dependence to be entirely lacking in local populations in a metapopulation, merely that, for whatever reason, populations are not necessarily regulated so efficiently that extinctions would be rare. The most likely cause for frequent local extinctions is simply that many local populations are small and hence vulnerable to stochastic extinction. Nonetheless, it is worth examining in some detail the question of local density dependence and metapopulation persistence, which has created confusion and controversy in the literature.

Density dependence and metapopulation persistence

Two obvious requirements for long-term persistence of an assemblage of extinction-prone local populations are sufficiently high colonization rate and some degree of asynchrony in local dynamics. If local dynamics were completely synchronous, the entire metapopulation would go extinct as soon as the least vulnerable population goes extinct. A more contentious requirement for long-term persistence is local density dependence. The striking and often quite erratic-looking variability of, especially, insect populations (Hanski 1990), and the failures to demonstrate significant density dependence with field data (den Boer 1987; Gaston and Lawton 1987; Stiling 1987), have led some ecologists (Andrewartha and Birch 1954; Thompson 1956; Schwerdtfeger 1958; Milne 1962; Murray 1979; Strong 1984; Wolda 1989; den Boer 1991) to question whether density dependence is really the ubiquitous phenomenon that theory implies (Royama 1992). Some ecologists, notably den Boer (1968, 1991), have taken refuge in the metapopulation concept by arguing that long-term persistence on a large spatial scale does not require local density dependence. Without density dependence, local populations will sooner or later go extinct but, according to this argument, such extinctions are balanced by recolonizations, and the metapopulation might persist for as long as required. This

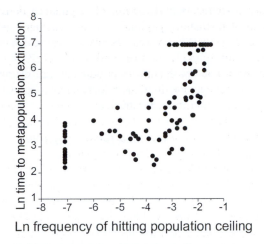

Fig. 2.1 Relationship between time to metapopulation extinction and the frequency of hitting the population ceiling per generation per population (a measure of density dependence) in a metapopulation model consisting of random walk local populations (Hanski *et al.* 1996a). A small number (0.001) was added to the frequency of hitting the ceiling to avoid taking logarithms of zeros (−1 on the horizontal axis corresponds to a frequency of 0.368, −6.9 corresponds to a frequency of 0). Data points represent model predictions for different parameter combinations. Note that there are no combinations leading to long persistence time with no or very little density dependence (the upper left-hand corner).

argument is, however, faulty. Either the metapopulation does persist, but then there is a substantial probability of some local populations exceeding any finite size, implying density dependence; or local populations never become large, and fluctuate without density dependence, but then their extinction rates are so high that the metapopulation does not persist for a long time (Fig. 2.1; Hanski *et al.* 1996a). The argument by which local populations cannot persist between finite positive limits without density dependence (Royama 1992) applies with equal force to metapopulations. Thus it is not true that metapopulation dynamics has 'solved' the deplorable argument about density-dependent population regulation, though it is true that it is possible to construct plausible examples in which a metapopulation persists with infrequent local density dependence (Hanski *et al.* 1996a). Such examples involve species with highly ephemeral populations, which agrees with the intuition of field ecologists: these are the populations that are generally thought to be good candidates for species with weak density dependence.

Although metapopulation dynamics of weakly regulated local populations might explain, to some extent, the past failures to detect density dependence with empirical data, this is not the full story. In recent years, the combination of more powerful statistical techniques (Pollard *et al.* 1987; Dennis and Taper 1994) and more comprehensive data sets (Hassell *et al.* 1989; Hanski 1990; Woiwod and Hanski

1992) has produced the consensus that density dependence is just as prevalent in natural populations as most ecologists have assumed (Hanski 1990; Godfray and Hassell 1992; Turchin 1995). What remains is a less exciting debate about semantics (Hanski *et al.* 1993b; Wolda and Dennis 1993; Wolda *et al.* 1994; Turchin 1995). At the same time, we now recognize that not all populations are strongly affected all the time by density-dependent interactions, and local extinctions of especially small populations might be common in many species.

Simple and complex dynamics, and population variability

Population size is affected by four population processes, natality and immigration, which increase population size, and mortality and emigration, which reduce population size. In the familiar models presented in ecology texts (Begon *et al.* 1996), migration is often omitted and the net effect of reproduction and mortality determines the rate of change in population size.

Some of the basic models, like the logistic model in continuous time:

$$\frac{dN}{dt} = rN\left(1 - \frac{N}{K}\right), \tag{2.1}$$

imply that a population reaches a constant equilibrium size, $\hat{N} = K$ in the case of the logistic model. The reality is of course less orderly. The processes affecting population size are stochastic, and hence changes in population size are more or less erratic. Rather than reaching a constant size, the population is expected to settle into a distribution of population sizes. If this distribution remains unchanged in time (stationary) the implication is that some density-dependent processes are acting upon population growth rate; the population is regulated (Turchin 1995). Depending on the strength of environmental perturbations and the strength and the type of the density-dependent processes, population size shows more or less variation around the average value. Generally, small ectothermic animals are more affected by environmental vagaries than are large endothermic animals, and because direct density dependence tends to be stronger in the latter (Sinclair 1989), it is no wonder that birds and mammals show significantly less variation in population size than do insects (Fig. 2.2). (There is a lively debate about the problems of measuring population variability, and how such problems may lead to biased results, but I believe that the above conclusion is nonetheless valid; Connell and Sousa 1983; Hanski 1990; McArdle and Gaston 1992; Gaston and McArdle 1993, 1994; Gaston and Blackburn 1995.) More variation in the demographic parameters generally means greater variability in population size and hence a greater chance of the population going extinct (Section 2.2).

The continuous-time logistic model has simple dynamics. Not all populations behave in the way predicted by this model even if we allow for stochasticity. Some populations show strikingly regular oscillations (Turchin 1990), and a few natural populations have been suggested to exemplify deterministic chaos (Turchin and Taylor 1992; Hanski *et al.* 1993a; Ellner and Turchin 1995). There is a plethora of

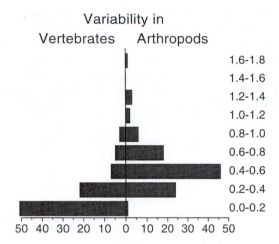

Fig. 2.2 Population variability in 91 species of terrestrial vertebrates (mammals, birds and lizards) and 99 species of terrestrial invertebrates (insects). Variability was measured by the standard deviation of log-transformed population density. (Based on Hanski 1990.)

models predicting complex dynamics, starting from very simple discrete-time models thought to be more appropriate than the corresponding continuous-time models for many natural populations (May 1976). In fact, practically any reasonable model with density dependence can lead to complex dynamics if the model involves time delays or explicit interactions among several species. Complex dynamics might amplify population variability and thereby increase the risk of population extinction. On the other hand, complex local dynamics would decrease synchrony in the dynamics of several local populations and thereby potentially contribute to metapopulation-level persistence (Ruxton 1996). The possibility of complex dynamics also changes the prediction that stronger and more consistent density dependence decreases variability by bringing the population faster towards the long-term average size following perturbations. Because stronger density dependence (non-linearity) might lead to complex dynamics (May 1976), stronger density dependence might increase variability. Figure 2.3 gives an intriguing example of apparently the opposite effects of the incidence of density dependence on population variability in moths and aphids. In moths, more consistent density dependence decreases, but in aphids it increases, population variability. A possible explanation of these contrasting relationships is that moths generally have simple dynamics, whereas the dynamics of aphids can be complex. The very high intrinsic rates of population increase in aphids (Dixon 1990) are consistent with this conjecture.

The Allee effect

Another phenomenon in local dynamics with potential significance to metapopulation dynamics is the Allee effect (Allee *et al.* 1949), reduced per capita growth rate

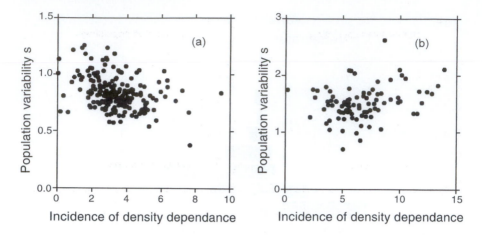

Fig. 2.3 Relationship between population variability, as measured by the standard deviation of log-transformed density, and the incidence of density dependence in moths (a) and aphids (b) in the UK. Incidence of density dependence measures the (transformed) probability of rejecting the null hypothesis of no density dependence. Note the different scales in the two panels. (Based on Hanski and Woiwod 1993a.)

at low population density. At low density, females may have reduced lifetime fecundity because it takes longer to mate, or they might fail to mate (Gilpin and Soulé 1986; Lande 1987; Dobson and Lyles 1989). Emigration rate may increase with decreasing population size (Kuussaari *et al.* 1996), and immigrants might avoid very small populations (the conspecific attraction hypothesis; Smith and Peacock 1990; Ray *et al.* 1991; Stamps 1991). In species with highly evolved sociality, low density may disrupt reproduction also for other reasons (Allee *et al.* 1949). In any case, reduced fecundity at low density may increase the risk of population extinction, with repercussions to metapopulation dynamics. A well-documented extinction of an isolated population in which the Allee effect was apparently involved is the disappearance of the middle-spotted woodpecker, *Dendrocopos medius*, from Sweden in 1982 (Pettersson 1985). Reduced mating success and increased emigration rate at low density lead to an Allee effect in the Glanville fritillary butterfly (Kuussaari *et al.* 1998; Fig. 2.4). The contrast between the more and less isolated populations in Fig. 2.4 also illustrates the effect of immigration on population size.

2.2 Population extinction

Over the past 30 years, population ecologists have constructed an extensive body of theory on population extinction (MacArthur and Wilson 1967; Leigh 1981; Goodman 1987a; Lande and Orzack 1988; Wissel and Stöcker 1991; Lande 1993;

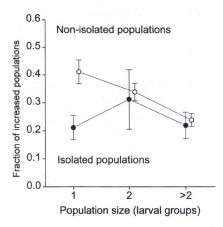
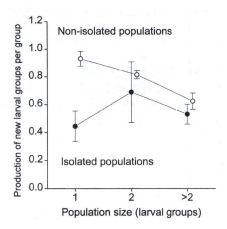

Fig. 2.4 Relationship between two measures of population growth rate and population size in isolated versus non-isolated populations of the Glanville fritillary butterfly. The values shown for the fraction of populations that increased and for the production of new larval groups per existing group are means for three years (1994–96), weighted by the numbers of local populations in each isolation and population size category in each two-year period. The bars show the standard errors for the means. (Based on Kuussaari *et al.* 1998.)

Mangel and Tier 1993a, 1994; Foley 1994, 1997; Haccou and Iwasa 1996). Another body of literature is devoted to the measurement of actual extinction rates (Diamond 1984; Lawton and May 1994). The purpose of this section is to present a brief overview of some key theoretical results. Empirical results on population extinction are discussed in Sections 8.2 and 8.3.

It should be made clear from the beginning that mathematical models, which typically deal with particular mechanisms of extinction in a stationary environment, such as demographic and environmental stochasticity, are not often very helpful in predicting the extinction of natural populations, because the actual causes of extinction often involve environmental changes, most notably habitat loss and change in habitat quality. Caughley (1994) has drawn attention to this contrast, making the point that although population biologists have traditionally worked in the realm of the small-population paradigm, assuming a stationary environment, what we really need is theory for the declining-population paradigm (for a critical comment see Hedrick *et al.* 1996). Metapopulation theory goes some way in this direction, because it deals explicitly with the consequences of habitat loss and fragmentation to long-term persistence (Section 4.4).

The single most important message from the extinction models is that the risk of population extinction increases with decreasing population size. Essentially, this is because with more individuals in a population it is more likely that some of them do relatively well even when most perform poorly. Focusing on environmental stochasticity, i.e. random environmental variation simultaneously affecting many individuals, Foley (1994, 1997) constructed a model of population dynamics in which population size performs a random walk between a reflecting upper boundary K, the population 'ceiling', and an absorbing lower boundary, population extinction (essentially the same model has been analysed by Lande 1993 and Middleton *et al.* 1995, and the ceiling model has been discussed in the ecological literature a long time ago by Milne 1957, 1962). In this model, density dependence occurs when population size hits the ceiling. Assuming that the finite growth rate is log-normally distributed with mean r and variance v, the expected time to population extinction is given by (Foley 1994):

$$T_e(n_0) = \frac{1}{sr}[\exp(sk)(1 - \exp(-sn_0)) - sn_0], \tag{2.2}$$

where $s = 2r/v$ and n_0 and k are the natural logarithms of the initial population size and the ceiling, respectively. If $n_0 = k$, eqn 2.2 simplifies to:

$$T_e(k) = \frac{K^s}{sr}\left[1 - \frac{1 + sk}{\exp(sk)}\right]. \tag{2.3}$$

Assuming further that $r > 0$ and sk is so large that the term in square brackets in eqn 2.3 is close to unity, we get the following simple asymptotic scaling of time to extinction with the population ceiling:

$$T_e(k) = \frac{K^s}{sr}. \tag{2.4}$$

This model assumes temporally uncorrelated (white) noise. A correction for temporally correlated noise (Turelli 1977; Foley 1994) is needed if the correlation time is long and the strength of noise (v) is large (Johst and Wissel 1997). This scaling of T_e with K will be used in the interpretation of the parameters of the incidence function model in Section 5.3 (Box 5.1).

Figure 2.5 shows how the expected time to extinction depends on the values of the ceiling and s, the ratio of the long-term growth rate and its variance (see also Lande 1993; Middleton *et al.* 1995; Ludwig 1996). Notice that when s is large, which corresponds to weak environmental stochasticity, T_e increases faster than linearly with K, but when s is small, corresponding to strong environmental stochasticity, T_e increases roughly logarithmically with K. The important message is that when environmental stochasticity is strong, even large populations have a substantial risk of extinction. Figure 2.5 also shows that T_e is small for small initial population sizes, such as typically occur after colonization, but T_e increases rapidly to an asymptotic value when the initial (propagule) size increases towards K (Ludwig 1996).

The extinction–area relationship

Making the plausible assumption that the expected size of an extant population increases with the area of the habitat patch, the population ceiling on the horizontal

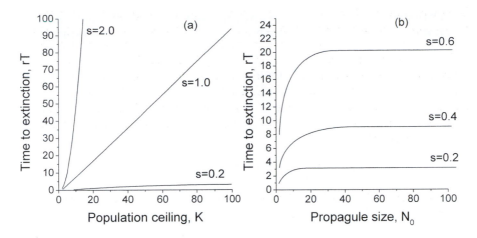

Fig. 2.5 Relationships between time to metapopulation extinction and population ceiling (a), and between time to extinction and propagule size (b), based on eqn 2.2. In (a) the propagule size (N_0) equals the ceiling (K), in (b) $K = 100$.

axis in Fig. 2.5 can be replaced by patch area; the probability of extinction decreases with increasing area. This is a relationship that can be used in metapopulation models (Section 5.3).

Extinction models assuming environmental stochasticity can be used to describe the decreasing risk of population extinction with increasing population size and increasing patch area, but they do not provide a mechanistic explanation, as these models do not explain why there is more or less temporal variation in population growth rate at particular spatial scales. I can think of three different scenarios which might yield the extinction–area relationship (Fig. 2.6). Extinction models typically assume Caughley's (1994) small-population scenario, an ideal population living in a habitat patch of uniform quality. When the patch becomes larger, the population ceiling (K) becomes larger, but v remains constant, because all individuals in the population are affected by exactly the same environment; there are just more individuals. In reality, however, temporal variation in the environmental conditions affecting demographic processes tends to have a spatial component, making stochasticity scale-dependent. In the changing-environment scenario large areas are spatially more heterogeneous than small ones, for instance they might include habitat types entirely lacking from small areas (this scenario has been referred to as the vegetation mosaic hypothesis; Short and Turner 1994). An important consequence of such heterogeneity is that when environmental conditions change, the essential resources for a particular species might disappear entirely from small areas, though perhaps only temporarily, whereas in large areas favourable conditions are more likely to remain somewhere (Ehrlich and Murphy 1987). The changing-environment scenario thus predicts that increasing area reduces extinction risk because increasing area means more habitat heterogeneity and hence reduced impact

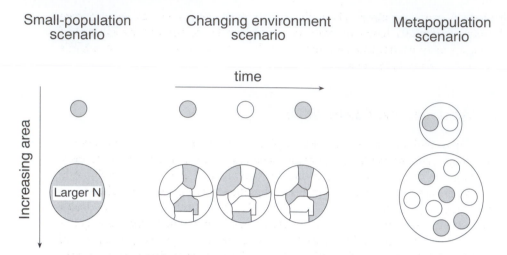

Small-population scenario

Changing environment scenario

Metapopulation scenario

time

Increasing area

Larger N

Fig. 2.6   Illustration of three ecological scenarios that might generate decreasing risk of population extinction with increasing area of suitable habitat.

of environmental stochasticity, and reduced variance (v) in population growth rate when calculated across the entire area (Goodman 1987a). The level of habitat heterogeneity depends on the kind of species. For instance, large vertebrates typically experience less habitat heterogeneity within some fixed area than insects. This means that the small-population scenario is relatively more relevant for large vertebrates, whereas the changing-environment scenario is more relevant for insects (Ehrlich and Murphy 1987). Section 8.3 gives some empirical examples. The third scenario is the metapopulation scenario, in which the key feature is that the 'population' within a larger area is not a single panmictic unit at all, but is instead structured into several more or less independently fluctuating local populations which together form a metapopulation (Holt 1992). In this case, though all local populations might have a high risk of extinction, the entire metapopulation survives possibly for a long time; and the larger the area and hence the greater the expected number of local populations, the longer the metapopulation is expected to survive (theory in Section 4.5, examples in Section 9.3).

The three hypotheses about the extinction-area relationship match the three explanations that Connor and McCoy (1979) put forward to explain the ubiquitous community-level relationship, the species–area curve (MacArthur and Wilson 1967). As Connor and McCoy (1979) named them, these are the sampling hypothesis (corresponding to the small-population scenario), the habitat diversity hypothesis (changing environment scenario), and the area per se hypothesis (metapopulation scenario). This correspondence is not so surprising when one recalls that extinction rate is one of the two processes on which basis MacArthur and Wilson (1967) predicted the species-area curve. One interesting observation however is that if, as seems likely, spatial heterogeneity generally reduces the risk of extinction, the habitat diversity hypothesis about the species-area curve involves an important dynamic component based on extinction rate, and hence the two hypotheses of habitat diversity (static) and area (dynamic) about the species-area curve cannot be disconnected from each other.

2.3 Migration and colonization

Migration of individuals away from their place of birth, or away from the previous breeding ground, and the factors affecting such movements, pose some of the more intriguing problems in population biology (Swingland and Greenwood 1983; Olivieri and Gouyon 1997; Section 6.1). Migration often involves an element of parent–offspring conflict (Hamilton and May 1977; P. D. Taylor 1988) and opposing selection pressures operating at the levels of genes, individuals and populations (Comins *et al.* 1980; Olivieri *et al.* 1995). In metapopulations persisting as a result of recurrent colonizations, a sufficiently high rate of migration is clearly necessary for long-term persistence. But given that migration often entails some cost to migrating individuals, a very high rate of migration might accelerate rather than delay metapopulation extinction (Hanski and Zhang 1993). A topical concern is the evolution of migration rate in increasingly fragmented landscapes (Section 6.1). In brief, migration is of fundamental significance to metapopulation biology, which to a large extent is nothing but the study of the population biological consequences of migration. This section presents a brief summary of the factors potentially affecting migration.

An individual is expected to migrate whenever moving away increases individual (or inclusive) fitness (Baker 1978), with the caveat that when the reproducing female rather than its offspring is in control of the migration decision, the parent–offspring conflict might be resolved to the advantage of the parent. In any case, it is clear that many factors affect the costs and benefits of migration, and it is unlikely that any one of the several hypotheses about the causes of migration will emerge as a clear winner. The following is a list of the local and metapopulation-level factors that are most likely to affect migration decisions.

Local factors

- Inbreeding avoidance. If breeding with close relatives reduces fitness (inbreeding depression), it might be advantageous to move away from the natal site, or to drive all or at least some of ones offspring away (Greenwood 1980; Dobson 1982; Shields 1982; Packer 1985; Pusey 1987). Inbreeding avoidance is a possible explanation of some evolved intrinsic migration tendencies, such as male-biased juvenile migration in mammals (Dobson 1982).
- Sib competition. It is bad enough to be forced to compete for limiting resources, but it is particularly bad if one has to compete with sibs, because this will reduce both individual and inclusive fitness. Avoidance of sib competition is thought to

be another intrinsic cause of migration, especially when the reproducing female is in control of the migration of her offspring (Hamilton and May 1977; Comins *et al.* 1980).

- High density: resource competition. High population density can increase emigration rate (Johnson 1969; Baker 1978), although, in contrast with the assumptions of many theoretical models, emigration is by no means always density-dependent (Hansson 1991). Density-dependent emigration is likely to be more frequent in species with relatively stable populations, for instance birds and mammals, than in insects, though good insect examples are not scarce either (Denno and Peterson 1995; Herzig 1995).

- Low density: conspecific attraction. Emigration can be elevated at very low densities (Kuussaari *et al.* 1996), most likely because of expected reduction in fitness due to a delay or entire failure to locate a mate in a sparse population or due to inbreeding depression. In some species with chronically low densities individuals have evolved an amazing capacity to locate a mate from long distances, usually with the help of olfactory signals. Satyrid moths include well-known examples. Such behaviour will lead to extensive migration of one sex, and one could even speculate that such species have evolved to be rare.

- Escaping imminent extinction. Species which live in ephemeral habitats have necessarily ephemeral local populations. Such species can only survive at the metapopulation level by having a migration rate sufficiently high to enable the continuous establishment of new populations (Brown 1951; Southwood 1962; Johnson 1969). Often emigration is triggered by adverse environmental conditions in the current habitat patch (Johnson 1969).

Metapopulation-level factors

Other factors affecting migration are not local but are rather properties of the entire patch network in which migration occurs:

- Temporal variance in fitness, when not completely correlated across the environment, tends to promote migration, essentially by the 'spreading-of-risk' principle (den Boer 1968)—an individual distributing its offspring among several more or less independently fluctuating populations will reduce the adverse impact of a poor year in the natal population (Gadgil 1971; Roff 1975; Levin *et al.* 1984; McPeek and Holt 1992; Section 3.3). It is usually assumed that temporal variance in fitness is caused by varying environmental conditions, but such variation could also be due to complex endogenous population dynamics (Holt and McPeek 1996).

- Spatial variance in fitness, as a rule, is not sufficient by itself to select for migration, because random movements would take more individuals from high-quality patches to low-quality patches than vice versa (Hastings 1983; Holt 1985).

- Cost of migration. Mortality during migration is the most direct cost of migration. Theoretical studies have clearly shown that migration costs greatly affect the level of adaptive migration (Hamilton and May 1977; Levin *et al.* 1984; Johnson and Gaines 1990; Olivieri and Gouyon 1997; Section 6.1). Several empirical studies have demonstrated elevated mortality during migration (Small *et al.* 1993; Larsen and Boutin 1994; Steen 1994; Ims and Yoccoz 1997), but often it is unfortunately very difficult to measure mortality during migration in empirical studies (Ims and Yoccoz 1997). Box 2.1 outlines one possible method that is particularly appropriate for metapopulation studies.

- Favourable conditions for migration. Movements of, especially, invertebrates are much affected by the prevailing environmental conditions (Johnson 1969). For instance, many butterflies are active only in sunny and warm weather, and hence in cloudy and rainy summers the level of migration remains at a low level.

Different species are affected by different mixes of the above factors. It seems probable that species which are more responsive to current environmental conditions, for instance population density, would have a greater chance of surviving in changing environments. However, if environmental changes occur primarily at the landscape level, as happens in habitat fragmentation, the evolutionary changes in migration rate are relatively slow (Section 6.1) in comparison with the demographic consequences of landscape-level environmental changes, which themselves occur with a delay. The dependence of migration rate on the prevailing environmental conditions might greatly affect metapopulation dynamics; colonizations of the more isolated habitat patches might occur primarily in the rare years when conditions are exceptionally favourable for migration.

Migration distances

Migration distances determine the scale of recolonization of empty habitat and thereby set the spatial scale of metapopulations. The simplest models of movement assume that individuals perform a random walk, and the models predict normally-distributed migration distances, with the variance increasing linearly with time and the coefficient of diffusion (Okubo 1980). (The distribution of migration distances is often called the redistribution (or dispersal) kernel in the modelling literature.) Empirical studies have however shown that the distribution of migration distances is typically leptokurtic (Dobzhansky and Wright 1943; Taylor 1978; Howe and Westley 1986; Johnson and Gaines 1990; Wilson *et al.* 1993), that is, more individuals have moved very short or very long distances than is predicted by the simple random walk (normal distribution). Leptokurtic distributions might indicate differences in individuals' migration tendencies (Dobzhansky and Wright 1943; Skellam 1951), but leptokurtic distributions might also reflect more complex movement behaviour than simple random walk in a homogeneous population of individuals (Okubo 1980). Perhaps the simplest and most natural way to obtain a leptokurtic distribution is to assume that migrating individuals have a constant probability of settling, which leads to a leptokurtic Laplace or back-to-back exponential distribution (Turchin and Thoeny 1993; Neubert *et al.* 1995). Ecologists

Box 2.1 Estimating the parameters of survival and migration in metapopulations

Individuals born to a metapopulation might stay all their life in the natal habitat patch or they might move on one or more times to a new patch during their life. Ecologists are interested in the rates of survival in habitat patches and during migration as well as in the rate and distances of migration. To measure these rates, Hanski *et al.* (1999) modelled a sample of individuals' capture histories obtained with a mark–recapture study conducted simultaneously in many populations in a metapopulation.

Assume that an individual has a constant probability, ϕ_p, of surviving one time unit (usually one day in insect studies) or until it emigrates (subscript p signifies survival in a habitat patch). The probability of emigrating from patch j, m_j, is assumed to scale with patch area A_j by the power function:

$$m_j = \eta A_j^{-\xi},$$

where $\eta > 0$ and $\zeta > 0$ are two parameters. Connectivity of patch j is measured by:

$$S_j = \sum_{k \neq j} \exp[-\alpha d_{jk}] A_k^{\zeta},$$

where d_{jk} is the distance between patches j and k and parameter $\alpha > 0$ determines the effect of distance on isolation as experienced by the study species. The probability, $\phi_{m,j}$, of surviving migration from patch j is assumed to increase with connectivity of patch j:

$$\Phi_{m,j} = \frac{S_j^2}{\lambda + S_j^2},$$

which formula adds one more parameter $\lambda > 0$ to the model (subscript m refers to survival during migration). The individuals that survive migration are distributed amongst all the target patches in proportion to the contributions that these patches make to S_j. Successful migration occurs within unit time, and the same individual may re-emigrate in the following time interval if still alive.

The model has five parameters, ϕ_p, η, ζ, α and λ, which describe the rates of survival and migration of individuals in a metapopulation. Assuming that an individual has been observed in habitat patches s_1, ..., s_n at times $t_1 < ... < t_n$, one can write an expression for the likelihood of the data for each individual, which leads to an expression for the likelihood of the combined data assuming independent individuals. Parameter estimates and their confidence intervals are obtained using a numerical optimization method (Hanski *et al.* 1999). The model is attractive for metapopulation studies because it allows us to tease apart mortality in habitat patches and mortality during migration, assuming that there are sufficient differences in the connectivities of the patches and assuming, of course, that the above model is otherwise appropriate for the study species. This model makes strong, but for many species realistic, assumptions about the effects of habitat patch area and isolation on movements. The incidence function model of metapopulation dynamics is based on comparable assumptions (Section 5.3).

working with vertebrates have found it natural to think about migration in terms of a search for a vacant territory. If p denotes the probability of locating a suitable territory before moving an additional distance corresponding to territory diameter, then the fraction of individuals moving a distance x in units of territory diameter is given by $p(1 - p)^x$ (Waser 1985; Buechner 1987; for elaborations of this model see Miller and Carroll 1989). The resulting geometric probability distribution is the discrete variant of the continuous exponential model. These models are consistent with the common assumption in metapopulation models and analyses of empirical data that the effect of isolation, and by implication the distribution of migration distances, is exponential.

The exact shape of the redistribution kernel is not expected to be critical for the study of established metapopulations, because their dynamics are dominated by short-distance movements. In contrast, the shape of the redistribution kernel is critical for the pattern of spread to a previously unoccupied region (Kot *et al.* 1996; Clark *et al.* 1998), for instance an empty habitat patch network. It is probably realistic to assume that the true distribution of migration distances has an even 'fatter' tail than the exponential distribution, for which reason a power function might be a preferable simple alternative to the exponential distribution (Hill *et al.* 1996; Thomas and Hanski 1997). Unfortunately, it is very difficult to measure the thickness of the tail in empirical studies. To find out the longest migration distances one might have to use indirect data, such as genetic markers (Turelli and Hoffmann 1991), or data on actual colonization rate of unoccupied habitat patches (Thomas and Jones 1993).

Empirical studies of migration distances have generally suffered from uneven sampling efficiency at different movement distances. Porter and Dooley (1993) used a simulation model that could be tailored for a particular spatial configuration of sampling points, such as habitat patches in a metapopulation, to derive correction factors for observed movement distances. They found that while the uncorrected distribution of movement distances was often well described by a simple model, such as the geometric model, the corrected distribution was typically more complex, possibly reflecting the influence of particular landscape structures on individual movement behaviour. In any case, it is imperative to correct for the bias introduced by distance-weighted sampling in any empirical study of migration distances. One approach is to model the actual capture histories of marked individuals in a set of local populations in a metapopulation (Box 2.1).

Colonization

The probability of successful colonization of an empty habitat patch depends on many factors specific to particular species and environments, but the propagule size, or the number of immigrants arriving at an empty patch, is likely to be a key factor. Figure 2.5 shows how the expected time to extinction increases rapidly with propagule size in the model described in Section 2.2 (eqn 2.2; for a comparable result based on a somewhat different model see Ludwig 1996). If we define successful colonization as the newly-established population surviving until it has reached the long-term expected size of extant populations, theory predicts a rather sharp effect of propagule size on colonization, which has also been demonstrated in several empirical studies (Sheppe 1965; Crowell 1973; Ebenhard 1987, 1991; Veltman *et al.* 1996; for an exception see Schoener and Schoener 1983). Inspecting how the various model parameters affect the probability of extinction in eqn 2.2, one gains further insight into the effects of species demography on colonization. Large r and small v (variance of r), in particular, increase the probability of successful colonization. Propagule size is not so significant when r/v is very large, or very small, which may explain the results of Schoener and Schoener (1983) on lizards (high colonization success) and Harrison (1989) on butterflies (low colonization success).

Some other factors not included in eqn 2.2 but likely to affect colonization success are the mode of reproduction (asexual reproduction advantageous; Jain 1976; Ebenhard 1991), the reproductive value of the colonizing individuals (MacArthur and Wilson 1967; Williamson and Charlesworth 1976), and niche width (generalists thought to be better colonizers than specialists; Grant 1970; Ehrlich 1986; Baur and Bengtsson 1987). When demographic stochasticity is the main cause of extinctions, a good colonizer is expected to have a large r via a large ratio of natality to mortality (MacArthur and Wilson 1967; Ebenhard 1991). Some empirical studies support this prediction (Ebenhard 1991; Veltman *et al.* 1996).

An important question for metapopulation dynamics is whether colonization involves an element of inverse density dependence, such that the per-immigrant probability of establishing a new population increases with propagule size up to some threshold size. Such inverse density dependence would further amplify the significance of propagule size in colonization, and would lead to a sigmoid relationship between the probability of colonization and propagule size. Empirical studies on this relationship would be most welcome.

3

Metapopulations of two local populations

This chapter is concerned with questions about migration among local populations that have no risk of stochastic extinction. How does migration affect population sizes, and how does it affect the type of dynamics? In the absence of population turnover it is sufficient, for many purposes, to consider the simplest case of a metapopulation consisting of just two local populations. The two populations and the respective habitat patches might or might not be similar. In source–sink metapopulations, there are substantial differences between the intrinsic growth rates of local populations.

3.1 The two-population metapopulation

Extending classical population models to not more than two local populations has the advantage that the resulting metapopulation model can often be analysed mathematically (Freedman and Waltman 1977; Holt 1985). This section is however restricted to a graphical analysis of a simple continuous-time model, with selected examples being used to make general points. The purpose is to give an intuitive feeling about the likely consequences of migration on local dynamics. The more complex dynamics that occur in discrete-time models are discussed in the following section.

A general two-population model takes the form:

$$\frac{dN_1}{dt} = g(N_1)N_1 - \gamma_{12}(N_1)N_1 + \gamma_{21}(N_2)(1 - \delta_1)N_2$$
$$\frac{dN_2}{dt} = g(N_2)N_2 - \gamma_{21}(N_2)N_2 + \gamma_{12}(N_1)(1 - \delta_2)N_1, \qquad (3.1)$$

where $g(N_i)$ gives the per capita rate of change of population i due to births and deaths, $\gamma_{ij}(N_i)$ is the per capita rate of emigration from population i to population j, and δ_i is the fraction of migrants dying during migration. As a standard and not too unrealistic specific example, let us assume that local dynamics are given by the logistic model and that emigration is density-independent and the same in the two populations. Equations 3.1 then become:

$$\frac{dN_1}{dt} = r_1N_1\left(1 - \frac{N_1}{K_1}\right) - mN_1 + m(1 - \delta)N_2$$
$$\frac{dN_2}{dt} = r_2N_2\left(1 - \frac{N_2}{K_2}\right) - mN_2 + m(1 - \delta)N_1, \qquad (3.2)$$

where r_i and K_i are the intrinsic rate of population increase and the carrying capacity of population i and m is the constant emigration rate.

For positive r_i and K_i, the two-population metapopulation has two equilibria, of which the one corresponding to metapopulation extinction, $(\hat{N}_1, \hat{N}_2) = (0,0)$, is unstable, and the one with $\hat{N}_1 > 0$ and $\hat{N}_2 > 0$ is stable ($\hat{N}$ denotes the equilibrium value). The pooled size of the two populations with migration is always equal to or less than their sum in the absence of migration, thus migration will reduce the overall metapopulation size (somewhat surprisingly, the opposite might happen when we change some model assumptions below). The pooled metapopulation size is reduced even if there is no mortality during migration, provided that there is a difference between the two carrying capacities (Fig. 3.1a)—more individuals move from the population with larger K to that with smaller K, where their per capita reproductive success is lower than in the natal population. Therefore, random migration of the type assumed in eqn 3.2 is not expected to evolve by natural selection (Hastings 1983; Holt 1985), implying that migration in natural populations has evolved in response to some forces not considered in this model, for instance temporally varying carrying capacities (Section 6.1). The equilibrium population size in the patch with smaller K might of course be higher with than without migration, even if there is substantial mortality during migration (Fig. 3.1a). Biased migration might produce the same result—increased density in the patch with lower emigration rate.

The logistic model assumes that the per capita growth rate decreases linearly with increasing population size. This is a convenient mathematical assumption, but biology might be different; for instance, density dependence might become strong only when population size is close to the carrying capacity, as would happen when growth is limited by space for individual territories. Such situations can be modelled by raising the term N/K in the logistic model to power θ, $(N/K)^\theta$. Assuming that θ is small (<1), which corresponds to weak density dependence near the carrying capacity, has the interesting consequence that the two populations connected by migration tend to become similar in size, even if the respective carrying capacities are quite different (Fig. 3.1b). With increasing mortality during migration, metapopulation size becomes reduced because of the generally reduced growth rate, and with high rate of mortality during migration the metapopulation might become extinct. The message from here is that non-linear per capita density dependence makes it ever harder to use observed population sizes or densities to draw inferences about the quality of the respective habitat patches. Assuming that in one population density dependence is weak near the equilibrium ($\theta < 1$), and that the other population has a large carrying capacity, produces the surprising result that the pooled size of the metapopulation may exceed the sum of the two carrying capacities, even if there is mortality during migration (Fig. 3.1c). This happens because one of the populations, the one with the greater carrying capacity, feeds large numbers of migrants to the other population, where weak density dependence allows population density to persist at a high level, much higher than the local carrying capacity. In other words, population sizes in this case are biased towards the

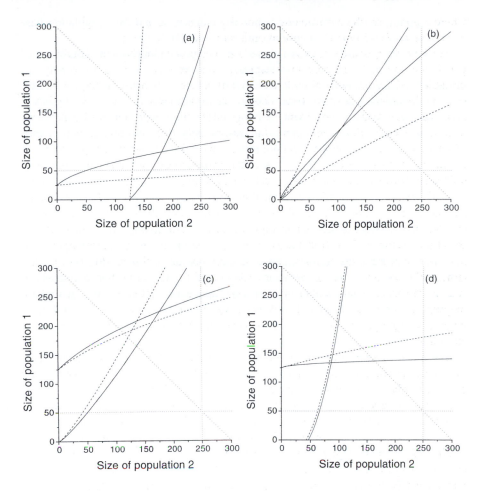

Fig. 3.1 Isoclines of the two-population metapopulation model. (The isocline for a population gives the combinations of population sizes for which there is no change in the size of the focal population. The intersection point of the two isoclines gives the combinations for which there is no change in the size of either population. This point is an equilibrium point. The thin dotted lines indicate the individual and pooled carrying capacities of the populations.) (a) Demonstrating how mortality during migration reduces metapopulation size: $K_1 = 50$, $K_2 = 250$, $r_1 = r_2 = 2$, $m_1 = m_2 = 1$, $\delta = 0$ (continuous line), $\delta = 0.9$ (broken line). (b) Weak density dependence near the equilibrium density ($\theta < 1$) tends to equalize population sizes: parameter values as in (a) but now $\theta_1 = \theta_2 = 0.1$ and $\delta = 0$ and 0.5. (c) Metapopulation size may exceed the sum of the two carrying capacities if density-dependence is weak ($\theta < 1$) in the population with smaller K: $\theta_1 = 0.1$, $\theta_2 = 1$, and $\delta = 0$ and 0.2. (d) Density-dependent emigration, $\zeta = 0.5$ and $\zeta = 1$ (continuous lines), and a difference in density-independent emigration rates, $m_1 = 0.3$ and $m_2 = 1$ (broken lines), have similar effects.

K of the population with stronger density dependence, and if this population has larger K metapopulation size might exceed $K_1 + K_2$ (Holt 1985).

So far we have assumed that emigration rate is density-independent. This is a fair assumption for many species, for instance for many insects, but especially in vertebrates emigration rate often increases with increasing density (Section 2.3). Let us model density-dependent emigration by replacing the term mN in eqn 3.2 by mN^ζ. If emigration is density-dependent in two similar populations, there are no new major consequences, but if emigration is density-dependent in one population and density-independent in the other, the equilibrium density is reduced in the former and increased in the latter (Fig. 3.1d). The result is roughly the same if density-independent emigration rate is higher in one population than in the other, which is, of course, exactly what density-dependent emigration in one of the populations tends to cause.

This simple model makes it plain that migration can greatly affect the sizes of local populations in a metapopulation. Some of the predicted consequences are at first surprising, for instance that metapopulation size might exceed the sum of the local population sizes in the absence of migration. The most important general message is that population density is an unreliable indicator of habitat quality in metapopulations with substantial migration. It is unfortunate that the measurement of migration rate in natural populations is generally very laborious (Ims and Yoccos 1997), making it hard to test model predictions with empirical data. However, given sufficient mark-recapture data from several populations in a metapopulation, it is possible to estimate the key parameters of individual survival and migration (Hestbeck *et al.* 1991; Nichols 1992; Nichols *et al.* 1993; Box 2.1).

3.2 Migration and complex dynamics

Some ideas are so obviously good that when the time is right the same idea occurs to many people at the same time. Here is an example. Take a simple discrete-time population model, like the Ricker model:

$$N(t+1) = N(t)\exp\left[r\left(1 - \frac{N(t)}{K}\right)\right], \tag{3.3}$$

where r and K set the growth rate and the equilibrium population size, respectively. The Ricker model has the desirable property that population size cannot become negative (it is less desirable, though not fatal for a simple deterministic model, that population size cannot become zero either). The behaviour of the Ricker model and other simple discrete-time population models has become familiar to ecologists thanks to pioneering research by May (1974, 1976) and May and Oster (1976) and 25 years of tutelage. The simple models motivated by population ecology have played a role in the development of non-linear dynamics more generally (Gleick 1987).

Now add to the discrete population model the metapopulation idea—local populations connected by migration. For simplicity, take two populations obeying the Ricker model and assume that they are connected to each other via density-independent random movements. I tried this exercise in 1990, and many other ecologists played the same or a similar game around the same time (McCallum 1992; Gonzales-Andujar and Perry 1993; Hastings 1993; Bascompte and Solé 1994). Serious mathematics has been done with these models (Gyllenberg *et al.* 1993, 1996 and references therein). My aim here is to describe some of the qualitative results and ecological implications.

Let us start with the observation I made in 1990. We know that discrete-time population models such as the Ricker model generate a range of dynamic behaviour depending on the value of the growth parameter r, which controls the degree of non-linearity (May 1976). The Ricker model has a stable equilibrium point when $r < 2$. For $2.000 < r < 2.526$ the model predicts a 2-point limit cycle, followed by more complex limit cycles and, when r exceeds 2.692, by chaotic behaviour (May 1974, 1976). For the purpose of an example, let us assume $r = 3$, which gives chaotic dynamics. The first 100 time steps in Fig. 3.2, obtained by numeric iteration of eqn 3.3, show the predicted dynamics in two uncoupled populations. At time 100, the two populations were connected, by allowing 30% of individuals to emigrate and by dividing the migrants equally among the two populations. Nothing special seems to happen in the dynamics of local populations, but note that there are now more consistent high-amplitude oscillations in the size of the metapopulation as a whole—migration has a synchronizing effect on local dynamics. The real surprise occurs, in this example, at time 346, when the dynamics change radically—local populations enter a 2-point limit cycle instead of being chaotic, and the two populations cycle out of phase, completely stabilizing the size of the metapopulation (Fig. 3.2). The model behaviour between time-points 100 and 346 in this example represents a transient, a period during which the dynamics have not yet settled to the attractor, out-of-phase 2-point limit cycle.

Gyllenberg *et al.* (1993) have conducted an exhaustive mathematical analysis of a similar model but assuming the logistic map rather than the Ricker model for local dynamics (the qualitative conclusions are the same for the two models; Gyllenberg *et al.* 1993, p. 47). Figure 3.3 shows a simplified summary of their findings for two similar local populations. With high growth rate and little migration, the dynamics of the metapopulation are complex (limit cycles or chaos), like the dynamics of isolated populations, though there is a tendency towards greater stability in the sense of decreased amplitude of oscillations. With somewhat higher migration rate, the sort of metapopulation stability depicted in Fig. 3.2 emerges, and chaotic dynamics may turn to 2-point limit cycles or other simple limit cycles. When migration rate becomes still higher, the dominant consequence is synchrony in local dynamics, without stability; with large r the two populations oscillate chaotically but in synchrony. For many parameter combinations, the model has alternative stable attractors, of which the most obvious example occurs in the double-hatched area in Fig. 3.3: here isolated populations show 2-point cycles, and in the metapopulation

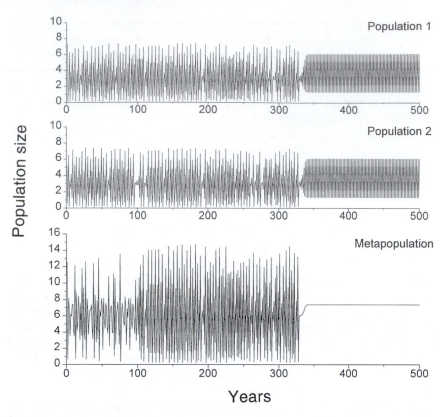

Fig. 3.2   An example of the dynamics of a metapopulation consisting of two connected local populations obeying the Ricker model. In both populations $r = 3$ and $K = 3$. The populations were initially unconnected, hence they oscillate independently and chaotically. At time 100, the populations were connected by assuming 30% emigration rate and an equal division of migrants among the two populations. See text for discussion.

both the in-phase and out-of-phase oscillations are stable attractors (Gyllenberg *et al.* 1993). These complex dynamics are not restricted to the specific model studied by Gyllenberg *et al.* (1993). For instance, Lloyd and May (1996) found similar dynamic behaviour in a multi-patch epidemic model.

　　One generally important question is the extent to which migration stabilizes local dynamics in a metapopulation consisting of many local populations. In the simplest case, with density-independent nearest-neighbour migration among identical local populations, the metapopulation has an equilibrium in which all local populations have the same density as they would have in isolation, and this equilibrium is stable if and only if the equilibrium in an isolated population is stable (Hassell *et al.* 1995; Rohani *et al.* 1996). In other words, migration does not make otherwise unstable populations stable, nor does it make stable populations unstable. However, this message should not be pushed too hard for field ecologists. The metapopulation

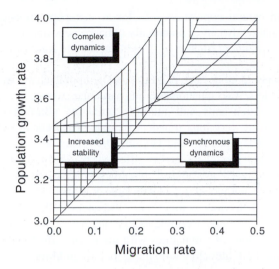

Fig. 3.3 Schematic depiction of how population growth rate and migration rate affect metapopulation stability and synchrony of local dynamics in a model of two coupled logistic maps. Complex dynamics (limit cycles and chaos) with little tendency towards metapopulation stability is retained in the upper left corner (large *r*, little migration). Synchronous dynamics are observed with high migration rate (horizontal shading), while intermediate migration rate has a strong stabilizing effect (vertical shading; simplified from the results of Gyllenberg *et al.* 1993).

might have other equilibria apart from that in which all local populations have the same density, and migration might influence the stability of these equilibria, as shown by the analysis by Gyllenberg *et al.* (1993) (see also Ruxton *et al.* 1997). On the other hand, the conclusion of Rohani *et al.* (1996) depends critically on model assumptions. For instance, if we assume realistically that some mortality occurs during migration, migration might have a strongly stabilizing effect on local dynamics (Ruxton *et al.* 1997). Migration from a permanent 'mainland' population (McCallum 1992) and migration from a population with low growth rate and hence stable dynamics might have a strongly stabilizing effect (Gyllenberg *et al.* 1993). Given the ubiquitous differences between growth rates in real populations, we could hence expect that, at least in the single-species context, migration has a tendency to eliminate chaotic dynamics in the habitat patches in which chaos would otherwise be most likely to occur (Doebeli 1995; Scheuring and Jánosi 1996). But to make things even more complicated, in other related models migration may generate chaos where simple dynamics would occur in the absence of migration (Ruxton 1993; Bascompte and Solé 1994), and inclusion of time delays in local dynamics can produce yet other surprises (Crone 1997). Simple density-independent migration can destabilize predator–prey dynamics (Rohani *et al.* 1996). Thus not many general conclusions have emerged about migration and the type of dynamics in meta-populations, and migration certainly makes it harder to characterize empirically the type of dynamics in natural populations. As far as metapopulation persistence is concerned, locally chaotic dynamics potentially enhance persistence by maintaining asynchrony in local dynamics (Allen *et al.* 1993; Bascompte and Solé 1994; Ruxton 1996).

If there are metapopulations that exhibit spatiotemporal chaos in the manner depicted in the models, two phenomena in particular would complicate their empirical study—supertransients and multiple attractors (Bascompte and Solé 1995). Supertransients are very long transient periods that occur in the iteration of spatially extended models (Hastings and Higgins 1994; Ruxton and Doebeli 1996). The implication for real metapopulations affected by stochasticity is that the dynamics might never reach asymptotic dynamics—and we might never know whether they will. With multiple attractors, the same ecological system might exhibit very different dynamic behaviour without any differences in the environment or in the traits of the species. A simple example of two alternative attractors is generated by single-species metapopulation dynamics with the rescue effect (Sections 4.2 and 12.3). In this example both attractors are simple point equilibria.

3.3 Source–sink metapopulations

Let us return to the two-population model in Section 3.1. What happens if we set the intrinsic growth rate negative in one of the populations? Clearly, if the populations were isolated, a population with $r < 0$ would go extinct, but when connected to another population with $r > 0$, the former population persists thanks to the flow of immigrants. To model such metapopulations, we have to replace the logistic model with the exponential model for the population with $r < 0$, otherwise we could end up having, meaninglessly, positive growth in the population with $r < 0$, when migration from outside has pushed population size beyond K.

The general notion of habitat-specific demography is an old one (Levene 1953), but ecologists' attention was drawn to it especially following Pulliam's (1988) paper, in which he popularized the terms source ($r > 0$) and sink ($r < 0$) population. Ecologists also refer to source and sink habitats, by which is meant habitat in which a particular species has, or would have, a source or a sink population. The source–sink scenario reinforces the point that the presence and the size of a population in a habitat patch might give an entirely misleading idea of the environmental conditions in that patch. Sink populations might occur at sites that are not included in the 'fundamental' niche of the species (Hutchinson 1957), hence, paradoxically, the 'realized' niche might become more extensive than the fundamental niche (Pulliam 1996). Depending on how fast the sink populations decline, how fast the source populations grow, and how extensive is migration between the two, the size of the metapopulation as a whole might be greater, or smaller, than the carrying capacity of the source population (Fig. 3.4; Holt 1985; Davis and Howe 1992). If emigration rate is higher from the source than from the sink, the size of the sink population might be greater than the size of the source population, but this may be unlikely in nature.

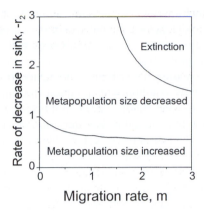

Fig. 3.4 Metapopulation size in a source–sink model compared with the carrying capacity of the source population. The source population grows logistically with $r_1 = 1$, the sink population declines exponentially with r_2 (vertical axis); the migration rate (m) is given on the horizontal axis. (Based on Holt 1985.)

There is some inconsistency in the use of the terms sink and source populations/ habitat patches in the literature. The most clear-cut definition of a sink population is based on negative intrinsic growth rate, as discussed above, implying that a sink population goes deterministically extinct in the absence of immigration. This definition is however different from that used by Pulliam (1988) and Pulliam and Danielson (1991), who defined sources and sinks on the basis of whether local births exceed deaths, or vice versa, at equilibrium. In the presence of emigration from and immigration to a population, local births are unlikely to exactly match deaths even over a long period, and therefore, in practice, by Pulliam's definition all metapopulations consist of sources and sinks. Furthermore, whether a particular population is a source or a sink would very much depend on which other populations it happens to be connected to, because the latter determine the level of immigration to the focal population. Watkinson and Sutherland (1995) have coined the term pseudo-sink to describe populations in which deaths exceed births at equilibrium, but which would decline to a positive new equilibrium, rather than to extinction, if they were cut off from other populations. Pulliam's definition is useful for population genetic purposes, because it draws attention to the question of which populations tend to be net importers of individuals (and genes) and which populations are net exporters. For population ecological purposes, the definition based on the sign of r is often preferable.

Another source of confusion arises from the habit of many ecologists to use the terms source and sink populations to describe large and small populations, respectively, because the latter often have a high risk of extinction. But small populations have a high risk of extinction for all sorts of reasons even though their expected growth rate would be positive (Section 2.2). It is preferable to use the terms mainland and island populations to describe metapopulations with much variation in patch size but not in patch quality. At the same time, we have to recognize that habitat patch size might affect the balance between births, deaths and migration. For instance, the per capita emigration rate often increases with decreasing patch size (Kareiva 1985; Turchin 1986; Back 1988; Kindvall 1995; Thomas and Hanski 1997), because of the increasing ratio of patch boundary to patch area, and hence with decreasing patch size the equilibrium population size might become zero, even though r is positive (Thomas and Hanski 1997). In this case r is not large enough to compensate for losses due to emigration. The question about the critical minimum patch size necessary for population survival has been extensively studied with diffusion models (Skellam 1951; Okubo 1980).

A related issue is how the relative abundances of source and sink habitats affect the growth rate, size and persistence of metapopulations. When migration rate is high and individuals move randomly between the two habitat types, the per capita growth rate in the metapopulation changes approximately linearly with the fraction of the source habitat. In contrast, when migration rate is low, most individuals occur in the source habitat, and the per capita growth rate becomes sensitively dependent on the fraction of the source habitat only when there is little of it left. Using an age-structured simulation model, Doak (1995) demonstrated a delay in the response of a grizzly bear population to declining amount of source habitat. This is worrying, because it would make difficult to observe the consequences of habitat degradation before a substantial change in the landscape has already occurred. The general message here is that migration and relative qualities and quantities of source and sink habitats have complex interactions, and no neat distinction between the consequences of habitat patch quality and patch size is possible (for a related theoretical study see Davis and Howe 1992).

One might be tempted to draw the conclusion that sinks are either irrelevant or detrimental to metapopulation persistence, but this does not need to be so. Gyllenberg *et al.* (1996) describe and analyse a model of a source and a sink, in which the source population shows chaotic behaviour. Assuming that small populations have a high risk of extinction for stochastic or deterministic reasons (Allee effect), the source population is liable to fluctuate to extinction, but it might be rescued if connected to a sink, or the site might become quickly re-colonized from the sink, especially if the rate of decline in the sink population is not fast. Sink and pseudo-sink populations might have an equally significant role also in cases where the source populations show great fluctuations in size because of sensitivity to environmental perturbations rather than because of complex endogenous dynamics (for an excellent example see Thomas *et al.* 1996; Section 8.4).

Taking into account environmentally induced stochastic fluctuations in population sizes, migration among independently fluctuating local populations might lead to another apparent paradox, a persisting metapopulation consisting entirely of true sink populations. This might happen because migration among independently fluctuating local populations enhances the overall growth rate in the metapopulation, essentially because the risk of poor reproduction in a particular population in a particular year is spread via migration among many independently fluctuating populations (the 'spreading of risk' concept of den Boer 1968). If λ_M and λ_L are the expected per capita reproductive rates in the metapopulation as a whole and in a local population, respectively, then in a metapopulation of k independently fluctuating local populations (Kuno 1981; Metz *et al.* 1983):

$$\lambda_M = \lambda_L + 0.5\left(1 - \sum_{i=1}^{k} w_i^2\right)c^2, \tag{3.4}$$

where w_i is the (fixed) fraction of immigrants reproducing in the ith local population and c is the coefficient of variation of r_{ij}, the per-capita growth rate in population i in year j (in this model, all individuals migrate). Clearly, it is possible that $\lambda_M > 0$ though $\lambda_L < 0$, and even if there is mortality during migration. Although in real metapopulations the positive effect of migration in enhancing growth rate is reduced by partial migration and by some degree of spatial synchrony in local dynamics (Hanski and Woiwod 1993b; Pollard and Yates 1993; Ranta *et al.* 1995; Sutcliffe *et al.* 1996), migration among stochastically fluctuating local populations adds another important element to the study of source and sink populations.

The source–sink population structure has outstanding consequences for interspecific interactions in multispecies communities (Holt 1984, 1997; Danielson 1991; Holt and Hassell 1993; Pulliam 1996; Stacey *et al.* 1997). The sink status of a population might be due to interspecific interactions, while migration from source populations might stabilize otherwise unstable interactions. Generally, the strength and even the nature of interspecific interactions becomes a function of the structure of the landscape, not only a function of species' intrinsic properties. Thus for instance the strength of interspecific competition might change with a change in landscape structure (Danielson 1992).

Evolutionary consequences

We should first ask why are there true sink populations at all? By definition, true sink populations persist because of immigration, but why should any individuals settle in inferior habitat? Two likely mechanisms are interference competition amongst animals, forcing subordinate individuals to move away from high quality habitat, and passive migration, which should be especially significant in plants and other sessile organisms (Holt 1993; Dias 1996). Recall also that metapopulations consisting primarily or entirely of true sink populations in stochastic environments might persist owing to risk-spreading among independently fluctuating local populations. Some species might thus persist in generally unfavourable environments—there simply are no dependable source habitats.

The source–sink metapopulation structure creates an asymmetry, a net flow of migrants from sources to sinks, and typically a higher abundance and fitness in the source. In this situation, evolution is likely to improve adaptations to the source habitat at the cost of performance in the sink habitat (Brown and Pavlovic 1992; Holt and Gaines 1992; Houston and McNamara 1992; Kawecki and Stearns 1993;

Kawecki 1995; Holt 1995, 1996). This might lead to what Holt (1995) calls 'niche conservatism'. Selection against deleterious mutations expressed only in the sink habitat might be expected to be weak (Kawecki 1995), which creates another barrier for the species to evolve to novel environmental conditions. How strong this barrier is depends on how low is the fitness of individuals in the sink habitat and how much migration there is between the habitat types (Holt 1995).

4

The Levins model and its variants

4.1 The Levins model and other patch models

The fundamental idea of metapopulation persistence in a stochastic balance between local extinctions and recolonizations of empty habitat patches is captured in the Levins (1969, 1970) metapopulation model. Considering the metapopulation as a population of local populations inhabiting an infinitely large patch network, Levins modelled the rate of change in metapopulation size, $P(t)$, measured by the fraction of patches occupied at time t. In the spirit of other simple population models, such as the logistic model, the Levins model assumes that all existing basic entities, which are here the local populations, are identical and hence have the same behaviour. Extinctions occur completely independently in different habitat patches, in other words the Levins model assumes completely asynchronous local dynamics. Being identical, the existing populations contribute equally to the pool of migrants, which is, however, not modelled explicitly. The migrants are spread out across the entire patch network, and hence they 'encounter' and colonize empty patches out of all patches in proportion to how many empty patches there are. Changes in $P(t)$ are then given by:

$$\frac{dP}{dt} = cP(1 - P) - eP, \qquad (4.1)$$

where c and e are the colonization and extinction rate parameters, respectively.

The Levins model gives a deterministic description of the rate of change in metapopulation size, though the model is implicitly based on stochastic local extinctions. The model assumes infinitely many habitat patches but also that colonization is not affected by distance. These two assumptions seem contradictory, as movements of most organisms are restricted in space, and hence not all patches in a large network are likely to be equally accessible from a given patch. However, what the model really assumes is that all patches are equally connected to other patches (not necessarily to all other patches), which is called the mean-field assumption and which is often a good approximation for a metapopulation at stochastic steady-state even if migration is distance-dependent (Nisbet and Gurney 1982; Durrett and Levin 1994). The mean-field approximation becomes dubious when local populations have a clumped distribution in space, either for environmental or population dynamic reasons, because then different populations are not likely to be equally connected to other populations (unless dispersal is not distance-dependent). To study such metapopulations one has to move from the spatially implicit patch models to the spatially explicit and spatially realistic models discussed in Chapter 5. The Levins model assumes, strictly speaking, infinitely many habitat patches, but in practice it provides a good approximation for networks of the order of 100 or more patches.

In the Levins model, all existing local populations are assumed to be equally large, which in a stochastic setting assumes that the existing populations settle to the same quasi-stationary distribution of population sizes. The Levins model assumes that on the time-scale of extinctions and recolonizations, local dynamics can be ignored. Drechsler and Wissel (1997) show that this assumption is well justified when the rate of immigration is sufficiently low in comparison with the intrinsic rate of population increase (r), and when the variance of the growth rate is less than the average growth rate minus emigration rate, $v < r - m$. Recalling that the ratio of the long-term population growth rate and its variance determines the scaling of the expected time to population extinction with the carrying capacity (Section 2.2), the above result makes good intuitive sense. When the variance (v) is small in relation to the mean ($r - m$), the size of newly-established populations quickly attains a quasi-stationary distribution, after which the extinction probability is constant and the persistence time is exponentially distributed (Mangel and Tier 1993b; Drechsler and Wissel 1997; Middleton and Nisbet 1997). The Levins model is most appropriate, as a simplified description of the dynamics, for metapopulations consisting of local populations satisfying these assumptions.

The equilibrium value of P is easily obtained by setting the right-hand side of eqn 4.1 equal to zero:

$$\hat{P} = 1 - \frac{e}{c}. \tag{4.2}$$

The dynamic behaviour of the Levins model is best illustrated by plotting the colonization and extinction rates against P (Fig. 4.1). It is apparent that the model may have at most one positive equilibrium point, the value of which increases with decreasing e/c. When the value of e/c is equal to or greater than one, the metapopulation goes extinct, $\hat{P} = 0$ (eqn 4.2). This is a fundamentally important result. Think about a patch network which has just been colonized, with only one local population in existence. P is therefore very small, and eqn 4.1 simplifies to $dP/dt = (c - e)P$, as the non-linear term cP^2 is negligibly small. Condition $e/c < 1$, or $c/e > 1$, implies that, during the lifetime of the original population, which is given by $1/e$, it has to produce at least one new population, via colonization, for the metapopulation to persist.

Eqn 4.1 can be rewritten in the form:

$$\frac{dP}{dt} = (c - e)P\left(1 - \frac{P}{1 - \frac{e}{c}}\right), \tag{4.3}$$

which has the same structure as the logistic model for a single population (parameterized as in, e.g., Berryman 1992). The difference $c - e$ thus gives the intrinsic rate of metapopulation increase, which is approached when P is small, whereas $1 - e/c$ is the equivalent of local 'carrying capacity', the stable equilibrium

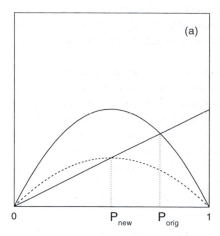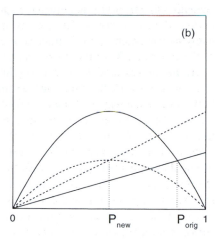

Fig. 4.1 A graphical illustration of the equilibrium in the Levins model. The colonization (parabola) and extinction rates (straight line) are plotted against P. The intersection points of the two lines are model equilibria. Panel (a) illustrates the expected consequence of patch removal, which reduces the colonization rate (modelled by reducing the value of c, broken line) and hence $\hat{P}$. Panel (b) shows the expected changes owing to a reduction in patch areas, which increases e and reduced c.

point towards which P moves in time. In the Levins model, density dependence is in the 'birth' rate (colonizations), and by definition P cannot exceed 1.

Metapopulations in dynamic patch networks

Many patch networks are dynamic in the sense that the habitat patches themselves might disappear and new ones might appear. Harrison (1994) among others has suggested that classical metapopulations might be especially common in successional habitats. Turnover of habitat patches increases the extinction rate of local populations and therefore reduces the fraction of occupied patches at equilibrium (eqn 4.2). Predator–prey metapopulations (Section 7.2) represent a particularly dynamic metapopulation situation, because not only is the patch network for the predator dynamic (the set of prey populations), but the predator–prey interaction itself may increase the rate of extinction of the prey population. Gyllenberg and Hanski (1997) present a thorough analysis of single-species metapopulation dynamics with independent patch dynamics.

Comparison with epidemiological models

The Levins model is analogous to basic epidemiological models, describing the spread of an infectious disease in a host population. Compared with a model of susceptible, infected and immune (recovered) host individuals (Anderson and May 1991, p. 122), the Levins model is simpler because vacated patches (infection lost) become immediately susceptible to recolonization (immune patches do not exist), and because susceptible (empty) patches do not disappear for reasons other than colonization (infection). If host individuals do not acquire immunity after their recovery, and if the host population remains stable, the epidemiological model (now called the SIS model) is in fact identical to the Levins model. The basic reproduction number of a parasite, R_0, is obtained from $R_0\hat{x} = 1$, where $\hat{x}$ is the fraction of host individuals susceptible at equilibrium (Anderson and May 1991, p. 17). In the Levins model, R_0 equals c/e, as the argument above revealed. As discussed above, in metapopulation models disappearance of habitat patches, comparable with infection-independent death of host individuals, is a justified addition if habitat patches may become unsuitable owing to, e.g., vegetation succession (Gyllenberg and Hanski 1997). In this case, the lost patches may be compensated for by the 'birth' of new patches elsewhere.

Instead of considering individual hosts as 'patches', one can consider groups of hosts as patches, such as villages, towns and cities in the case of human infections. Grenfell and Harwood (1997) apply this perspective to measles metapopulation dynamics. They discuss the 'critical community size', below which measles tends to die out between epidemics; this is analogous to the concept of 'minimum viable population' or 'minimum viable metapopulation size' (Section 4.5). Town area plays a key role in measles dynamics, just as the habitat patch area plays a key role in the metapopulation dynamics of free-living organisms (Chapter 5).

The epidemiological theory has a much longer pedigree than the ecological metapopulation theory (Nee *et al.* 1997). The threshold theory of Kermack and McKendrick (1927) demonstrated that an epidemic outbreak could only spread into a host population if the number of susceptible individuals exceeds a critical minimum number. Even earlier, Ross (1909) had alluded to the threshold condition from the opposite angle, while discussing the conditions for eradicating malaria by reducing the numbers of mosquitoes. This perspective has a more sinister counterpart in the metapopulation literature, in the form of the threat that habitat destruction poses to the persistence of endangered species (Sections 4.4 and 4.5). There is scope for a more penetrating comparison of epidemiological models with metapopulation models than can be accomplished here. Explicit comparisons might well reveal overlooked research opportunities in both fields (Grenfell and Harwood 1997). Recent studies of spatial dynamics in plant pathogens have successfully combined concepts, types of analysis and models from the epidemiological literature and from the metapopulation literature (Burdon *et al.* 1995; Thrall and Antonovics 1995).

Qualitative predictions for fragmented landscapes

The two parameters of the Levins model are most naturally interpreted in terms of the biology of the species, as the species' colonization ability and proneness to local extinction. But an alternative interpretation can be formulated in terms of the structure of the fragmented landscape. Think about a patch network which is modified by removing a fraction of the patches but leaving the rest unchanged. The extinction rate is not affected by this change, but the colonization rate is, as there are now fewer local populations and empty patches; the level of patch connectivity is reduced, which can be modelled by reducing the value of c. Habitat destruction thus leads to a decrease in the fraction of the remaining patches that are occupied (Fig. 4.1). Alternatively, no patches might be completely destroyed, but the patch areas might be reduced. Such a reduction is expected to increase the extinction rate, because small patches will tend to have small populations with a high risk of extinction (Section 2.2). At the same time, the colonization rate per empty patch is also reduced, because the smaller patches (populations) will create a weaker colonization pressure on empty patches. Hence reduction in patch areas is modelled by increasing e and reducing c; the net result is again a reduction in the fraction of occupied patches (Fig. 4.1). Clearly, the metapopulation is predicted to disappear before all the habitat is destroyed; the metapopulation goes extinct when the remaining patches are so far apart, and so small, that the condition c/e ($= R_0$) > 1 fails to be met. Despite its extreme simplicity, the Levins model has thus been helpful in enabling us to develop a key insight about the population dynamic consequences of habitat destruction and fragmentation. A more quantitative analysis is presented in Section 4.4.

Mainland–island metapopulations

In the Levins model, all patches are identical and the metapopulation survives in a colonization–extinction equilibrium. This is not even approximately true for all metapopulations. Another extreme type is called the mainland–island metapopulation structure (Harrison 1991), with one or more very large populations—the mainland—with a negligible risk of extinction. The remaining (island) populations run a high risk of extinction in their small habitat patches. Colonization of empty patches is now enhanced by migration from the mainland,

$$\frac{\mathrm{d}P}{\mathrm{d}t} = (c_m + cP)(1 - P) - eP, \tag{4.4}$$

where c_m is the colonization rate per empty patch from the mainland. In the extreme case, when all colonizations are due to migration from the mainland, $c = 0$ and the equilibrium value of P is given by $\hat{P} = c_m/(c_m + e)$. The species will not go permanently extinct from the network of small patches as long as there is some migration from the mainland, which guarantees the establishment of new populations however small is P. Assuming that there are R species on the mainland, each island has $R\hat{P}$ species, on average, at equilibrium. Such a multispecies version of eqn 4.4, assuming independent dynamics in the species, is the basis of MacArthur and Wilson's (1967) dynamic theory of island biogeography.

4.2 The rescue effect and alternative equilibria

In an important paper published in 1977 Brown and Kodric-Brown introduced the concept of rescue effect into the metapopulation literature. By the rescue effect, they referred to increasing population size and hence decreasing risk of extinction with increasing rate of immigration. The term 'rescue effect' is widely used in the literature but often in somewhat different meanings. I first discuss the phenomenon itself and the mechanism creating it, then describe the consequences for metapopulation dynamics.

Brown and Kodric-Brown (1977) envisioned the rescue effect in the context of the dynamic theory of island biogeography (MacArthur and Wilson 1967), which assumes the mainland–island metapopulation structure. Brown and Kodric-Brown assumed that constant migration from the mainland increases the expected size of an island population and hence reduces its risk of extinction. Migration also has the effect of reducing the apparent extinction rate between two censuses at times t and $t + \tau$. This happens when immigration leads to the establishment of a new population before time $t + \tau$ following extinction after time t. I call this the pseudo-rescue effect. The pseudo-rescue effect leads to underestimation of true population turnover rate, and more so the longer the interval τ (Diamond and May 1977; Russell *et al.* 1995).

Turning to metapopulations without a mainland, we still have both the true and the pseudo-rescue effect for individual populations, only with the difference that now immigration rate varies temporally, depending on how many habitat patches within migration range from the focal population happen to be occupied. With increasing size of the metapopulation (more occupied patches), there is more immigration to a particular patch and hence a stronger rescue effect, just as would happen if an island were moved closer to the mainland. With increasing immigration rate, the local population sizes are expected to increase, generating a positive relationship between the fraction of occupied patches and the average size of local populations. This argument ignores the negative effect of emigration on population size. But considering the consequences of migration for population extinction in real metapopulations with differences in population sizes, this does not really matter, because most migrants originate from large populations, where the losses increase only slightly the risk of extinction. In contrast, many migrants end up in small populations, where they may significantly reduce the risk of extinction. This sort of situation cannot be analysed in the context of the Levins model with equally large local populations, hence I describe below a metapopulation model with differences in population sizes.

Let us assume that the risk of local extinction is greater in small than large populations, and that the effect of emigration in increasing the risk of extinction of large populations is negligible in comparison with the positive effect of immigration in reducing the risk of extinction of small populations. Let us extend the Levins model by making a distinction between small and large local populations, though retaining the assumption that all habitat patches themselves are identical. Let E, S and L denote the fractions of patches that are empty, have a small population and have a large population, respectively ($S + L = P$ in the Levins model). The following equations describe the rates of change in these fractions (Hanski 1985):

$$\frac{dE}{dt} = e_S S - cLE$$

$$\frac{dS}{dt} = cLE + e_L L - e_S S - rS - mLS$$

$$\frac{dL}{dt} = rS + mLS - e_L L. \tag{4.5}$$

An empty patch turns to a patch with a small population after colonization from large populations (parameter c). Migration from small populations is omitted for simplicity without a qualitative effect on the results (Hastings 1991). Extinctions of small populations generate empty patches (e_S). A small population may grow large (r) and a large population may become small (e_L). The model assumes that a large population does not become extinct without first becoming a small population. Finally, and what is really important here, migration from large populations into small populations increases the rate of change of small populations to large ones (m).

Alternative equilibria

When there is no rescue effect ($m = 0$), the dynamics predicted by eqn 4.5 are qualitatively the same as the dynamics predicted by the Levins model (Fig. 4.2a). With the rescue effect ($m > 0$), a new dynamic feature emerges, because for a range of parameter values the model has three equilibria, of which two are stable and one is unstable (Fig. 4.2b, for $c = 0.11..0.20$ in this example). When this occurs, we cannot predict the size of the metapopulation without knowing the past history of perturbations, in other words without knowing on which side of the unstable equilibrium in Fig. 4.2b the metapopulation is located. Alternative equilibria occur in this model because of the effect of immigration on local dynamics—the rescue effect.

The same conclusion can be demonstrated with the more complex structured models described in the following section. In fact, eqn 4.5 represents the simplest possible metapopulation model in which local populations are structured by their size. The following is an intuitive explanation of alternative stable equilibria in these models. When the metapopulation is small, and only a small fraction of patches is occupied, a large fraction of migrants is 'wasted' in the empty patches which they failed to colonize and in the small populations that happened to go extinct. In contrast, when most patches are occupied and local populations are larger, such loss of migrants is smaller. Therefore, the per-individual growth rate in the metapopulation as a whole may be negative when the metapopulation is small, but becomes positive when the metapopulation size exceeds a threshold value. In very large metapopulations, the growth rate again becomes negative, and the size of the metapopulation settles at a positive equilibrium. The local rescue effect as originally described by Brown and Kodric-Brown (1977) turns to a metapopulation-level rescue effect—a metapopulation may perform the trick of rescuing itself.

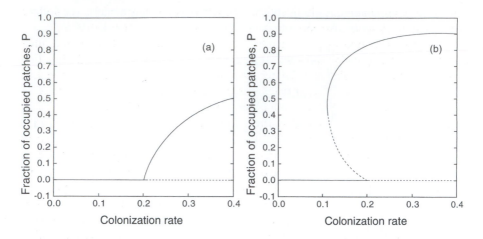

Fig. 4.2 Fraction of occupied habitat patches (small and large populations put together) against the colonization rate (c) in the simple structured model, eqn 4.5. In panel (a), there is no rescue effect ($m = 0$), in panel (b) there is strong rescue effect ($m = 0.5$). The continuous and broken lines represent stable and unstable equilibria, respectively. Other parameter values: $e_S = 1.0$, $e_L = 0.02$ and $r = 0.1$.

4.3 Structured metapopulation models

Chapter 3 described simple metapopulation models in which migration among populations greatly affected local dynamics. In the Levins model and other unstructured patch models, migration is consequential only in the establishment of new populations after local extinction. Both approaches can be defended for the study of particular questions in particular types of metapopulation. When local dynamics occur on a fast time-scale and migration rate is low, it is reasonable to ignore the effect of migration on local dynamics and to focus on extinction–colonization dynamics. In the other extreme, when the local growth rate is low in comparison with migration rate, the impact of migration on local dynamics may be substantial and should not be ignored.

Structured metapopulation models attempt to merge the two types of situation and combine high migration rate with significant population turnover. This combination might seem implausible, and related biological and technical arguments have been raised against this modelling approach. Harrison (1994) has suggested that genuine metapopulation persistence in a balance between extinctions and colonizations is an improbable condition, because it requires that migration rate is high enough to prevent metapopulation extinction, but low enough not to create a large continuous, even if not entirely panmictic, population. In reality, however, although the expected migration rate may be high, large numbers of immigrants do not necessarily arrive at every patch in every generation. And even though there is substantial immigration, newly-established populations run a high risk of immediate or virtually immediate extinction. It should also be noted that although populations in some patch networks might indeed be so large and located so close to each other that there is no population turnover at equilibrium, the same system might function as a classical metapopulation when perturbed to a small size with only a few extant populations. And, finally, many field studies have demonstrated the presence of empty habitat despite a high migration rate (Sections 9.1 and 11.5).

The related technical trouble is the following. The models discussed in this section are deterministic and formulated in continuous time, and they assume implicitly that the migrants are spread equally among the habitat patches. If there is enough migration to influence local dynamics, there should be no completely empty habitat. One solution is to turn to stochastic models. Lande *et al.* (1998) have recently analysed a stochastic metapopulation model for finite patch networks. Val *et al.* (1995) have analysed deterministic models in which all habitat is always occupied. Encouragingly, these models make the same key predictions as the models described below, which are more straightforward, structured analogues to the Levins model and which allow for empty patches via a limited rate of recolonization after extinction. The presentation here omits all mathematical steps, for which the reader should consult Gyllenberg and Hanski (1992) and Hanski and Gyllenberg (1993). For a more general mathematical framework see Gyllenberg *et al.* (1997).

A structured model

Consider first the dynamics of one local population. In the absence of migration, the rate of change in population size n is given by:

$$\frac{\mathrm{d}n}{\mathrm{d}t} = g(n), \tag{4.6}$$

where g is some function of population size, for instance the logistic function. Let the number of migrants searching for a new patch at time t be $D(t)$, and let α denote the rate at which migrants arrive at a patch. The rate of emigration, which might depend on population size, is given by $\gamma(n)$. Finally, as the local dynamics themselves have no stochastic element, we add explicitly a term giving the rate of population extinction, which might depend on population size, $\mu(n)$.

The purpose here is not to model the changes in the sizes of individual populations, but instead to model changes in the size distribution of local populations, denoted by $p(t, n)$. $p(t, n)$ is a normalized population size distribution, such that when $p(t, n)$ is integrated over all sizes n, we obtain the fraction of occupied patches $P(t)$ (Gyllenberg and Hanski 1992). Given the above assumptions, changes in $p(t, n)$ are given by the partial differential equation (for $n > 1$):

$$\frac{\partial}{\partial t} p(t, n) + \frac{\partial}{\partial n} ([\, g(n) - \gamma(n) + \alpha D(t)]\, p(t, n)) = \mu(n) p(t, n). \tag{4.7}$$

The number of migrants outside the patches increases because of emigration from the existing populations and it decreases because of immigration and death during migration. Denote the death rate by δ. The rate of change in $D(t)$ is given by:

$$\frac{\mathrm{d}D(t)}{\mathrm{d}t} = -(\alpha + \delta)D(t) + \int\limits_1^\infty \gamma(n)p(t, n)\mathrm{d}n. \tag{4.8}$$

The integral gives the sum of emigrants originating from the currently occupied patches. We also need an equation for the change in the number of occupied patches due to successful colonization. The colonization rate is given by $\beta\alpha$, where β sets the rate of successful colonization:

$$[g(1) - \gamma(1) + \alpha D(t)]p(t, 1) = \beta\alpha D(t)[1 - P(t)]. \tag{4.9}$$

Finally, we have to specify some initial size distribution for local populations, $p(0, n)$, and the initial number of migrants, $D(0) = D_0$.

Equations 4.6 to 4.9 describe a lot of biology but they also comprise a complicated system of non-linear equations. Although limited mathematical analysis is possible (Gyllenberg and Hanski 1992; Gyllenberg *et al.* 1997), it is hard to extract many useful ecological results and predictions without making some further simplifying assumptions. One general conclusion that can be drawn concerns the shape of the population size distribution. If local dynamics occur on a fast time-scale, most populations are expected to be large. The Levins model represents a limiting case in which all existing populations are of the same size. On the other hand, if local dynamics are slow, most populations are expected to be small (Gyllenberg and Hanski 1992). Hanski *et al.* (1996a) and Middleton *et al.* (1995) have described comparable stochastic models, which demonstrate that the skew in the population size-distribution is especially sensitive to the ratio of the long-term growth rate over its variance (Lande 1993; Foley 1994; Section 2.2). It is worth stressing that the model as described here assumes similar habitat patches. This assumption can be relaxed, and one can assume either fixed differences in the sizes (or quality) of habitat patches (Hanski and Gyllenberg 1993), or a dynamic landscape with changes in patch quality, including patch disappearances and appearances (Gyllenberg and Hanski 1997).

A simplified model

A more tractable model can be constructed by returning to the Levins model assumption of fast local dynamics. Viewed from the time-scale of metapopulation dynamics, a newly-established population is assumed to grow quickly to local equilibrium. The difference from the Levins model is, however, that the equilibrium population size is now affected by immigration and emigration.

As there are no differences between local population sizes in this model, the partial differential equations simplify to ordinary differential equations. Furthermore, as the equilibrium local population size $\hat{n}$ is not a dynamic variable, it can be eliminated from the model by use of the equation:

$$g(\hat{n}) - \gamma(\hat{n}) + \alpha D = 0. \tag{4.10}$$

The number of migrants is related to the fraction of occupied patches by (Gyllenberg and Hanski 1992):

$$D = \frac{\gamma(\hat{n})P}{\alpha + \delta}, \tag{4.11}$$

and it also can be eliminated as a dynamic variable. What we then have left is a single equation for P. To study a specific example, let us assume that local population growth is given by the logistic model, $rn(1 - n/K)$, and that emigration (m) and extinction (e) rates are density-independent. With these assumptions we arrive at the model (Hanski and Zhang 1993)

$$\frac{dP}{dt} = \beta\alpha'mK\left(1 - \frac{m[1 - \alpha'P]}{r}\right)P(1 - P) - eP, \tag{4.12}$$

where $\alpha' = \alpha/(\alpha + \delta)$ is defined as the fraction of migrants that survive migration and land in a habitat patch. The metapopulation persists at equilibrium when the following conditions are met (Hanski and Zhang 1993)

$$m < \frac{r - \sqrt{rB}}{1 - \alpha'} \qquad \text{if } m > \frac{r}{1 + \alpha'}$$

$$\frac{r - \sqrt{r^2 - rB/\alpha'}}{2} < m < \frac{r + \sqrt{r^2 - rB/\alpha'}}{2}, \qquad \text{if } m \leq \frac{r}{1 + \alpha'}$$

where $B = 4e/\beta K$. These conditions are illustrated in Fig. 4.3. A necessary condition for metapopulation persistence is that emigration rate m exceeds a threshold value, which gives the minimum amount of migration needed to generate sufficient recolonizations to compensate for extinctions; and that m is simultaneously below another threshold value, which is set by the requirement that local populations must grow fast enough to compensate for the losses due to migration. Figure 4.3 shows graphically how m, e, α and βK affect metapopulation persistence. Note that with increasing extinction rate the range of emigration rates compatible with meta-population persistence decreases. The equilibrium value of P is given by:

$$\hat{P} = 0.5 - \frac{r - m \pm \sqrt{A^2 - rB}}{2\alpha'm}, \tag{4.13}$$

where $A = r + m(\alpha' - 1)$. The equilibrium size of local populations is a function of $\hat{P}$:

$$\hat{n} = K\left(1 - \frac{m}{r}[1 - \alpha'\hat{P}]\right). \tag{4.14}$$

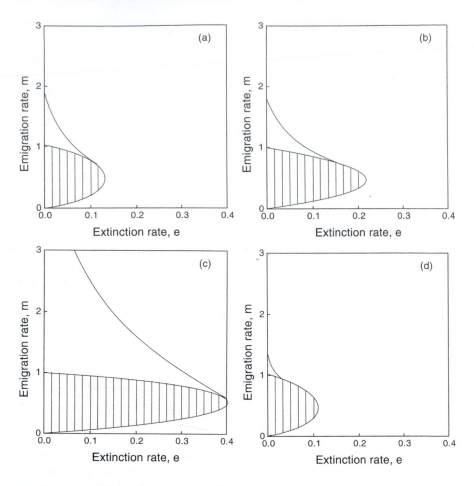

Fig. 4.3 Combinations of emigration rate (m) and extinction rate (e) that enable metapopulation persistence. There is one stable positive equilibrium point in the shaded region, and two alternative stable equilibria above the shaded region but below the upper line, one of which is the positive equilibrium. Parameter values: $r = 1$, (a) $\alpha' = 0.5$, $\beta K = 1$; (b) $\alpha' = 0.5$, $\beta K = 2$; (c) $\alpha' = 0.8$, $\beta K = 2$; and (d) $\alpha' = 0.2$, $\beta K = 2$.

Figure 4.4 illustrates how the equilibrium patch occupancy ($\hat{P}$) and local abundance ($\hat{n}$) depend on emigration rate: the latter always decreases, but the fraction of occupied patches first increases, then declines, with increasing m.

For a range of parameter values, the model predicts alternative stable equilibria, one of which corresponds to metapopulation extinction (Figs 4.3 and 4.4). Alternative equilibria are most likely when the impact of immigration on local dynamics is great, which occurs when m/r and α' are large. However, this result should be interpreted with some caution, as the model assumes distinct time-scales

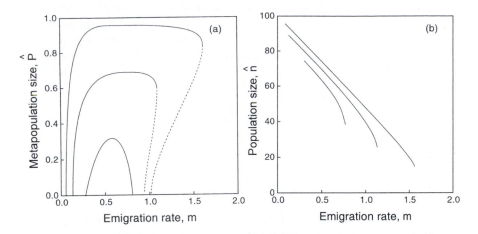

Fig. 4.4   Effect of emigration rate (m) on metapopulation size ($\hat{P}$, fraction of patches occupied) and the size of local populations ($\hat{n}$) at equilibrium. The three lines are for three different values of the extinction rate, $e = 0.1$, 0.05 and 0.01 (from down to top). Other parameter values: $r = 1$, $\alpha' = 0.5$ and $\beta K = 1$. The dashed lines in panel (a) represent unstable equilibria.

of local and metapopulation dynamics, and hence that local populations 'jump' to the local equilibrium population size $\hat{n}$ immediately after colonization. Alternative stable equilibria are more probable in this model, like in the simple structured model in the previous section (eqn 4.5), when extinction risk decreases with increasing population size (Hanski and Zhang 1993), allowing migration from large populations to enhance the survival of small populations—the rescue effect. This is not a severe restriction, as extinction risk practically always decreases with increasing population size in real metapopulations (Section 2.2). Extra complexity occurs if we assume differences in the areas of the habitat patches, which are the rule in real metapopulations. Large patches enhance metapopulation persistence and, when combined with the rescue effect, might generate up to four equilibria, including two positive stable equilibria (Hanski and Gyllenberg 1993).

4.4 Habitat destruction and metapopulation extinction

The Levins model contributes a critical insight into the mechanisms of species extinction in fragmented landscapes. With increasing fragmentation, the density of habitat patches decreases and hence the rate of establishment of new populations decreases. Eventually, and before all suitable habitat has been destroyed, the colonization rate drops below the threshold level necessary to compensate for extinctions and the metapopulation goes extinct (Fig. 4.1).

May (1991), Nee and May (1992), Lawton *et al.* (1994) and Nee (1994) have modified the Levins model by parameterizing the process of habitat destruction (following Lande 1987). Assume that fraction $1 - h$ of the patches is permanently destroyed. The colonization rate is reduced because the density of empty but suitable patches available for colonization is reduced from $1 - P$ to $h - P$, and we obtain the model:

$$\frac{dP_{tot}}{dt} = cP_{tot}(h - P_{tot}) - eP_{tot}. \tag{4.15}$$

In this variant of the Levins model, the fraction of occupied patches, P_{tot}, is calculated as the fraction of all patches, including those that have been destroyed. This is related to the P in the original model as $P = P_{tot}/h$. Substituting P for P_{tot} in eqn 4.15 gives eqn 4.1 with the colonization parameter ch, demonstrating that we have just two ways of parameterizing the same model, but now making clear the assumption as to how habitat destruction affects colonization rate.

At equilibrium, the fraction of empty patches out of all patches is given by:

$$h - \hat{P}_{tot} = \frac{e}{c}. \tag{4.16}$$

Thus the fraction of empty patches remains constant as long as the metapopulation does not go extinct, which happens when $h < e/c$ (Fig. 4.5). By multiplying the fraction of empty patches by the original number of suitable patches, this result can also be expressed in terms of the number of empty but suitable patches, which is predicted to remain constant. Hanski *et al.* (1996b) referred to this result as the

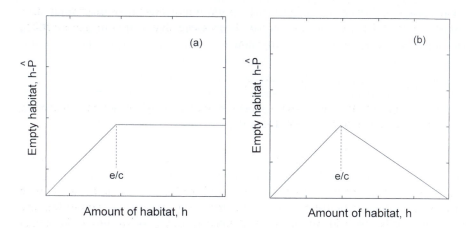

Fig. 4.5 (a) A schematic illustration of the Levins rule, eqn 4.16, which gives the relationship between the fraction of empty patches at equilibrium ($h - \hat{P}_{tot}$) and the fraction of suitable patches (h). (b) The analogous result with the rescue effect ($\omega > 0$), eqn 4.18. Note that in this case the fraction of empty patches is not independent of h.

Levins rule: 'A sufficient condition for metapopulation survival is that the remaining number of habitat patches following a reduction in patch number exceeds the number of empty but suitable patches prior to patch destruction.' The Levins rule is attractive because it gives an estimate of the minimum amount of suitable habitat necessary for long-term metapopulation persistence from the very limited information of the number of empty patches in a landscape in which the metapopulation survives; no detailed knowledge of local nor metapopulation dynamics is required (Nee 1994).

The Levins rule has substantial heuristic value, but it should not be used as a management tool, because in practice the minimum amount of suitable habitat might be greatly affected by factors not included in the simple model (eqn 4.15). Firstly, the Levins rule is based on a deterministic model assuming a large number of patches. In real patch networks with often a small number of patches, stochasticity will increase the probability of metapopulation extinction (Section 4.5). Secondly, the Levins rule assumes homogeneous mixing of local populations among the suitable patches (Lawton *et al.* 1994), which is often violated due to a clumped distribution of patches. The probability of colonization of a particular patch then depends strongly on which particular other patches are occupied. I return to this question in Section 5.1 while discussing spatially explicit metapopulation models. Thirdly, the Levins rule assumes that the metapopulation has reached an equilibrium, which is a dubious assumption for metapopulations in recently fragmented landscapes (Section 12.5). For the metapopulation to be at equilibrium requires close tracking of a changing landscape, which is possible only when the environment changes slowly or the population turnover rate is high. The latter in turn implies high immigration rate, but if immigration rate is high, it becomes hard to defend the Levins model assumption that metapopulation dynamics occur on a slow time-scale in relation to the time-scale of local dynamics (Section 4.1). This leads to the fourth reason for distrusting strict interpretation of the Levins rule—it is based on a model that ignores the rescue effect, the impact of immigration on local dynamics.

Habitat destruction with rescue effect

The following variant of the Levins model includes, though in a non-mechanistic manner, the rescue effect (Hanski 1983a),

$$\frac{dP_{tot}}{dt} = cP_{tot}(h - P_{tot}) - e(1 - \omega P_{tot})P_{tot}, \tag{4.17}$$

where ω is an extra parameter setting the rate of decline of the extinction rate with increasing P_{tot} (the justification for this assumption is that average population size tends to increase with increasing P_{tot}). If there is no rescue ($\omega = 0$), the model reduces to eqn 4.15. With the rescue effect, eqn 4.16 is replaced by:

$$h - \hat{P}_{tot} = \frac{(e/c)}{1 - \omega \frac{e}{c}}[1 - \omega h]. \tag{4.18}$$

The fraction of empty patches is no longer independent of h but it increases with decreasing h (Fig. 4.5). As in the Levins model, the metapopulation goes extinct if $h < e/c$, but here the fraction (or number) of empty patches at equilibrium cannot be used to estimate the amount of habitat necessary for metapopulation survival.

In this model migration occurs among populations in the same metapopulation. Immigration from outside the metapopulation, for instance from a permanent mainland population, would further reduce the fraction of empty patches and would lead to an underestimate of the amount of suitable habitat necessary for long-term persistence if migration from the mainland would be cut off (Pagel and Payne 1996). Naturally, as long as there is some migration from outside, the metapopulation will not go permanently extinct regardless of the amount of habitat destruction.

Other factors

Other factors apart from the rescue effect are likely to affect the relationship between the fraction of empty patches and the amount of suitable habitat. For instance, let us relax the Levins model assumption that all patches are equally large or have the same quality. Keeping the patch size or quality distribution fixed, the fraction of empty patches at equilibrium, $\hat{E}$, remains constant when h changes, provided that e (extinction risk) is the same in all occupied patches (Gyllenberg and Hanski 1997). In this case, the equilibrium density of empty patches is independent of patch quality. If, however, c and e, are, respectively, increasing and decreasing functions of patch quality, which are plausible assumptions, the overall fraction of empty patches will increase with increasing h (Gyllenberg and Hanski 1997). If habitat destruction is associated with a decrease in the average quality of patches, as might often be the case, the above trend can be reversed, and $\hat{E}$ might now increase with decreasing h (Gyllenberg and Hanski 1997). The quality of the empty patches, which represent the 'resource' limiting metapopulation growth (Nee 1994), plays a decisive role here. Gyllenberg and Hanski (1997) formulated the following general principle: '$d\hat{E}/dh > 0$ if and only if the value of an average (randomly chosen) empty patch is lower than the value of an average empty patch after habitat destruction.' The 'value' of a patch is here measured by

$$\int_{(0,\,\infty)} \frac{c(x)}{e(x)} \psi(dx),$$

where x is a measure of the intrinsic patch quality, c and e are the colonization and extinction rates, and ψ is the patch quality distribution (Gyllenberg and Hanski 1997).

The effect of habitat subdivision

A fundamental question about persistence of species in fragmented environments concerns the effect of habitat subdivision on extinction risk. In other words, given a fixed amount of suitable habitat, is time to metapopulation extinction longer, or shorter, when the habitat is subdivided into a larger number of smaller pieces? Quinn and Hastings (1987) showed with a model without recolonization that subdivision is likely to increase persistence time in networks of independent habitat patches if extinctions are due to environmental stochasticity. In this case, even large populations have a substantial risk of extinction, and spreading the extinction risk among several, albeit smaller, populations tends to increase the overall persistence time. The opposite conclusion emerges when extinctions are caused by demographic stochasticity, which poses a serious threat to small populations only; now subdivision is detrimental. In qualitative terms these conclusions extend to metapopulations with recolonization.

Several theoretical studies have subsequently identified a number of factors, apart from the relationship between local extinction risk and patch area, which affect the likely outcome of habitat subdivision. These factors include spatially correlated local dynamics (Gilpin 1988; Hanski 1989; Harrison and Quinn 1990) and the effect of connectivity on recolonization (Burkey 1989; Bascompte and Solé 1996; Section 10.2). Furthermore, whether habitat subdivision increases or decreases the probability of extinction during some time interval T might well depend on the value of T itself (Quinn and Hastings 1987). In summary, no universally valid answer exists to the question about habitat subdivision and (meta)population persistence, except perhaps that, as concluded by Fahrig (1997), in most realistic scenarios the decisive factor is not subdivision but the pooled area of suitable habitat. Still, it would be wrong to conclude that the spatial configuration of suitable habitat makes no difference at all (Fahrig's (1997) individual-based model is likely to underestimate the extinction rate of small populations and hence underestimate the adverse effect of habitat fragmentation). We need to ask more specific questions than the general question about habitat subdivision. The most important reason for developing spatially realistic metapopulation models (Chapter 5) is to develop a flexible modelling framework in which to predict the consequences of habitat loss and fragmentation for particular species living in particular fragmented landscapes.

4.5 Minimum viable metapopulation size

Numerous empirical studies have confirmed the prediction (Section 2.2) that the expected lifetime of a population increases with its current size (Section 8.2). Provided that the environment does not change greatly (unrealistic as this assumption might be), very large populations are expected to last for so long that no conservation measures are called for, whereas very small populations are likely to go rapidly extinct. The minimum viable population (MVP) size is intended to be an estimate of the minimum number of individuals in a population which has a good chance of surviving for some relatively long period of time, for instance 95% chance of surviving for at least 100 years (Soulé 1980, 1987; Lande 1988b). In the case of metapopulations consisting of small and hence extinction-prone local populations, an analogous concept of minimum viable metapopulation (MVM) size may be defined as the minimum number of interacting local populations necessary for long-term persistence. Additionally, one has to consider the minimum amount of suitable habitat (MASH) necessary for metapopulation persistence, discussed in the previous section, because not all suitable habitat is simultaneously occupied at the stochastic equilibrium. In the previous section, the consequences of habitat loss were modelled using simple deterministic models. An obvious shortcoming of this approach is that when only a few local populations remain the metapopulation might go extinct by chance, due to extinction–colonization stochasticity (Hanski 1991a), which is analogous to demographic stochasticity in local populations (May 1973).

Gurney and Nisbet (1978) and Nisbet and Gurney (1982) have analysed a stochastic version of the Levins model with a finite number of habitat patches and local populations (see also the analysis in Frank and Wissel 1998). Assume that there are H identical habitat patches, of which Q are occupied and E are empty. The probability of successful colonization in an infinitesimally small time interval is proportional to the product QE, as in the Levins model, and the corresponding probability of extinction is proportional to Q. This model yields the following approximation for the expected time to metapopulation extinction, T_M (Gurney and Nisbet 1978):

$$T_M = T_L \exp\left[\frac{H\hat{P}^2}{2(1 - \hat{P})} \right],$$
(4.19)

where T_L is the expected time to local extinction and $\hat{P}$ is the fraction of occupied patches at a stochastic steady state ($=Q/H$). If one defines long-term metapopulation persistence as T_M exceeding 100 times T_L, eqn 4.19 leads to the following condition for reasonably large H (Gurney and Nisbet 1978):

$$\hat{P}\sqrt{H} \geq 3.$$
(4.20)

For example, if there are 50 habitat patches, eqn 4.20 says that the colonization and extinction rates must be such that $\hat{P} > 0.42$ for the metapopulation to persist for longer than 100 times T_L.

These results were derived from a stochastic Levins model with identical habitat patches, no spatial structure and no rescue effect. Hanski *et al.* (1996b) have explored numerically the validity of eqn 4.20 in a spatially realistic metapopulation model, the incidence function model (Section 5.3), in which these restrictions can be relaxed. The incidence function model relates the stationary probability of patch occupancy to patch size and isolation from the existing local populations. Let us first consider a hypothetical network of 100 equally large patches located randomly within a square area. Colonization probability is modelled either as an exponential or sigmoid function of the expected number of migrants arriving at a patch. The former assumption corresponds more closely with the assumptions of the Levins model but the latter is often more realistic (Hanski 1994a). The model includes the rescue effect and distance-dependent migration (Hanski *et al.* 1996b).

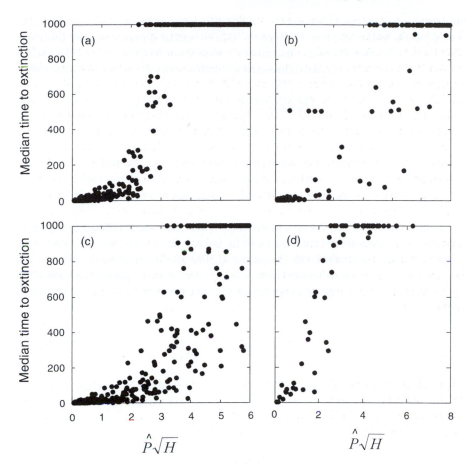

Fig. 4.6 Relationship between median time to metapopulation extinction, T_M, and the product $\hat{P}\sqrt{H}$ in simulations of the incidence function model (the median $T_L = 3.3$ in these examples; details in the text and in Hanski *et al.* 1996b). Panel (a) is for a hypothetical network of equally large patches, assuming exponentially increasing probability of colonization with the number of immigrants arriving at an empty patch and distance-dependent migration. (b) As (a) but with sigmoid colonization probability (eqn 5.12). (c) With spatially correlated environmental stochasticity. (d) Results for real patch networks inhabited by the Glanville fritillary butterfly. (Based on Hanski *et al.* 1996b.)

The results in Fig. 4.6a suggest that eqn 4.20 is a reasonably good approximation for exponential colonization probability. With sigmoid colonization probability there is more variation in the results (Fig. 4.6b), apparently because now there is more variation in the fraction of occupied patches at stochastic steady state (Hanski *et al.* 1996b). Variance in P is greatly increased by adding regional stochasticity (spatially correlated environmental stochasticity) to the model, which also substantially shortens metapopulation lifetime (Fig. 4.6c). The latter can be understood as a result of reduction in the 'degrees of freedom' due to spatial correlation. The following quantity can be thought of as the equivalent number of independent habitat patches among the total of H patches influenced by regional stochasticity (Chesson 1998):

$$H_{\text{effective}} = \frac{H^2}{\sum\limits_{j=1}^{H} \sum\limits_{i=1}^{H} \rho(j, i)},$$

where $\rho(j,i)$ is the Pearson correlation in the stochasticity affecting patches j and i, assumed to decrease to zero as the distance between patches j and i increases. Clearly, if the dynamics are completely independent, $H_{\text{effective}} = H$, otherwise $H_{\text{effective}} < H$.

In a second set of simulations, we used the incidence function model to iterate the dynamics of the Glanville fritillary butterfly in real patch networks using parameter values estimated with field data (Hanski *et al.* 1996b). To examine the effect of the number of patches on metapopulation survival, we created smaller patch networks by deleting patches from the periphery of the original network, thus keeping patch density relatively constant. In this example, with large variance in patch areas (Hanski *et al.* 1996b), metapopulations tended to survive for a long time when, roughly, $\hat{P} \sqrt{H} > 2$ (Fig. 4.6d). The reason for the enhanced survival in comparison with the other examples in Fig. 4.6 seems to be variance in patch areas and hence in local population sizes: metapopulation lifetime is increased by the low risk of extinction of the largest local populations (annual extinction probability was ≈ 0.3 for median-sized patches compared with <0.01 for the largest ones).

It should be noted that the attributes of the species and the patch network, as reflected in the value of $\hat{P}$, are combined with the size of the network (H) in eqn 4.20. Rearranging eqn 4.20 as $\hat{P}H = \hat{Q} > 3\sqrt{H}$ suggests that MVM is given, in this model, by $3\sqrt{H}$. Because MVM is a function of H, MVM and MASH cannot be evaluated independently. Somewhat paradoxically, the more habitat patches there are in the network (H), the greater must be the expected number of extant populations ($\hat{Q}$) for the metapopulation to persist. The minimum number of local populations for long-term metapopulation persistence, assuming a species which is expected to occupy all the patches, is roughly 10, though the approximation (eqn 4.20) becomes poor when H becomes small. The point to appreciate is that even a large number of habitat patches is not sufficient for long-term metapopulation persistence if these patches are thinly spread across a large area, so thinly that the expected fraction of occupied patches is less than $3H^{-1/2}$.

5

Spatially explicit approaches

The patch models described in the previous chapter are based on the assumption that all populations are equally connected—called the mean-field approximation—hence these models deal with space only implicitly. With patch models, we can explore the metapopulation dynamic consequences of reduced average patch density, but not, for instance, the consequences of habitat destruction in any specific part of a hypothetical or real landscape. This chapter is concerned with modelling approaches, commonly referred to as spatially explicit models, which account for the actual spatial locations of populations. I have made a distinction between spatially explicit and spatially realistic models (Hanski 1994b), the latter including information about the actual geometry of fragmented landscapes and hence allowing modelling of real metapopulations.

Spatially explicit and spatially realistic approaches are used when it seems that the spatial locations of individuals in populations or local populations in metapopulations critically underpin the phenomena of interest. For instance, if one is interested in spatial pattern-formation due to population dynamic processes one cannot ignore spatial locations (but see e.g. Levin and Pacala 1997). Examples discussed in this book include the neighbourhood competition model and diffusion–reaction models (Section 1.2), cellular automata and related models applied to the study of habitat fragmentation (Section 5.1), and coupled-map lattice models of predator–prey dynamics (Section 7.3). In the more applied context, when models are used to study real meta-populations in particular fragmented landscapes, a spatially realistic model is needed, especially if the aim is to make quantitative predictions. Apart from Section 5.1 on cellular automata-type spatially explicit models, this chapter deals with spatially realistic models. This emphasis reflects my aim in this book to keep the gap between modelling and empirical studies as narrow as possible.

Four kinds of spatially realistic model are discussed: a simple spatially realistic extension of the Levins model (Section 5.2), the incidence function model (Section 5.3), state transition models (Section 5.4), and n-population simulation models (Section 5.5). The incidence function model is given much space because it represents my own contribution to the literature (Hanski 1994a) and will be used extensively in Part III of this book. I also expect that the incidence function model has potential applications in spatially extended population viability analyses (Section 10.6). The state transition models represent a related empirically-based modelling approach. The n-population simulation models are straightforward extensions of the two-population models discussed in Section 3.1. These models appeal to many field ecologists, because one may incorporate any amount of biological detail in simulations. Finding the most useful balance between generality and biological realism in model structure is not an easy task, but when the purpose of the modelling is to make accurate predictions, and given the constraint of a fixed amount of information about the metapopulation to be modelled, at least we know how to evaluate the performance of different models. In practice, unfortunately, the problem often is that there are not enough data to test model predictions rigorously. An alternative approach that has not yet been sufficiently explored is to generate simulated data with some sufficiently complex model, to obtain a realistic sample from the simulated data, and to arrange contests among competing models based on such data.

5.1 Cellular landscapes

A general and computationally attractive spatially explicit approach to spatial dynamics depicts a landscape divided into a lattice, with each of the identical lattice cells having four nearest neighbours. The cells have two or more possible states, one of which is an empty cell (no population). An ecologist can readily devise a set of reasonable rules that could be used to determine the change of the state of a cell. Below I describe one example. In the mathematical literature, these models are referred to as interacting particle system models, and much of the relevant literature for ecologists has recently been reviewed by Durrett and Levin (1994). Models with deterministic rules for the change in the state of cells are known as cellular automata, but in the ecological literature this term is commonly used also for models with stochastic rules.

To appreciate the most general dynamic properties of a basic lattice model, consider the following model from Durrett and Levin (1994). Cells are classified either as occupied or empty, that is, they have two possible states like the patches in the Levins model. In each time step there is a probability e that an occupied cell becomes empty. If cell x survives, it sends out a propagule to cell y with probability $\gamma(y - x)$, which typically depends on the distance between x and y. Cell y will be occupied at the next time step if it did not become empty or if one or more propagules was sent to y. As a specific example, let us assume that $\gamma(y - x) = c$ if y is one of the four nearest neighbours of x, otherwise $\gamma(y - x) = 0$. As one could intuitively expect, for any given value of c there is a critical value of e, say $e_{crit}(c)$, such that if $e > e_{crit}(c)$ the metapopulation goes certainly extinct, but if $e < e_{crit}(c)$ it may avoid extinction. If we change the model assumptions and assume that propagules are dispersed over longer distances, up to R neighbouring cells, then the metapopulation persists if (Durrett and Levin 1994):

$$(1 - e)Rc > e. \tag{5.1}$$

The left-hand side of the inequality gives the mean number of offspring produced by a surviving cell, which must be greater than the probability of death for the metapopulation to persist (note that a cell survives with probability $1 - e$, and a propagule is sent to R cells with probability c). Inequality 5.1 gives an analogue of the threshold condition for metapopulation persistence in the Levins model (Section 4.1).

Spatial dynamics of rare species

One of the types of rare species in the classification of Rabinowitz *et al.* (1986) is a species which has a restricted distribution but which might be abundant where it occurs. In other words, the species only occurs in a small fraction of the suitable habitat. In the Levins model, a species fits this description if the extinction rate is high and the colonization rate is low, but not so low as to lead to metapopulation extinction. However, the Levins model and other spatially implicit models imply a random pattern of spatial distribution. Below I describe a lattice model with assumptions motivated by ecological considerations, and use the model to explore the spatial dynamics of rare species.

Let us assume that each occupied cell produces n 'offspring', where n is Poisson-distributed with mean λ. The offspring perform independently a random walk from the natal cell, settle into a cell with probability P and continue the random walk with probability $1 - P$. If the cell in which an offspring settles is empty the cell becomes occupied, whereas an already occupied cell just remains occupied. A fraction μ of all cells is hit by a disturbance in each time step, causing the extinction of the corresponding population. The disturbances are clumped in space in the following manner. Randomly select the first cell to destroy (it may be occupied or not). Then select d other cells, and destroy the one which is closest to a cell that has already been destroyed. Continue until fraction μ of the cells has been destroyed. If $d = 1$ destruction is random; with larger values of d destruction is increasingly clumped. Following the destruction of the cell population, the cell itself remains suitable for recolonization, hence there is no habitat loss in this model, only destruction of populations.

Let us consider four contrasting situations, for which the model was iterated on a 256 by 256 lattice. All four parameter combinations generate a rare species in the sense that, at steady state, when the model has been run for a long time, only *ca.* 10% of the cells are simultaneously occupied. The four cases represent species with short-distance versus long-distance migration (large versus small P) and spatially random versus spatially aggregated disturbance (small versus large d). The spatial patterns are described by three scale-dependent measures, which are plotted in Fig. 5.1 against the size of lattice subdivisions, called sites below, starting from single cells and including sites of 2^n cells up to $n = 8$ (2^8 is the entire lattice, 65 536 cells). The three measures are 'distribution', which gives the fraction of occupied sites, that is, sites with at least one occupied cell; 'abundance', which is the fraction of occupied cells at occupied sites; and 'vagility', which is the standard deviation of abundance at particular sites during the model simulation, and is thus a measure of the permanence of site occupancy.

The point of this exercise is that rare species can exhibit very different spatial patterns and dynamics. The extreme of long-distance migration and random disturbance is an analogue of the spatially implicit Levins model. In this case, most larger sites are occupied, though abundance is low: few occupied cells per occupied site (Fig. 5.1). Vagility is also low at larger sites, as most sites are occupied most of the time. At the other extreme are species with short-range migration and aggregated disturbance. For these only a small fraction of even large sites is occupied, for instance only 25% of the sites with an area of 1024 cells are occupied in Fig. 5.1. In

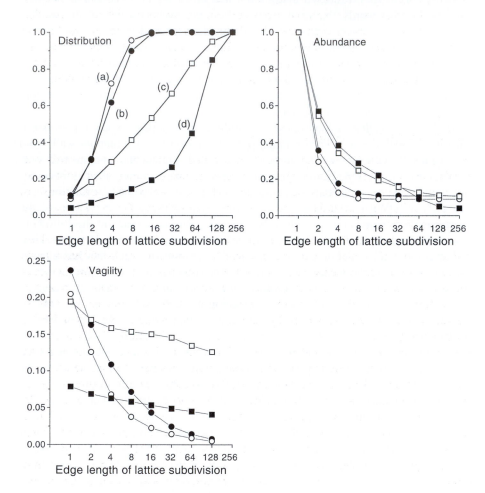

Fig. 5.1 Scale-dependent distribution, abundance and vagility in the lattice model described in the text. Parameter values: (a; long-distance migration, random disturbance), $\lambda = 0.8$, $P = 0.01$, $d = 1$; (b; short-distance migration, random disturbance), $\lambda = 1.0$, $P = 0.45$, $d = 1$; (c; long-distance migration, aggregated disturbance), $\lambda = 1.2$, $P = 0.01$, $d = 21$; and (d; short-distance migration, aggregated disturbance) $\lambda = 1.6$, $P = 0.45$, $d = 21$.

these species, abundance is relatively high and vagility is also high. Such species show dense population aggregates that slowly drift in space. The species is absent from a particular part of the lattice for a long time, but might later make a temporary appearance in high abundance. Note that the pattern of distribution, and especially the pattern of abundance, is largely determined by the scale of disturbance, less by the range of migration.

Spatial dynamics that seemingly agree with the predicted 'nomadic' spatial behaviour have been detected in plant species in species-rich tropical forests (Kwan and Whitmore 1971; Condit 1995, 1996) and in limestone grasslands (van der Maarel and Sykes 1993). Pest outbreaks and other disturbances might be spatially correlated, and the resulting clumped distributions tend to persist because of limited dispersal (Condit 1996). Communities including such species never reach an equilibrium locally, a viewpoint advocated by Hubbell and coworkers for tropical trees though based on a somewhat different model (Hubbell and Foster 1986).

Habitat destruction revisited

In the previous model there was spatially aggregated destruction of populations but no destruction of the habitat itself. Let us now return to the question of habitat destruction, which was discussed in Section 4.4 in the context of spatially implicit patch models. What difference does it make to keep track of exactly where the habitat is destroyed?

The simplest spatially explicit scenario of habitat destruction involves random removal of suitable habitat and does not lead to conclusions really different from those based on the spatially implicit models. Assume that a randomly selected fraction $1 - h$ of lattice cells is destroyed, while fraction h remains suitable for occupation. $1 - h$ thus represents habitat loss, while the degree of habitat fragmentation can be characterized by various indices (Gustafson and Parker 1992), for instance by the size of the largest continuous patch divided by the total area of suitable habitat (Bascompte and Solé 1996). The value of this parameter equals 1 as long as so few cells are destroyed that the remaining ones are all connected to each other. With increasing habitat destruction, the suitable habitat becomes rather suddenly fragmented into isolated patches, at a point when around 40% of the cells have been destroyed (the percolation theory gives the exact value of 0.4072; Schroeder 1991). At this point, the ratio of the largest patch to pooled habitat area plummets towards a small value (Fig. 5.2). The degree of connectivity itself does not, however, accurately portray the response of a metapopulation to habitat destruction, as there is no metapopulation dynamic component in this 'landscape model'. Nonetheless, making the realistic assumptions that migration across non-habitat is limited and that small populations are prone to go extinct, the highly non-linear change in habitat connectivity is likely to lead to metapopulation extinction at some critical level of habitat destruction, as also predicted by the Levins rule (Section 4.4).

In the real world, habitat destruction seldom occurs in the random manner depicted in Fig. 5.2—it is likely to occur non-randomly. It is not difficult to see that the spatial pattern of habitat loss can make a big difference, and, therefore, that the Levins rule (Section 4.4) cannot be generally used to assess the consequences for

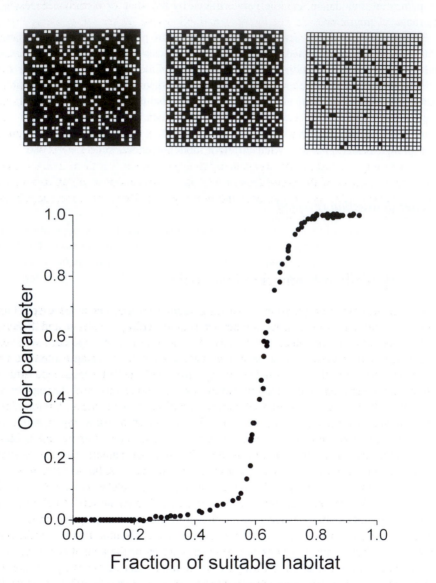

Fig. 5.2 Habitat loss and fragmentation caused by randomly removing cells from a regular lattice. The three upper panels show three levels of habitat destruction (increases to the left). The lower panel gives the ratio of the size of the largest connected patch over the total area of suitable habitat against the fraction of suitable habitat. (Based on Bascompte and Solé 1996.)

metapopulation persistence. If habitat is removed in large chunks, the populations within the remaining (large) patches of suitable habitat might remain largely unaffected (McCarthy *et al.* 1997). If habitat loss is strongly aggregated and leaves a well-connected network of fragments, the metapopulation might survive for a long time, as long as the remaining network is so large that extinction–colonization dynamics (Section 4.5) do not increase extinction risk. Other things being equal, non-random loss of habitat is generally less detrimental to metapopulation persistence than random loss (Dytham 1995; Moilanen and Hanski 1995), though if destruction leaves many completely isolated small fragments, it is possible that metapopulation persistence is harmed more than by more random loss of habitat (Bascompte and Solé 1996). Alternatively, habitat may be removed by reducing the size of each patch without reducing their number, corresponding to an increase in e and a decrease in c in the Levins model (Section 4.1). In this case, the fraction of empty patches increases with habitat destruction (Fig. 4.1), and the Levins rule is not valid. In reality, habitat destruction is likely to involve several of these elements. For accurate prediction of the consequences of habitat destruction one has to take the spatial pattern of destruction into account by using spatially realistic metapopulation models, to which we now turn.

5.2 A spatially realistic simple model

Lattice models are appropriate for studying general questions about metapopulation dynamics and for more general studies of spatial ecology (Tilman and Kareiva 1997), but they are not necessarily ideal for modelling metapopulations in real fragmented landscapes, especially if the landscape is highly fragmented. If the habitat patches are relatively small in comparison with the total landscape area, it might be sufficient and more economical to treat the patches themselves rather than grid cells which they consist of as the basic spatial units. To achieve that, consider the following changes to a lattice model. First, instead of assuming a lattice of identical contiguous cells, let the cells be scattered in the midst of unsuitable habitat, as would happen if a large fraction of the cells were permanently eradicated (Fig. 5.3a). Second, let the cells, which we now call habitat patches, be of different sizes, and let us draw circles instead of squares to represent the patches (Fig. 5.3b). Third, the rule of colonization is changed; an occupied patch now inflicts a 'colonization pressure' on all empty patches, with the pressure declining exponentially or otherwise with distance (Fig. 5.3c). And finally, let us assume that the extinction probability of an existing local population is a function of the area of the respective patch. Patch areas and isolations thus play the fundamental role of setting the rates of extinction and colonization, which are reflected in the pattern of patch occupancy in relation to patch area and isolation (Fig. 5.3d).

With these assumptions we can construct a simple variant of the Levins model for a finite number of R patches with known areas and spatial locations. Following

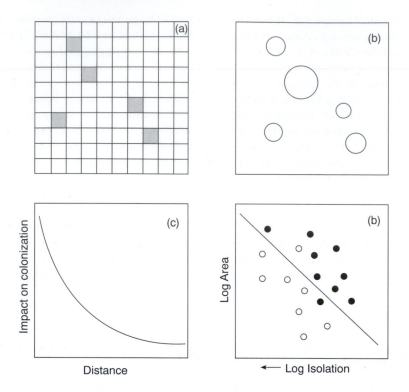

Fig. 5.3 The four steps by which an interacting particle system model is transformed into a spatially realistic patch model. (a) Only a fraction of lattice cells are suitable for occupancy. (b) These cells, now called habitat patches, may be of different sizes, and are here drawn as circles instead of squares. (c) Each occupied patch imposes an exponentially declining 'colonization pressure' on empty patches. (d) Information on the effects of patch area and isolation on occupancy can be used to parameterize spatially realistic models (filled dots are occupied patches, open dots are empty patches).

Hanski and Gyllenberg (1997; Section 7.5), denote by p_i the probability that patch i is occupied. The rate of change in p_i is given by:

$$\frac{\mathrm{d}p_i}{\mathrm{d}t} = C_i(t)[1 - p_i] - e_i p_i, \tag{5.2}$$

where $C_i(t)$ sets the colonization rate and e_i sets the extinction rate in patch i. Below, I assume that $e_i = 1/A_i$, where A_i is the area of patch i, on the assumption that large patches tend to have large populations with small risk of extinction (Section 2.2). The equilibrium value of p_i is then given by:

$$\hat{p}_i = \frac{\hat{C}_i A_i}{\hat{C}_i A_i + 1}, \tag{5.3}$$

where $\hat{C}_i$ is the equilibrium value of $C_i(t)$. Making the assumption that the contributions of the existing populations to the colonization of patch i depend on their sizes and distances from patch i, we obtain:

$$C_i(t) = c \sum_{j \neq i}^{R} \exp[-\alpha d_{ij}] p_j(t) A_j, \qquad (5.4)$$

where c is a colonization parameter, α sets the migration range of the species, and d_{ij} is the distance between patches i and j. We are interested in the equilibrium values $\hat{p}_i$, which can be calculated numerically for the set of R patches with known areas and spatial coordinates (needed to calculate the d_{ij} values), by substituting eqn 5.3 into eqn 5.4, finding the equilibrium values $\hat{C}_i$ by iteration, and finally substituting these values into eqn 5.3. The equilibrium fraction of occupied patches, $\hat{P}$, is the sum of the equilibrium probabilities $\hat{p}_i$. Note what is different here from the Levins model: a finite number of patches, patch-specific colonization and extinction rates, and hence patch-specific long-term probability of occupancy.

Neighbourhood habitat area

There is a simple and practical way of characterizing the degree of habitat fragmentation in the context of the model expressed by eqn 5.2. I first define the connectivity of patch i as:

$$\Gamma_i = \sum_{j \neq i}^{R} \exp(-\alpha d_{ij}) A_j. \qquad (5.5)$$

Note that the value measured does not depend on the occurrence of the species in the landscape, it reflects the properties of the landscape, with the necessary caveat that the spatial scale, as reflected by the value of α, must be selected to be appropriate for particular species. The neighbourhood habitat area for the entire patch network is then defined as:

$$H_n = \frac{\Sigma\Sigma\exp(-\alpha d_{ij})A_i A_j}{\Sigma A_i} = \frac{\Sigma(A_i^2 + A_i\Gamma_i)}{\Sigma A_i}. \qquad (5.6)$$

The rationale for this measure is that the pooled habitat 'availability' in pairs of patches increases as the areas of the patches increase and with decreasing pair-wise distance. The double sum for the different pairs of patches is divided by the pooled habitat area to express H_n in units of area. After some calculation we obtain:

$$H_n = \bar{A} + \bar{\Gamma} + \frac{\text{Var}(A)}{\bar{A}} + \frac{\text{Cov}(A, \Gamma)}{\bar{A}}. \qquad (5.7)$$

The neighbourhood habitat area of a patch network thus increases with increasing average patch area, with increasing average connectivity, with increasing variance of

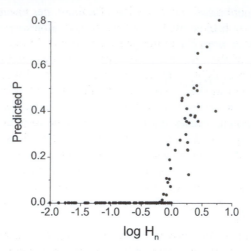

Fig. 5.4   The equilibrium fraction of occupied patches (P) against the logarithm of the neighbourhood habitat area (H_n) in 127 patch networks. Parameter values used in the calculations: $\alpha = 2$ and $c = 10$.

patch areas and with increasing covariance between patch areas and connectivities, which in practice increases with patch aggregation. Note that the average connectivity and hence H_n increase with increasing number of patches in the network, up to a point where the network is much bigger than the migration range of individuals.

We can now return to the question about metapopulation dynamic consequences of habitat destruction. The value of H_n can be used to characterize the loss and fragmentation of habitat in a particular patch network. For the same network, the equilibrium fraction of occupied patches can be calculated with eqns 5.3 and 5.4, for chosen values of α and c. Figure 5.4 gives an example of the 127 real patch networks that are potentially inhabited by the Glanville fritillary butterfly in Finland (Section 11.1). With decreasing neighbourhood habitat area, the fraction of occupied patches is expected to decrease. Ultimately a threshold value is reached and the metapopulation goes extinct.

5.3 The incidence function model

The simple model in the previous section incorporates the spatial structure of a finite patch network into a metapopulation model, representing in this respect a substantial improvement over the Levins model and other spatially implicit patch models. In this section, I describe another spatially realistic model, the incidence function model (IFM), which is built upon similar reasoning to the model expressed by eqn 5.2, but with the following differences. Firstly, the IFM is a discrete-time stochastic patch model, a metapopulation-level extension of a first-order linear Markov chain model for an individual habitat patch. Gyllenberg and Silvestrov (1994) and ter Braak *et al.* (1998) relate the IFM to the general theory of Markov chain models. Secondly, the scaling of the extinction risk with patch area, and the relationship between colonization probability and immigration rate, are described with more flexible expressions than was used in eqn 5.2. And thirdly, the IFM is constructed to enable parameter estimation with field data and thereby to facilitate the application of the model to specific real metapopulations. The IFM was inspired by Diamond's (1975) original concept of the incidence function, describing the generally increasing probability of island occupancy with increasing area or species richness (for examples see Gilpin and Diamond 1976, 1981; Diamond 1979; Diamond and May 1981). Diamond's research suggested a mainland–island IFM (Hanski 1992, 1993), which was subsequently generalized to metapopulations without an external mainland (Hanski 1994a,c).

Consider first a single habitat patch *i*. Assuming that patch *i* is presently empty, there is a constant probability C_i of recolonization in unit time. Similarly, if patch *i* is presently occupied, there is a constant probability E_i of extinction in unit time. Only one turnover event, extinction or colonization, is allowed per time step. With these assumptions, the stationary (long-term) probability of patch *i* being occupied, which is called the incidence J_i, is given by

$$J_i = \frac{C_i}{C_i + E_i}.$$ (5.8)

The same result is valid for an analogous continuous-time model (see eqn 5.3).

In the case of mainland–island metapopulations with no interactions amongst the local (island) populations, the model expressed by eqn 5.8 can be lifted to the level of a metapopulation without extra assumptions, as in this case the colonization and extinction probabilities are indeed patch-specific constants. However, if there is no external mainland, and if colonizations are due to migration from the occupied patches, connecting several populations into a metapopulation leads to apparent conflicts with the assumptions of the Markov chain model for one patch. Firstly, the value of C_i is not constant, as assumed in the model, but it will vary depending on which other patches happen to be occupied and hence function as sources of colonists. However, what really matters is what happens at the stochastic steady state, to which eqn 5.8 refers. It is reasonable to assume that at the steady state all patches retain a relatively constant patch-specific level of connectivity to the existing populations. Secondly, the model assumes independent dynamics in different patches. This assumption is not met either, because two nearby patches are affected by largely the same populations and hence their colonization rates are correlated. In practice, barring pathological examples, spatial non-independence is not likely to be a severe problem at the steady state and it will be ignored here (but see Section 12.1). Finally, with no external mainland, a finite metapopulation consisting of extinction-prone local populations will ultimately go extinct, and hence metapopulation extinction is the only true steady state. Before going extinct, the metapopulation may, however, settle for a very long time in a quasi-stable positive steady state, which is here called 'steady state' for short.

Assumptions about extinction and colonization probabilities

In the spirit of focusing only on the effects of patch area and isolation on metapopulation dynamics, let us make the following assumptions about how patch area affects extinction and how isolation affects colonization. The aim is to make sufficiently general assumptions to accommodate a range of possible metapopulation structures in the same modelling framework and thereby to facilitate the application of the model to real metapopulations. To keep the model simple, let us assume that the extinction probability E_i depends only on patch area A_i, because the extinction probability generally depends on population size (Section 2.2) which in turn is usually given by a simple linear (Kindvall and Ahlén 1992; Eber and Brandl 1996) or power function (Hanski *et al.* 1996c) of patch area. The generally much weaker effect of isolation on extinction is ignored at first for simplicity, though it will be included below via the rescue effect. The following relationship between E_i and A_i can be justified both on empirical (Hanski 1994a) and theoretical grounds (Box 5.1):

$$E_i = \min\left[\frac{e}{A_i^x}, 1\right], \tag{5.9}$$

Box 5.1 Parameters of local extinction in the incidence function model

In the extinction model of Lande (1993), Foley (1994) and Middleton *et al.* (1995), the asymptotic extinction risk for positive long-term growth rate (r) and large population ceiling (K), under the assumption that extinctions are caused by environmental stochasticity, is given by (Section 2.2):

$$E \approx \frac{sr}{K^s},$$

where $s = 2r/v$ and v is the variance of the growth rate. Assuming a power function relationship between patch area and the population ceiling, $K = DA^\zeta$, where D is the density per unit area, the parameters of the extinction model are related to the parameters of the IFM (eqn 5.9) as $s = x/\zeta$ and $r = eD^s/s$. Parameter x in the IFM thus reflects the effective strength of environmental stochasticity (r/v), which increases as the value of x decreases.

Although the correspondence between x and s is the most useful result that can be gleaned from this comparison, I illustrate here how, in principle, the parameters of local dynamics can be inferred from the parameters of the IFM (Hanski 1998). As an example, I use the metapopulation of the American pika (*Ochotona princeps*) studied by Smith (1974, 1980) and Smith and Gilpin (1997) in a large network of small habitat patches in California. The parameters of the IFM were estimated as described in Section 12.4, giving $x = 1.28$ and $e = 0.0046$. Moilanen *et al.* (1998) found that $K = 22.5A^{0.74}$ (area measured as the length of the patch boundary, as suggested by pika biology). From these values we may calculate the values of r and its variance, $r = 0.19$ and $v = 0.54$. These values are consistent with the low rate of reproduction in the pika, which produce only one small litter per year.

where e and x are two parameters. Note that the extinction probability equals unity for patches smaller than or equal to a minimum patch area, A_0, given by $e^{1/x}$.

The colonization probability C_i is an increasing function of the numbers of immigrants (M_i) arriving at patch i in unit time. In the case of mainland–island metapopulations, with a permanent mainland population as the sole or main source of colonists, it is reasonable to assume that:

$$C_i = \beta \exp(-\alpha d_i), \qquad (5.10)$$

where d_i is the distance of patch (island) i from the mainland and α and β are two parameters. For common species, which recolonize a minimally isolated patch (d_i close to zero) without a delay, eqn 5.10 may be further simplified by setting $\beta = 1$. For metapopulations without an external mainland, M_i is the sum of individuals originating from the surrounding populations. Taking into account the sizes of and distances to these populations, we may assume that (Hanski 1994a):

$$M_i = \beta S_i = \beta \sum_{j \neq i}^{R} \exp(-\alpha d_{ij}) p_j A_j, \qquad (5.11)$$

where p_j equals 1 for occupied and 0 for empty patches, d_{ij} is the distance between patches i and j, and α and β are two parameters as in eqn 5.10. The sum in eqn 5.11 is denoted by S_i for short. If there are no interactions among the immigrants in the establishment of a new population, C_i increases exponentially with M_i. Often, however, the probability of successful establishment of a new population depends on propagule size in a more complex manner (Schoener and Schoener 1983; Ebenhard 1991), in which case the following sigmoid relationship is justified:

$$C_i = \frac{M_i^2}{M_i^2 + y^2}, \qquad (5.12)$$

where y is an extra parameter. Note that when eqn 5.11 is substituted into eqn 5.12, only the parameter combination y/β remains, which I denote simply by y below. There is also another justification for eqn 5.12. The IFM is a patch model in which a newly-established population grows instantly to the local carrying capacity. This is a convenient but a potentially troublesome assumption especially for larger patches. By using eqn 5.12 we can somewhat alleviate the consequences of this assumption and take into account, though in a phenomenological manner, the possibility of many newly-established populations rapidly going extinct without having had time to grow to the local carrying capacity. Substituting eqns 5.9 and 5.12 into eqn 5.8 gives an expression for the incidence as a function of patch area and isolation. This is the key idea in the IFM. The expression for the incidence enables us to parameterize, at least in principle, a model of spatio-temporal metapopulation dynamics using spatial data only, a snapshot of patch occupancies.

It is useful to add at this point one more ingredient into the model, the rescue effect (Section 4.2), which enables us to apply the model to metapopulations with relatively high rate of migration and hence high rate of population turnover. I use the following argument to incorporate the rescue effect into the IFM. Recall that a maximum of one change in patch state may occur in one time step, but assume now that if population i is well connected to other populations at time t, its size has become elevated (although this is not modelled explicitly) and hence the extinction probability E_i between t and $t+1$ is reduced. For simplicity, I assume that the extinction probability is reduced to $(1 - C_i)E_i$ (to reiterate, note that this does not assume extinctions followed by colonizations in the time interval from t to $t+1$). With this assumption, eqn 5.8 becomes

$$J_i = \frac{C_i}{C_i + E_i - C_i E_i}. \tag{5.13}$$

Inserting the above assumptions about how patch areas and isolations affect extinctions and colonizations into eqn 5.13, we obtain the following expression for the incidence of patch i:

$$J_i = \frac{1}{1 + \frac{ey}{S_i^2 A_i^x}}, \tag{5.14}$$

which is valid for patches greater than A_0.

One could make some other reasonable assumptions about the functional forms of E_i and C_i, and one may include in the model the effects of other patch attributes apart from area and isolation (Moilanen and Hanski 1998; Section 12.1). The essential point is that with such assumptions eqn 5.8 is transformed into a model which can be fitted to empirical data to estimate the values of the model parameters. Having done that, one may proceed to use the estimated parameter values to simulate the dynamics of the species in the original or in some other patch network, to study and to make predictions about the transient and steady state metapopulation dynamics. Parameter estimation is discussed at length in Section 12.1, where the IFM is applied to real metapopulations. Section 12.1 is self-contained, and the interested reader might want to skim through it at this point. The best way to demonstrate the potential value of the incidence function approach is via a hypothetical example; this is given in Section 5.6, following the description of two other spatially realistic modelling approaches in the following two sections.

5.4 State transition models

Instead of modelling the states of local populations in a metapopulation, as in lattice models and in the IFM, one can model state transitions, extinction and colonization events. This approach seems to have important advantages in comparison with the IFM. First, given that state transitions are modelled statistically, for instance using logistic regression (below), it is straightforward to incorporate any empirically-observed effects of habitat quality or the surrounding landscape into models of extinction and colonization probabilities. Although this can also be done in the IFM, the procedure is more complex (Section 12.1). Second, modelling state transitions seems to make no assumption about steady state, unlike eqn 5.8, and hence, in principle, model parameters can be estimated from non-equilibrium metapopulations. The snag here is that such parameter estimates are likely to be badly biased, because the model does not account for any delay in metapopulation change—which is likely to occur in non-equilibrium situations in changing environments.

A practical drawback of state transition models is the large amount of data needed to parameterize the model; one has to accumulate at least tens of turnover events for parameter estimation, which restricts the application to metapopulations with high turnover rate. Most metapopulation studies have accumulated data for a few years only. With such limited data, useful information can be extracted using both the spatial pattern of patch occupancy and the observed turnover events. This can be accomplished with the IFM (Section 12.1) but not with logistic regression-based state transition models.

Verboom *et al.* (1991), Kindvall (1996a) and Sjögren Gulve and Ray (1996) have used logistic regression to construct state transition models for metapopulation dynamics. Empirical data on observed extinction and colonization events were first modelled with multiple logistic regression. Given the parameter values, probabilities of extinction and colonization could then be calculated, for each patch and time step, and for any patch network and initial state of patch occupancies, assuming that the values of the explanatory variables in the logistic models are known for each patch. Verboom *et al.* (1991) applied the model to a nuthatch metapopulation, Kindvall (1996a) to a bush cricket and Sjögren Gulve and Ray (1996) to a frog metapopulation. I return to state transition models in Section 5.6, where this approach will be compared with the IFM using simulated data.

5.5 *n*-population simulation models

The key word in this section is not simulation. With few exceptions (Durrett and Levin 1994; Levin and Pacala 1997), all results of all spatially explicit and realistic metapopulation models are based on simulation. The key word is structured, populations structured by their size and spatial location, enabling local populations to have their own dynamics, including the effect of location-dependent immigration on dynamics. These models include *n*-population simulation models, an extension of the two-population model (Section 3.1), and coupled-map lattice models, which introduce an explicit description of local dynamics in the setting of lattice models.

The philosophy of structured simulation models is brute force, which is taken to its extreme in individual-based models. If there are *n* local populations in the metapopulation, local dynamics in each population are simulated separately. Migration is modelled explicitly, often with substantial detail about the presumed behaviour of migrating individuals. The aim is to maximize the realism and accuracy of model predictions, generality is of less concern. Thus the *n*-population simulation

Table 5.1 Attributes of *n*-population simulation models as implemented in the computer programs ALEX, RAMAS/Space and VORTEX (according to Lindenmayer *et al.* 1995)

Attribute	ALEX	RAMAS/Space	VORTEX
Density dependence	Limited	Yes	Limited
Demographic stochasticity	Yes	Yes	Yes
Environmental stochasticity	Yes	Yes	Yes
Regional stochasticity	Partial	Yes	No
Allee effect	No	Yes	Yes
Density-dependent migration	Yes	Yes	No
Distance-dependent migration	Yes	Yes	No
Genetic variability	No	No	Yes
Social population structure	No	No	Yes

models are typically used when one wants to make predictions about the dynamics of a particular metapopulation in a particular fragmented landscape. It is, however, questionable whether sufficient data will ever be available for more than a handful of species to parameterize complex simulation models rigorously. Detailed modelling of migration is a particularly troublesome component in these models (Kareiva *et al.* 1997).

Several generic computer programs are available for simulating *n*-population metapopulation dynamics, including ALEX (Possingham and Noble 1991), VORTEX (Lacy 1993) and RAMAS/Space (Akçakaya and Ferson 1992). Representative examples of their application to particular metapopulations include Lindenmayer's work on Leadbeater's possum, an Australian marsupial (Lindenmayer *et al.* 1993; Lindenmayer and Possingham 1995), the study by Lahaye *et al.* (1994) of the Californian spotted owl, the modelling by Akçakaya *et al.* (1995) of the helmeted honeyeater, and the study by Akçakaya and Atwood (1997) of the California gnatcatcher. For more examples see Lindenmayer *et al.* (1995), who also present a comparison of the main features of the three simulation programs mentioned above (Table 5.1). These models are application-oriented, written and used primarily for metapopulation viability analysis. Generic problems in their use for the intended purpose include many untested assumptions about local dynamics and migration, difficulty of parameterizing the model for particular species, and the great difficulty of testing model predictions. This third problem is, of course, shared with all models that purport to predict long-term dynamics of species on large spatial scales.

Coupled-map lattice models (Kaneko 1992, 1993) are theory-oriented structured simulation models. These models assume a grid of identical cells, like the models in Section 5.1, but now each cell is occupied by a local population with its own explicit dynamics. In these models, space and time are discrete, but population size is represented by a continuous variable. The local populations are coupled by migration, for instance by assuming that a constant fraction of individuals in a cell emigrates to the four nearest-neighbour cells in each time step. Coupled-map lattice models have been used to study the role of migration in general in spatial dynamics

Table 5.2 Parameter values of the simulation model estimated for three butterfly species (Hanski and Thomas 1994)

Parameter	*Melitaea cinxia*	*Hesperia comma*	*Plebejus argus*	Description
r	2.0	1.5	2.0	Intrinsic growth rate
θ	0.1	0.2	1.0	Strength of density dependence
K	0.2	0.1	0.1	Carrying capacity (ind m^{-2})
m	0.3	0.1	0.1	Emigration rate
α	1.0	1.0	2.0	Migration distance parameter
δ	0.25	0.4	0.5	Mortality during migration
e_s	20	32	32	Environmental stochasticity
e_l	0.006	0.008	0.02	Catastrophic stochasticity

(Section 3.2) and, for instance, host–parasitoid metapopulation dynamics, which are discussed in Sections 7.2 and 7.3.

A model of butterfly metapopulations

As an example of the *n*-population simulation approach I use here a model which was constructed for butterfly metapopulations (Hanski and Thomas 1994; Hanski *et al.* 1994), although the model could be applied just as readily to many other species with comparable life histories. This model has eight parameters, of which five describe local dynamics and the remaining three describe migration and colonization (Table 5.2). The model assumes a network of habitat patches which can have different areas, and which have their distinct spatial positions, but no other features of landscape geometry nor other landscape properties are included, for instance the shape of the patches is ignored.

Local dynamics in an occupied patch are modelled with an extension of the Ricker model:

$$N_i(t+1) = N_i(t)\exp[\, r(1 - (N_i(t)/K_i)^{\theta})], \qquad (5.15)$$

where $N_i(t)$ is population size in patch i in year (generation) t, r is the intrinsic growth rate, K_i is the carrying capacity of patch i, defined as the equilibrium number of individuals in the absence of migration, and θ sets the strength of density dependence near equilibrium (density dependence increases with θ). This model is deterministic, yet we wish to model metapopulation dynamics with population turnover. Rather than turning to a stochastic model of local dynamics, I make the more phenomenological assumption that the probability of local extinction in patch i in year t is given by:

$$E_i(t) = 1 - \frac{(1 - e_l)N_i(t)^2}{e_s^2 + N_i(t)^2}, \qquad (5.16)$$

where e_s is roughly the size of a population which has a 50% chance of surviving until next year, and e_l gives the asymptotic per-year probability of extinction in large populations. e_s and e_l may be interpreted as reflecting the levels of demographic/environmental and catastrophic (population size-independent) stochasticities, respectively. In reality, of course, large populations typically become small before going extinct, but this is not explicitly modelled here.

Emigration is assumed to be density-independent, fraction m of individuals leaving their natal population in each year. Immigration is similarly assumed to be density- and patch size-independent, though these assumptions could be easily changed at the cost of one or two extra parameters. The numbers of emigrants from patch i are distributed to the other patches using the formula:

$$M_{ij}(t) = mN_i(t)\exp(-\delta d_{ij}) \frac{\exp(-\alpha d_{ij})}{\sum_{j \neq i} \exp(-\alpha d_{ij})}, \tag{5.17}$$

where $M_{ij}(t)$ is the number of individuals moving from patch i to patch j in year t, α is the constant of an exponential distribution of migration distances, and $\exp(-\delta d_{ij})$ is the fraction of individuals surviving migration over distance d_{ij}. An empty patch is colonized if $\Sigma M_{\cdot j}(t) > 0.5$, in which case the new population is started with $\Sigma M_{j(t)}$ individuals, truncated to the nearest integer value.

Parameter estimation and applications

Even if there are only eight parameters in this model, it is hard to estimate their values with independent data, which highlights a general problem with complex simulation models. Hanski *et al.* (1994) obtained rough estimates of K, m, α and r for the Glanville fritillary butterfly, but there were no data suitable for estimation of the values of δ (mortality during migration), θ (strength of density dependence), e_s and e_l, the two parameters setting the rate of local extinction. The values of these four parameters were estimated indirectly by running many model simulations with different sets of parameter values. The results were characterized by five criteria, evaluated at the stochastic steady state: the number of occupied patches, the effects of patch area and isolation on occupancy, and the effects of patch area and isolation on population density in the occupied patches. The simulation results were compared with the observed results for one year, and those parameter values were selected that minimized the difference between the observed and predicted values (Hanski *et al.* 1994). This sort of approach has problems. If the number of parameters to be estimated is large, several parameter combinations might produce approximately equally good results. In the butterfly example (Hanski *et al.* 1994), it was reassuring that the parameter values estimated indirectly were consistent with the known biology of the species (Table 5.2).

Hanski and Thomas (1994) tested the model with data on the silver-spotted skipper butterfly (*Hesperia comma*), for which data were available from two independent patch networks in South England (Thomas and Jones 1993). In one network with 94 patches the skipper was widespread and seemed to occur at or close to the steady state. In another network with 75 patches, the skipper had declined during the early part of this century, and in a survey conducted in 1982 it was found in one large and in two small patches only. Subsequently habitat quality improved due to increased grazing by rabbits, and the species started to spread back into the network, which was judged to be entirely suitable for the species by the early 1980s (Thomas and Jones 1993). Model parameters were estimated from the first network, and the dynamics were predicted for the second network (Hanski and Thomas 1994). We assumed that initially the skipper was restricted to the three patches in which it was actually observed in 1982, and we predicted the number of patches occupied after 9 years (generations), when another field survey was conducted (1991) and 21 patches were found to be occupied (Fig. 5.5). In all but one of 100 replicate simulations the metapopulation was successfully established from the three initial populations, and on average 8.6 patches were occupied after 9 years (median 9, minimum 1, maximum 11). This is significantly less than 21, the observed number of occupied patches. In the model, 14.4 patches were occupied after 50 years, which is still significantly less than the observed number after 9 years, though there was a good agreement between the prediction and observation in terms of which patches were occupied (Fig. 5.5). A comparable application of the IFM to these data predicted that 11.4 populations would exist, on average, after 9 years (Hanski 1994a).

There are two likely reasons for the discrepancy between the predicted and observed numbers of occupied patches. First, the metapopulation in the network that was used to estimate parameter values was perhaps not quite at steady state in 1991, but might have been expanding (Hanski and Thomas 1994). This would result in parameter values underestimating the species' colonization potential. Second, the model with exponentially distributed migration distances might underestimate long-distance migration, which is critical for metapopulation expansion as well as for the spread of species in general (Kot *et al.* 1996; Clark *et al.* 1998). Though this is hard to establish with mark–recapture studies, recent work suggests that the redistribution kernel for the silver-spotted skipper might indeed have a 'fat' tail (Hill *et al.* 1996). In the network in which expansion occurred, three patches had been colonized by 1991 although they were predicted to have zero probability of colonization even after 50 years (Fig. 5.5). These patches were isolated by up to 8.7 km from the nearest occupied patch, indicating substantial potential for long-range migration and colonization (it is, of course, also possible that these populations were simply missed in the 1982 survey). These results demonstrate that it is difficult to make accurate predictions about transient dynamics, but models can be useful in predicting the pattern of spread (Fig. 5.5) and in assessing whether a metapopulation is likely to succeed in colonizing a particular patch network. Hanski and Thomas (1994) describe other examples on the silver-spotted skipper and other butterfly species.

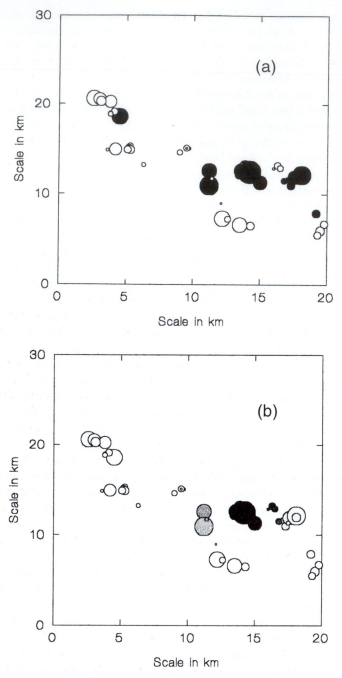

Fig. 5.5 Observed and predicted distribution of the silver-spotted skipper butterfly in a 75-patch network in southern England. In 1982, the species occurred in only one large and two small patches. In the following 9 years the butterfly expanded its distribution, and had reached the patches shown as black dots in panel (a) by 1991. Panel (b) shows the distribution predicted by the simulation model, with the predicted probability of patch occupancy indicated by the shading (large probabilities indicated by dark shading). The relative size of the habitat patch is given by the size of the symbol. (Based on Hanski and Thomas 1994.)

5.6 Comparison of modelling approaches

In this section I return to the IFM and state transition models and apply them to simulated data generated by the eight-parameter simulation model described in the previous section. To have a realistic example, I used the 50-patch real network of dry meadows described by Hanski *et al.* (1994) for the Glanville fritillary butterfly. The habitat patches vary in size from 12 m^2 to 2.7 ha and they are located within an area 5 by 5 km^2. I generated data on patch occupancy in this network using the parameter values given in Table 5.2. As these values represent the best estimates for the Glanville fritillary butterfly (Hanski *et al.* 1994), this example is as realistic as possible.

A snapshot of patch occupancies in the simulation results was sampled after 200 years. In this sample, 42 of the 50 patches happened to be occupied. I fitted eqn 5.14 to these data as explained in Section 12.1, and obtained the following parameter estimates: $x = 0.41$, $y = 3.91$ and $e = 0.063$ (independent estimates of y and e were obtained by assuming a minimum patch size $A_0 = 12$ m^2, which was the observed minimum size of occupied patches; the exact value of A_0 affects the predicted turnover rate but has little effect on the steady state metapopulation size). I assumed $\alpha = 1$ in eqn 5.11, a value which was suggested by mark–recapture data for this species (Hanski *et al.* 1994). The exact value of α is not critical, because there is strong correlation between the values of α and y, which both affect the colonization probabilities (Hanski 1994a).

In the output of the simulation model, representing the 'true' dynamics in this example, the fraction of occupied patches was $P = 0.87$ (standard deviation 0.042 in data for 500 years). Using the estimated parameter values, I simulated the IFM as explained in Section 12.1 for the same length of time. In these results, P was 0.83 (sd 0.048). Comparing the fraction of time that each individual patch was occupied during the 500-year period, Fig. 5.6 shows that there was no great systematic difference between the true values and the values predicted by the IFM, though the model somewhat overestimated turnover rate (Fig. 5.6b). A more stringent test can be made by running both the simulation model and the IFM for another patch network using the same parameter values as above. For this purpose, I selected another 50-patch network from our large data base on the Glanville fritillary butterfly (Section 11.2). In this case, the fraction of occupied patches was $P = 0.47$ (sd 0.11) and $P = 0.57$ (sd 0.08) in the outputs of the simulation model and the IFM, respectively. The predicted incidences were somewhat too high for patches with true incidence smaller than 0.5 but not very small (Fig. 5.6c). The predicted turnover rate was again somewhat overestimated (Fig. 5.6d).

Another test of the model is to compare the true and the predicted transient behaviour after a perturbation. To do this, I simulated both models in the previous 50-patch network using the same initial condition of three occupied patches. In both cases, two of 10 replicates went extinct in 50 years, but there were also differences in the true and predicted trajectories (Fig. 5.7). In the predictions of the IFM, the surviving metapopulations tended to reach a greater size than in the output of the

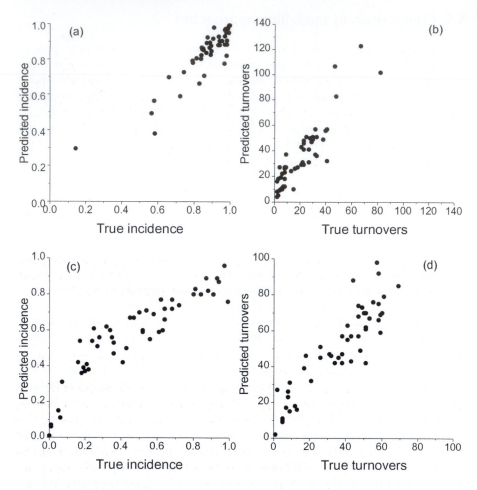

Fig. 5.6   (a) Comparison of true (simulation model) and predicted (IFM) patch incidences, which give the fraction of time that each patch was occupied during a period of 500 years. (b) The true and predicted numbers of local extinctions in the 50 patches. Panels (c) and (d) give the same information as (a) and (b) but for another 50-patch network, using the same parameter values as for the first network.

simulation model. Such discrepancies are not surprising in view of the simplicity of the IFM and the very limited amount of data used for parameter estimation. Given the intrinsic stochasticity of metapopulation dynamics, it clearly matters in which particular year the data for parameter estimation were collected, and more accurate results could be obtained if data were available for parameter estimation for several years (Hanski *et al.* 1996c; Section 12.1). Note that independent replicates of the

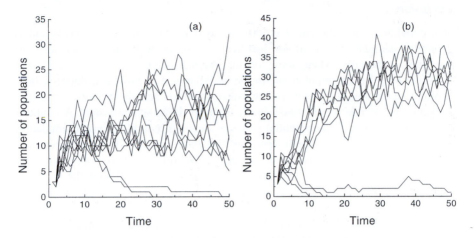

Fig. 5.7 Ten replicate trajectories of metapopulation size (number of extant populations) during a 50-year period with each run started with the same three patches being occupied. Panel (a) gives the results for the simulation model, panel (b) for the IFM.

'true' dynamics also show much variability (Fig. 5.7) and place a definite limit on the accuracy of any model prediction.

These model-generated data can be used to compare the performance of the IFM and a state transition model based on logistic regression (Section 5.4). The minimum amount of data needed to parameterize a state transition model consists of two consecutive snapshots of patch occupancies, which yield a number of extinction and colonization events. In the present example, I sampled the simulated metapopulation in years 199 and 200. As it happened, no colonization events occurred during this transition, and the number of extinctions was only two. I hence used data collected in years 197 to 200 (three transitions) to have more data. The number of colonization events was still only two and the number of extinctions was seven. Logistic regression models were nonetheless parameterized with these (inadequate) data, by regressing the observed extinction and colonization events against the logarithm of patch area and the index of patch isolation, S_i (eqn 5.11). The dynamics of the metapopulation were then simulated in the same patch network from which the logistic regression model was estimated. The result was a rapid extinction of the entire metapopulation, apparently because data for parameter estimation were collected during four years when the metapopulation happened to decline, from 47 to 42 patches occupied. A logistic regression model parameterized with data from such a period is liable to extend the temporary trend to the future, and the predictions are clearly unreliable. Another obvious problem in this particular example is the limited amount of turnover data used, but in practice much more data are seldom available.

Conclusion

The *n*-population simulation models, with all the detail that one may so easily add to a complex simulation, might be justified for some well-studied species and landscapes, especially when some of the habitat patches are large and have potentially long-lasting populations. In the latter case, it is, however, doubtful how important an element the metapopulation-level analysis actually provides, and it is practically impossible to test model predictions for large spatial scales. A counter-example demonstrating the potential value of detailed individual-based simulation models is the forest stand model developed by Pacala *et al.* (1996 and references therein). This spatial and mechanistic model successfully predicts both population and community-level dynamics in multispecies forest stands. Although the model of Pacala *et al.* (1996) is not a metapopulation model as such, it could be applied to landscapes with fixed spatial heterogeneity and patchiness. Tree dynamics might be particularly amenable to individual-based modelling, especially if one is able to commit the amount of resources that Pacala *et al.* had for field work.

In the case of highly fragmented landscapes with numerous small populations in small habitat fragments, which is the classical metapopulation scenario, populations turn over faster and testing model predictions becomes more realistic. At the same time, the effects of habitat patch area and isolation on extinction and colonization rates become increasingly prominent. To model such metapopulations one might do better with the much simpler incidence function and state transition models than with complex simulation models; the former models make fewer assumptions and have fewer parameters without losing the flexibility to describe the key elements of colonization–extinction dynamics. Of course, one should not have too high expectations of simple models, but given the stochasticity that is inherent in the dynamics of real metapopulations, and the nearly inevitable environmental changes that occur during any longer period of time, perhaps the best we can ever achieve is robust semi-quantitative predictions. Comparing the IFM and state transition models, only the former can use spatial information on patch occupancy for parameter estimation, which is a big advantage in practice. The cost of this possibility is the assumption that the metapopulation is close to a stochastic steady state. If this assumption cannot be made, there is no alternative but to collect data for many years to parameterize the models with data on population turnover.

6

Metapopulation genetics and evolution

The spatial structure of populations influences population dynamics but also the dynamics of genes in populations. An obvious but important example is increased genetic differentiation with increased population fragmentation (Hastings and Harrison 1994). The reverse—gene dynamics driving spatial structure—can also occur—as in the local adaptation of herbivores to individual host plants, or groups of host plants, which may generate or amplify spatial population structure (Mopper and Strauss 1998). A thorough discussion of metapopulation genetics and evolution is beyond the scope of this volume (for reviews see Barton and Clark 1990; Olivieri *et al.* 1990; Hastings and Harrison 1994; Harrison and Hastings 1996; Antonovics *et al.* 1997; Barton and Whitlock 1997; Olivieri and Gouyon 1997). Genetics cannot be completely ignored even here, however, as it is possible that the genetic composition of local populations and the entire metapopulation has some immediate consequences on population dynamics and ecology. Some questions of particular significance in this context include: Does reduced genetic variation and the expression of deleterious alleles due to inbreeding in local populations increase the risk of local population extinction? Does metapopulation structure hinder local adaptation and hence possibly increase extinction rate? How fast can species' migration rates evolve in an increasingly fragmented landscape, and what difference does this make to metapopulation dynamics? And what information about migration rates can be gleaned from the study of allele frequencies?

6.1 Genetics, migration and colonization

Broad comparative studies have convincingly demonstrated that migration tendency is especially high in species living in ephemeral habitats (Brown 1951; Southwood 1962; Roderick and Caldwell 1992; Roff 1994; Denno *et al.* 1996), which is a necessary condition for their long-term persistence. Detailed studies of wing size in polymorphic insects, such as water striders, have shown that the selective advantage of short-winged versus long-winged individuals is related to habitat permanence (Vepsäläinen 1974; Järvinen and Vepsäläinen 1976; Harrison 1980; Roff 1986; Kaitala 1988), although wing size polymorphism in, e.g., water striders has also a phylogenetic ingredient (Andersen 1993). Considering populations within meta-populations, if emigrants from or immigrants to a habitat patch represent a non-random genetic sample of individuals in the metapopulation as a whole, migration (gene flow) will change local allele frequencies. Assuming that variation in migration tendency, or ability, is to some extent heritable, for which there is much evidence for many taxa (Wilson 1995; Dingle 1996; Roff 1996), immigration and emigration themselves will cause a change in the migration rate in a population. Two testable predictions follow. First, if emigration greatly exceeds immigration, as is likely to happen in isolated populations, migration selects for reduced migration rate via selective loss of the most migratory individuals. Second, if immigration exceeds emigration, which might happen in small populations located close to large ones, the population is expected to show a particularly high migration rate (the more general question about the evolution of migration rate at the metapopulation level is discussed in the next section).

The first prediction is supported by data from very isolated populations— flightless birds (Hesse *et al.* 1951; Carlquist 1966) and insects (Brinck 1948; Gressit 1964) on oceanic islands, and flightless insects on mountain tops (Darlington 1943; Hanski 1983b; Roff 1990). Of greater concern to us is whether, in comparisons of conspecific populations, reduced emigration rate can be detected in populations living in the more isolated habitat patches in fragmented landscapes. Not much is known about this, but a few studies on butterflies seem to provide supporting evidence. Dempster (1991) and Dempster *et al.* (1976) conducted a morphometric study of museum specimens of the swallowtail butterfly (*Papilio machaon*) and the large blue butterfly (*Maculinea arion*) from isolated populations in the UK. He found a reduction in the ratio of thoracic width to thoracic length, suggestive of reduced mobility, coinciding with an increase in population isolation. In another morphometric study on a third British butterfly, the silver-studded blue (*Plebejus argus*), Thomas *et al.* (1998) reason that as populations in small patches receive relatively more immigrants than populations in large patches, individuals in small patches should exhibit a greater migration tendency. They found that the relative thorax size of butterflies decreased, while the relative abdomen size increased, with increasing habitat patch size. These results were interpreted as reflecting greater allocation of resources to flight muscles and flight capacity in butterflies in the smaller patches.

From the perspective of metapopulation dynamics, the important issue is to what extent possible evolutionary changes in mobility modify the connectivity of populations. This might clearly happen—if isolated populations exhibit especially low migration rate, they are effectively even more isolated; and if well-connected populations evolve a high rate of migration, they become effectively better connected than might be inferred from the physical connectivity of the populations. At the same time, it becomes apparent that for better understanding of possible evolutionary scenarios, we have to shift the focus from individual populations to metapopulations as a whole. For instance, in the study of Thomas *et al.* (1998), populations in small patches have shorter life-times and are hence younger, on average, than populations in large patches, which might be an even more important cause of elevated migration rate in small populations than the balance between emigration and immigration during the life-time of the populations.

Evolution of migration rate in metapopulations

Evolution of migration rate is an important topic in evolutionary ecology (Van Valen 1971; Comins *et al.* 1980; Levin *et al.* 1984; Olivieri and Gouyon 1985, 1997), on which much empirical work has been conducted especially with plants (Venable 1979; Venable and Lawlor 1980; Olivieri *et al.* 1983) and insects (Southwood 1962; Dingle *et al.* 1980; Harrison 1980; Roff 1986; Roderick and Caldwell 1992). In the present context, the interesting question is what difference does colonization–extinction dynamics make to evolution of migration rate at the metapopulation scale and how does that evolution feed back to metapopulation dynamics.

Let us first consider a simple hypothetical scenario, in which there are no local extinctions and no fluctuations in local population sizes, but migration is costly or there are fixed spatial differences in habitat suitability and hence in population sizes. In this case migration is selected against, and the evolutionary stable (ES) migration rate is zero (Comins *et al.* 1980; Olivieri and Gouyon 1997), because the expected fitness of an average migrant is necessarily lower than the expected fitness of a resident. For a positive migration rate to evolve there must be some mechanism which increases the expected fitness of a migrant. The most likely mechanism is temporal variance in population sizes, and ultimately in fitness, of which population turnover represents an extreme case. The possibility that a migrant arrives at an empty site, or in a small population which has the potential to become larger, greatly boosts the expected fitness of a migrant.

It is both natural and useful to consider explicitly two levels of selection on migration rate and other life-history traits in metapopulations—selection within established populations and selection among populations (Van Valen 1971; Olivieri and Gouyon 1997). Within each population, a gene reinforcing migration should decline in frequency, because more copies are likely to leave the population than to arrive, with the exception mentioned above of small populations located close to large ones. In contrast, at the metapopulation level selection reinforces migration tendency, and more so the greater the temporal variance in fitness. Olivieri and Gouyon (1997) have termed the two antagonistic levels of selection a metapopulation effect.

The metapopulation effect makes the testable prediction that the frequency of migratory genotypes should decline with the age of the population. Olivieri and Gouyon (1997) (see also Roff 1990) review the yet scanty evidence, which seems to be largely supportive. To take two examples, Peroni (1994) compared five populations of the red maple from early successional environments with five populations from late successional environments. Populations from the early successional environments showed a slightly but significantly lower wing-loading ratio (samara mass/samara area) than populations from the late successional environments. Cody and Overton (1996) measured dispersal-related morphological traits in several species of Asteraceae on small islands off Vancouver Island, British Columbia. They estimated the ages of the plant populations by using long-term records of the island populations. They found that in *Lactuca muralis*, for which

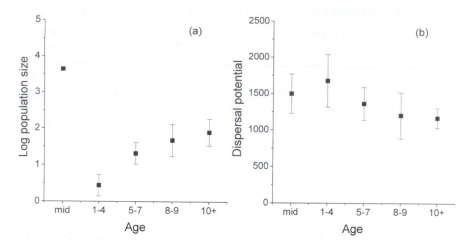

Fig. 6.1 Comparison of population density (a) and seed dispersal potential (b) in *Lactuca muralis* on small islands off Vancouver Island, British Columbia. Dispersal potential was measured by the ratio of pappus volume to achene volume. Note that density increased with island age, while seed dispersal potential was highest in the youngest populations, then declined to a level lower than in the mainland in island populations older than 5 years. (Based on Cody and Overton 1996.)

extensive data were available, a measure of seed dispersal potential was highest in the youngest populations, but it subsequently declined in only a few generations to a level lower than in the mainland populations (Fig. 6.1). A line of research awaiting empirical studies is comparison of populations living in small versus large habitat patches in the same metapopulation. Populations in small patches have a higher risk of extinction, hence they are on average younger, than populations in large patches, and one could expect the frequency of migratory genotypes to be higher in small than in large patches. Similarly, sink populations should show a comparable excess of migratory genotypes in comparison with source populations.

With increasing loss and fragmentation of natural habitats, the extinction risk of local populations is expected to increase and the colonization rate is expected to decrease. Are these changes likely to be amplified, or to be counteracted, by selection on migration rate in metapopulations? Increased fragmentation is likely to increase the cost of migration (mortality during migration) and hence the relative strength of within-population selection against migration. On the other hand, reduced population sizes increase the extinction rate, which will increase temporal variance in fitness, which will select for increased migration rate at the metapopulation level. As local dynamics are necessarily faster than metapopulation dynamics, we might expect that a sudden fragmentation first enhances local selection against migration, while the metapopulation-level selection for increased migration rate occurs with a delay, in the course of establishment of populations consisting of individuals with the highest migratory tendency. Therefore, in the short-term evolution is likely to hasten the demise of a population in a fragmented landscape, whereas the long-term effects, assuming that the metapopulation survives to experience any long-term effects, are more specific and difficult to predict (Leimar and Norberg (1997) have recently demonstrated these effects with a simple genetic model). One reason why 'naturally rare' species are likely to perform much better in highly fragmented landscapes than 'new rare' species is that the former, but not the latter, have had sufficient time to experience metapopulation-level selection for increased migration rate. And, of course, the species which did not evolve a sufficiently high migration rate are already extinct.

Genetics of colonizers

Life-history traits other than migration rate can be affected by metapopulation dynamics, and these traits, in turn, might affect the dynamics. Examples include dormancy in plants and diapause in animals (Olivieri and Gouyon 1997), and allocation of resources to reproduction and survival (Kawecki 1993; Ronce and Olivieri 1997), as in the case of annual versus perennial life histories in plants (Crawley and May 1987). The following example is from our studies on the Glanville fritillary butterfly, described at length in Part III. This example relates closely to population turnover, the hallmark of classical metapopulation dynamics.

The Glanville fritillary has two larval host plants in the Åland islands, SW Finland, *Plantago lanceolata* and *Veronica spicata*. A distinct east–west gradient exists in the relative abundances of the two host plants, and this gradient is paralleled by a gradient in genetically determined host plant preference by ovipositing females within a distance of 30 km (Kuussaari 1998). Such small-scale adaptation to the regionally more abundant host plant raises an interesting question about metapopulation dynamics. Assume that butterflies in a particular region have evolved a preference for *V. spicata* (or vice versa). Host species composition varies from patch to patch, and though most patches in certain regions are dominated by *V. spicata*, some patches have only *P. lanceolata*. Given the host preferences in female butterflies, we might ask whether the host plant composition of the habitat patches influences extinction–colonization dynamics.

The answer is yes. Habitat patches that are dominated by the host plant to which the metapopulation in a particular region has adapted have a substantially higher rate of colonization than patches dominated by the less preferred host plant (Fig. 6.2). In one out of three years, there was a comparable difference also in the extinction rate, which was higher in populations with the 'wrong' host plant. The effect of female preference on population turnover is probably caused by the effects of female preference on emigration from and immigration to small habitat patches rather than on larval performance (M. Singer, personal communication). In this example, colonization probability depends on a genetically determined trait of colonizers, which is selected in the same direction or in different directions at the level of local populations and at the level of the metapopulation. Other studies have elucidated

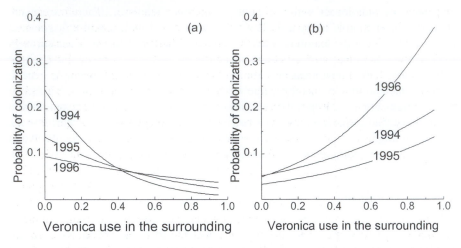

Fig. 6.2   The probability of colonization of an empty habitat patch by the Glanville fritillary butterfly as a function of the fraction of larvae on *Veronica spicata* (as opposed to *Plantago lanceolata*) in the surroundings of the focal patch. The focal patches have been divided into two classes, those with *Plantago* only (a) and those with relative *Veronica* cover >70% (b; there are only few patches with *Veronica* only). Note that the probability of colonization is higher when the patch is surrounded by populations using the same host plant which predominates in the focal patch. This 'colonization effect' is explained by genetically determined female oviposition preference influencing both host use in the surroundings and the probability of colonization of the focal patch. The lines are based on logistic regression, which also included the effect of isolation to existing populations on colonization probability (here set at a constant value; I. Hanski and M. Singer, unpublished).

how local adaptation and gene flow promote genetic variation in metapopulations (Karban 1989; Wade 1990; Antonovics *et al.* 1994) and how local adaptation can be influenced by gene flow (Holt 1995; Dias 1996; Pulliam 1996; Mopper and Strauss 1998). The Glanville fritillary example suggests that gene flow and the establishment of new populations can be influenced by regional adaptation.

6.2 Genetics and extinction

The literature on conservation biology has for a long time referred to inbreeding depression and reduced genetic variance as potential contributory causes of population extinction (Frankel and Soulé 1981; Soulé 1987; Frankham 1995). Inbreeding is defined as increased incidence of mating with a relative in comparison to random mating. In the case of spatially structured populations, inbreeding does not require active mate selection, as random mating within local populations leads to inbreeding relative to the metapopulation as a whole. Inbreeding increases the frequency of homozygous loci in offspring, which is often associated with reduced fitness (Ralls *et al.* 1988; Thornhill 1993; Saccheri *et al.* 1996), most likely because of the expression of deleterious recessive alleles (Charlesworth and Charlesworth 1987; Crow 1993). Reduced genetic variance due to drift is not expected to cause a reduction in fitness in the short term, but in the long term it might lower the rate of adaptive evolution and thereby increase the risk of extinction in a changing environment (Fisher 1930). At present, however, major environmental changes are so rapid that the distinction between 'short term' and 'long term' loses significance.

For some time, the consensus has been that small populations are generally likely to go extinct so rapidly because of demographic and environmental stochasticity that genetics is unlikely to make a substantial difference (Lande and Barrowclough 1987; Lande 1988b; Nunney and Campbell 1993; Caro and Laurenson 1994; Caughley 1994). More recent theoretical studies have stressed the high rate of fixation of slightly deleterious mutations in small populations, which can reduce fitness (population growth rate) and thereby increase the risk of extinction even in populations with relatively large effective sizes, up to several hundreds (Gabriel *et al.* 1991; Noordwijk 1994; Lande 1995; Lynch *et al.* 1995). So far, these studies have considered isolated populations, and it remains to be seen what are the predictions for metapopulations. The empirical support is also largely lacking; one recent 50-generation study on *Drosophila* failed to find the predicted greater reduction in fitness in small compared with large populations (Gilligan *et al.* 1997).

Empirical evidence for inbreeding depression in captive animal populations is widespread (Ralls *et al.* 1988; Bryant *et al.* 1990; Thornhill 1993; Saccheri 1995), but the general relevance of these results to natural populations has been questioned. The same can be said about Frankham's (1995) interesting analysis of extinction rate in laboratory lines of two *Drosophila* species and the house mouse (*Mus musculus*), which showed higher extinction rates in inbred than in outbred lines. In these studies, the rate of inbreeding did not affect the extinction risk as expected (Lande 1988b), but the number of studies and lines analysed was small.

Studies of plant populations have frequently reported reduced fitness in small populations (Ellstrand and Elam 1993; Lamont *et al.* 1993; Ouborg and van Treuren 1994; Prober and Brown 1994; Heschel and Paige 1995). One of the more comprehensive examples is that on *Gentiana pneumonanthe* in the Netherlands (Oostermeijer *et al.* 1994, 1995; Oostermeijer 1996a,b). This species has become endangered in many parts of Europe due to habitat loss and fragmentation. Enzyme electrophoretic studies revealed loss of genetic variation from small isolated populations, in which plants showed a marked reduction in several fitness components (Oostermeijer *et al.* 1994). A significant difference was found between non-flowering (juvenile) and flowering (adult) individuals in allozyme heterozygosity, which Oostermeijer *et al.* (1995) interpreted as reflecting selection against the more homozygous (inbred) individuals in the population. Keller *et al.* (1994) in their study of a sparrow population on an island found that the more inbred individuals had a higher risk of mortality than less inbred individuals during a stressful period.

Few field studies have attempted to test directly the hypothesis that inbreeding depression increases the risk of extinction of natural populations (Jiménez *et al.* 1994; Keller *et al.* 1994; Newman and Pilson 1997). This question has major practical implications. If inbreeding depression increases the risk of population extinction, a possible management strategy is to cross small inbred populations by transfer of individuals; crossing of inbred populations should improve fitness (heterosis). The few studies conducted so far have produced mixed results (Ouborg and van Treuren 1994; Oostermeijer *et al.* 1995).

Inbreeding and extinction in the Glanville fritillary butterfly

The Glanville fritillary metapopulation described in Part III consists of a large number of mostly very small local populations. Local populations have a high risk of extinction for many reasons, including demographic and environmental stochasticity and parasitism (Section 11.2). The caterpillars live in large sib-groups, and often there is just one group of full sibs in a population. Mating among close relatives must therefore occur commonly in the Glanville fritillary metapopulation. Inbreeding reduces heterozygosity and if inbreeding depression increases the risk of extinction we would expect, other things being equal, that reduced heterozygosity would be associated with an elevated risk of population extinction. Saccheri *et al.* (1998) tested this prediction by genotyping a sample of butterflies from 42 local populations, of which seven went extinct in one year. The populations that went extinct were characterized by small size, they were isolated (small numbers of butterflies in the neighbouring populations, hence no rescue effect), and the habitat patch had a low density of nectar flowers (which increases emigration and reduces immigration; Kuussaari *et al.* 1996). Our previous studies have demonstrated that such populations tend to have a high risk of extinction (Section 11.2). However, in addition to these ecological factors, the level of heterozygosity also made a highly significant contribution to the model explaining the observed extinctions (Fig. 6.3). Inbreeding affects several fitness components in the Glanville fritillary, including egg hatching rate, the weight of post-diapause larvae, pupal period (inverse relationship) and adult longevity (Saccheri *et al.* 1998). Other studies on captive (Brakefield and Saccheri 1994) and wild butterfly populations (Ebenhard 1995) have found reduced egg hatching rate in inbred populations. Just a single round of brother–sister mating, which must occur commonly in the small populations in the Glanville fritillary metapopulation, was sufficient to reduce egg hatching rate by *ca.* 30% (Saccheri *et al.* 1998). These results suggest that the Glanville fritillary metapopulation maintains a high genetic load. Apparently, even strong selection against deleterious recessives exposed by inbreeding in local populations does not lead to rapid purging, most likely because of high probability of fixation of slightly deleterious alleles in small populations and because of gene flow among neighbouring local populations. Exactly how the effects of mutation, selection, inbreeding and migration blend in metapopulations remains an exciting challenge for theoretical and empirical research. Meanwhile, the general message of these

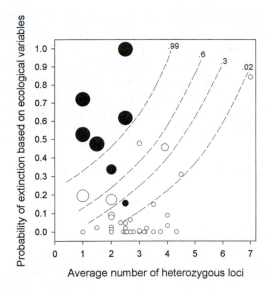

Fig. 6.3 The predicted probability of local extinction due to ecological factors (vertical axis, from logistic regression) and the average number of heterozygous loci in 42 populations of the Glanville fritillary butterfly. The ecological factors include population size, isolation and the availability of nectar flowers. The size of the symbol gives the predicted probability of extinction by a logistic model incorporating both the ecological factors and the average number of heterozygous loci. The model predicts well the observed seven extinctions (black dots), and heterozygosity makes a highly significant contribution to this model. (Based on Saccheri *et al.* 1998.)

results is that the effect of inbreeding on local extinction should not be ignored especially in species with a highly fragmented population structure.

6.3 Loss, or gain, of genetic variation?

Local populations tend to lose underlying genetic variation due to drift, hence small populations tend to have less genetic variation than large ones (Soulé 1976; Frankham 1996). The variance-effective size of a population gives the size of an idealized population in which drift causes the genetic variance to decline at the same rate as in the focal population. Haldane (1939) derived the approximation for the variance-effective size:

$$N_{e(v)} = \frac{4N}{V_k + 2},$$ (6.1)

where N is (constant) census population size and V_k is the variance in the number of gametes produced by the parents to the next generation (Kimura and Ohta 1971;

Nagylaki 1992). Variance-effective size thus depends on population size and the variance in the reproductive success among individuals. Metapopulation structure might affect variance-effective size by reducing total population size, for instance due to habitat loss, and by increasing, or reducing, the variance in the reproductive success of individuals. Population turnover in particular might greatly reduce the effective size, and more so the larger the local populations that turn over, the higher the extinction rate, and the smaller the number of individuals that establish new populations (propagule size), especially if drawn from a single extant population (Hedrick and Gilpin 1997). Population turnover therefore most likely reduces the total amount of genetic variation in the metapopulation as a whole (Whitlock and Barton 1997). Nonetheless, the Glanville fritillary metapopulation does not have markedly reduced genetic variation (Palo *et al.* 1995; Saccheri *et al.* 1998), despite rampant population turnover (Section 11.2), probably because the sizes of local populations are mostly very small, often just one or a few sib-groups of larvae, and the rate of gene flow is relatively high.

Though drift tends to reduce genetic (and phenotypic) variance in small populations, additive genetic variance may actually increase in population bottlenecks due to epistatis and dominance: non-additive variance may become converted to additive variance in inbred populations (Avery and Hill 1977; Bryant *et al.* 1986; Whitlock *et al.* 1993; Whitlock 1995). Whitlock and Fowler (1996), working on a large number of inbred *Drosophila* lines, have shown that there can be substantial chance variation in genetic variance after a bottleneck, which is especially significant for metapopulations with many local populations.

In the classical 'island' model of Sewall Wright (1931, 1932), all populations have a constant effective size N_e. A fraction m of individuals in each population emigrates per generation and is replaced by the same number of immigrants drawn at random from the common pool of migrants. At equilibrium, the genetic variance among local populations in the metapopulation is given by the coefficient:

$$F_{st} \approx \frac{1}{4N_e\,m + 1}. \tag{6.2}$$

Clearly, genetic differentiation of local populations decreases with the number of migrants, $N_e m$ (if there are differences among the sexes in the migration rate, m should be replaced by an 'effective' migration-rate parameter m_e; Berg *et al.* 1998). The level of gene flow that is sufficient to prevent substantial population differentiation is surprisingly low. Maruyama (1970, 1971) has worked out the exact relationship between differentiation and gene flow in a 2-dimensional stepping stone model, in which each population exchanges migrants with four neighbouring populations. In this model, noticeable local differentiation is possible only if $N_e m$ is less than unity, in other words if the populations exchange less than one individual per generation on average. This model, however, assumes constant population sizes, and it leaves open the question about the consequences of classical metapopulation dynamics to population differentiation.

Equation 6.2 indicates that, in principle, gene flow and hence migration (m) in a metapopulation can be estimated indirectly with data on genetic differentiation of local populations (F_{st}). Two recent metapopulation studies have used eqn 6.2 to study the correspondence between ecological and genetic measures of spatial population structure. Lewis *et al.* (1997) studying the butterfly *Plebejus argus* used direct estimates of N_e and m to calculate F_{st} and obtained a value (0.024) similar to the one estimated with genetic data (0.014). Appelt and Poethke (1997) used an estimate of F_{st} (0.056) for the grasshopper *Oedipoda caerulescens* to calculate $N_e m$ and thereby the probability of recolonization of an empty patch (0.77), which was very similar to the colonization probability (0.76) estimated with the incidence function model (Chapter 5.3). These results are encouraging, but they must be interpreted with caution, because other studies have found great discrepancy between direct and indirect estimates of m (for a review see Slatkin 1994). The use of eqn 6.2 makes many assumptions (Hastings and Harrison 1994; Slatkin 1994), most notably that the metapopulation is close to demographic and genetic equilibria, that genetic differentiation is due to drift only (neutral genetic markers), and that eqn 6.2 applies to the often complex spatial structures in real metapopulations (Slatkin and Barton 1989). Rather than relying exclusively on either ecological or genetic information, one should attempt to collect all possible information about spatial population structure and dynamics and base the final judgement on all available information. If ecological studies reveal only restricted migration but genetic studies indicate extensive gene flow, the explanation might be temporally varying migration and gene flow. Or the metapopulation might not have attained a genetic equilibrium after a regional population collapse and recovery, such as exemplified by the maps in Fig. 11.7, or after a recent range expansion.

Extinction–colonization dynamics and genetic differentiation of populations

Extinction–colonization dynamics generally reduce genetic variation in a metapopulation, as was discussed above, but how does population turnover affect population differentiation? Several recent studies have examined the combined effects of gene flow and extinction–colonization dynamics on genetic differentiation among local populations (Slatkin 1977; Wade and McCauley 1988; Whitlock and McCauley 1990; McCauley 1993; Whitlock *et al.* 1993; Barton and Whitlock 1997). These studies demonstrate that extinction–colonization dynamics can both enhance or reduce genetic differentiation, relative to the case of stable local populations. For gene flow m much less than unity, Whitlock and McCauley (1990) derived the following condition for increased genetic differentiation due to extinction–colonization dynamics:

$$k < \frac{2Nm}{1-\varphi} + 0.5, \tag{6.3}$$

where k is the number of individuals establishing a new local population (assuming diploid individuals), N is the equilibrium size of local populations (assumed to be

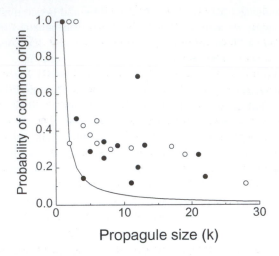

Fig. 6.4 Relationship between ϕ_i and k_i in experimental data on the Glanville fritillary butterfly. Open circles represent originally empty habitat patches, black dots are patches into which butterflies were released ('occupied' patches). The line gives the theoretical minimum value of ϕ_i, obtained when all immigrants originate from different source populations (data from Kuussaari *et al.* 1996).

reached in one generation following establishment; but see Ingvarsson 1997), and ϕ gives the probability that two gene copies (alleles) in the group of k individuals establishing a new local population are drawn from the same source population. The two key factors determining the effect of extinction–colonization dynamics on genetic differentiation are thus the pattern of gene flow, as summarized by ϕ, and the number of individuals establishing a new population, k, in comparison with gene flow among established populations, given by Nm.

Figure 6.4 gives an empirical example of the ϕ values from an experiment on the Glanville fritillary in which nearly 1000 marked butterflies were released into 16 habitat patches within a system of 65 previously empty patches (Kuussaari *et al.* 1996). In this network of small patches located within a small area (1.6 km^2), there was much migration, m being around 0.5. Figure 6.4 gives the patch-specific values of ϕ_i against the number of immigrants k_i, with a line giving the theoretical minimum value of ϕ, obtained when all immigrants originate from different populations (note that even in this case $\phi > 0$ because each diploid individual brings two gene copies from the same population). In these results, ϕ is around 0.3, demonstrating an intermediate situation between a 'migrant-pool' model, in which migrants are drawn randomly from the metapopulation, and a 'propagule-pool' model, in which all immigrants to a patch come from the same source population. The intermediate case is likely to be common in nature (Whitlock 1992). Using allozyme and other molecular data to estimate F_{st} among recently established and older populations of the plant *Silene alba*, and a direct estimate of k (4.2 individuals), McCauley *et al.* (1995) derived indirectly the value of 0.73 for ϕ. Ingvarsson (personal communication; Ingvarsson *et al.* 1997) estimated $\phi = 0.78$ for a beetle species living in small habitat patches.

In Fig. 6.4, the number of immigrants arriving at empty and occupied patches was similar, $k \approx Nm$, suggesting that there was no strong conspecific attraction nor repulsion. With $k \approx Nm$ condition (6.3) is satisfied, and extinction–colonization dynamics should enhance genetic differentiation among local populations. However, with as much migration as observed in this experiment the absolute level of differentiation will remain practically zero, which is likely to be a common situation for species with high turnover rate and hence high gene flow in general (Slatkin 1985; Harrison and Hastings 1996; below). The role of extinction–colonization dynamics in amplifying population differentiation is likely to be greater in plants than in animals, because generally in plants $k < Nm$ due to gene flow in pollen transfer, which can occur only among existing populations.

Whitlock and McCauley (1990) have derived the following expression for the level of genetic differentiation amongst populations with extinction–colonization dynamics:

$$F_{st} = \frac{[1 - e + e\varphi(1 - 1/2k)]/2N + e2k}{1 - [1 - e + e\varphi(1 - 1/2k)](1 - m)^2(1 - 1/2N)}, \tag{6.4}$$

where e is the extinction rate (all extinct populations are assumed to become recolonized immediately). Whitlock (1992) and Ingvarsson *et al.* (1997) have estimated the parameter values of eqn 6.4 for two mycophagous beetle species living in patchy habitats, the fungus beetle *Bolitotherus cornutus* and the smut-feeding *Phalacrus substriatus*. In both cases, the predicted F_{st} values, 0.067 and 0.070, agreed remarkably well with the observed values, 0.059 and 0.077.

The prediction that extinction–colonization dynamics increase genetic differentiation can be tested by comparing the level of differentiation among newly established populations, $F_{st}^{(new)}$, with the level of differentiation among older populations, $F_{st}^{(old)}$. Assuming that $N \gg k$, the following approximate result holds (McCauley *et al.* 1995):

$$F_{st}^{(new)} = \frac{1}{2k} + \varphi\left(1 - \frac{1}{2k}\right)F_{st}^{(old)}. \tag{6.5}$$

When ϕ is small, the level of differentiation among newly established populations is primarily determined by the propagule size, k. In contrast, when ϕ is large, there is little mixing at colonization and extinction–colonization dynamics amplify the existing genetic differentiation in the metapopulation (for $\phi = 1$ eqn 6.5 simplifies to $F_{st}^{(new)} = F_{st}^{(old)} + (1 - F_{st}^{(old)})/2k$). A few empirical studies have compared values of $F_{st}^{(new)}$ and $F_{st}^{(old)}$ (Table 6.1). Two studies on *Silene* plants and two studies on unrelated mycophagous beetles showed significantly more differentiation among young than old populations. The exception is Dybdal's (1994) study on a tidepool-

Table 6.1 Comparisons of F_{st} values for recently established and older populations in the same metapopulation

Species	Young populations		Old populations		P	Reference
	n	F_{st}	n	F_{st}		
Silene alba	12	0.197	11	0.126	0.05	McCauley *et al.* 1995
Silene dioica	13	0.057	30	0.030	0.05	Giles and Goudet 1997
Bolitotherus cornutus	NA	0.112	NA	0.040	0.005	Whitlock 1992
Phalacrus substriatus	32	0.090	10	0.059	0.03	Ingvarsson *et al.* 1997
Tigriopus californicus	6	0.013	6	0.037	0.02	Dybdal 1994

inhabiting marine copepod, in which older populations were more differentiated. This result might be due to the older tidepools being generally more isolated (Dybdal 1994; Giles and Goudet 1997), which would tend to reduce immigration rate and hence enhance population differentiation.

6.4 The shifting balance theory

Sewall Wright's (1932) hypothesis about adaptive evolution through a 'shifting balance' between the evolutionary forces in spatially structured populations has remained an influential idea for the past 60 years even in the absence of any substantial empirical support. A fundamental premise of the shifting balance theory is that populations might evolve to multiple locally stable states, which create the 'adaptive landscape' of fitness peaks with disparate heights. This is a reasonable assumption, as multilocus genetic models are typically capable of generating many equilibria (Feldman 1989; Barton and Whitlock 1997). One empirical approach to finding out how frequently different populations actually are located at different 'peaks' would be to measure the extent of outbreeding depression—crosses between populations at different peaks should exhibit reduced fitness. Outbreeding depression has been observed in some plant metapopulations (Burt 1995), but more empirical studies are needed (Barton and Whitlock 1997).

The shifting balance process has itself been divided into three phases (Wright 1932; Barton and Whitlock 1997). In the first phase, local populations cross from the domain of one equilibrium (adaptive peak) to the domain of another equilibrium due to random fluctuations in allele frequency. In the second phase, natural selection drives the local population to the new peak. And in the third phase, populations occupying different adaptive peaks compete with each other, which is expected to facilitate the spread of the fitter adaptive peaks into the metapopulation (species) as a whole. Theory for the first two phases of the shifting balance process is well understood, but the third phase has received more attention only recently (Barton and Whitlock 1997).

An apparent difficulty with the shifting balance process relates to the narrow limits within which migration rate must lie for the process to continue. The probability of crossing to the domain of a new peak is generally proportional to W^{2N}, where W is the mean fitness of the population in the adaptive valley compared with the original adaptive peak (Barton and Rouhani 1987). Thus a shift is more likely in a small population and when the 'valley' is not deep. However, a third important factor is the level of gene flow into the population—Nm immigrants per generation reducing the probability of a shift by a factor of 2^{-4Nm} (Lande 1979; Barton and Rouhani 1991). Substantial gene flow thus practically eliminates peak shifts; for $Nm = 1$ the above factor is 0.06, for $Nm = 2$ it is 0.004. In the third phase, the fitter peaks are assumed to outcompete the less fit ones via gene flow. However, gene flow tends to bias peak shifts to the advantage of the more common peaks, and when gene flow exceeds a critical value, roughly equal to one migrant per generation, a rare adaptive peak cannot spread because of gene flow from the more common but less fit peak (Barton and Whitlock 1997).

Several deviations from the assumptions of the simple island model, which is the basis of the above conclusions, might increase the chance that the shifting balance process will work. First, the fitter peaks might send out more migrants than the other peaks (Wright 1932; Phillips 1993), which would facilitate the third phase. Second, migration rate might vary spatially or temporally, which would enable all three phases of the shifting balance process to occur. Migration rate is indeed likely to vary from place to place, largely because the degree of isolation of local populations varies, both within metapopulations and, which is probably more significant for the shifting balance process, from the edge of the species range towards the centre. Migration is often likely to occur among nearby populations only, which is advantageous for the shifting balance process (Barton and Rouhani 1993). In many taxa, substantial temporal variation in migration rate is caused by year-to-year variation in environmental conditions, and one can imagine variation at longer time scales due to exceptional events. Shifts in the geographical ranges of species will generate spatio-temporal variation in migration rate as well as in the presence of populations. Range shifts are probably more extensive and more common than previously thought. During glacial periods range shifts have been dramatic, possibly creating favourable conditions for the shifting balance process. Nonetheless, despite these various mechanisms that could 'save' the shifting balance process, the consensus is now emerging that it is unlikely to be important in adaptive evolution (Barton and Whitlock 1997; Coyne *et al.* 1997).

6.5 Extinction–colonization dynamics—summary

The shifting balance theory might not represent the grand recipe of adaptive evolution, but it is unquestionable that spatial structure and dynamics of populations have many fundamental consequences for their evolutionary change. For instance, extinction–colonization dynamics may facilitate the conversion of non-additive into additive genetic variance (Whitlock *et al.* 1993; Hastings and Harrison 1994). One general consequence of the increasing attention to spatial structure of populations is the closing of the gap between population ecology and genetics. Though I am the last ecologist to phrase all ecological problems in genetic and evolutionary terms, genetic variation among and local adaptation of populations forces us to reconsider the role of genetics in population dynamics and ecology. The reverse is true even more generally: acknowledging the spatial structure of populations injects a huge amount ecology into population genetics and evolutionary biology.

More specifically, we might ask whether ecologists primarily interested in the ecological dynamics of metapopulations should nonetheless include genetic and evolutionary components in their research projects. Do the dynamics of local extinction and recolonization significantly hinge on the genetic composition of local populations? Several results reviewed in this chapter suggest that they might often do so. For instance, the role of inbreeding depression in contributing to population extinction has been discussed for a long time, but empirical work has not really taken advantage of the opportunities that metapopulations consisting of numerous small local populations provide. Small local populations have a high risk of extinction for all sorts of reasons, but inbreeding depression might often be one of them, as the Glanville fritillary example suggests. Extinction–colonization dynamics is likely to reduce genetic variation in metapopulations, but if the metapopulation consists of small populations with relatively high migration rates, the reduction might not be great. An interesting open question is the dynamics of slightly deleterious mutations in metapopulations. If local populations are small, the probability of local fixation is high, but what is the probability of regional fixation given local extinctions and distance-dependent migration and colonization? Is metapopulation 'meltdown' due to fixation of deleterious alleles more, or less, likely than the meltdown of a similar-sized local population (Lande 1995; Lynch *et al.* 1995)? Evolution of migration rate in increasingly fragmented landscapes is an area of metapopulation biology where important empirical discoveries can be made.

The finding that female oviposition preference greatly influences the colonization rate of empty habitat patches in the Glanville fritillary butterfly (Fig. 6.2) was a big surprise to me. In this case extinction–colonization dynamics is likely to amplify rather than oppose local selection, which might not be an uncommon situation in metapopulations. Local adaptation of herbivores and other parasites to host individuals (Mopper and Strauss 1998) might tend to bring the extinction–colonization dynamics to a halt—except if there is reciprocal evolution in the host and the parasite, which might lead to accelerated spatial dynamics (Frank 1997). Yet another interesting scenario occurs in source–sink metapopulations, where there is an asymmetric gene flow possibly leading to 'niche conservatism' (Holt 1995)—local adaptation in source habitats at the cost of adaptation to sink habitats (Kawecki and Stearns 1993; Holt 1995, 1996; Kawecki 1995). In all these cases local adaptation influences spatial dynamics and spatial dynamics influence local dynamics.

7

Interacting metapopulations and metacommunities

Two or more species can live in the same patch network without any interactions among the respective metapopulations. More interestingly, interspecific interactions might affect the rates of migration, extinction and colonization, in which case interspecific interactions modulate metapopulation dynamics of the species concerned. Extinction rates might become elevated, but local interactions between the species may also enhance metapopulation persistence by increasing asynchrony in local dynamics. An important recurring theme in the dynamics of two or more interacting metapopulations is the creation and maintenance of aggregated spatial distributions of species in the absence of any environmental heterogeneity. In single-species metapopulations, where extinctions are due to external perturbations, increased migration rate is generally beneficial for long-term persistence in facilitating colonization, with the caveat that too much is too much: massive emigration and high mortality during migration may increase extinction rate so much that metapopulation persistence is reduced. In contrast, in interacting metapopulations with extinctions at least partly due to interspecific interactions, increased migration rate may have a destabilizing effect in increasing synchrony (Harrison and Taylor 1997). In this chapter, I discuss interspecific competition and predator–prey dynamics in the metapopulation context, covering both basic models and selected empirical studies. The last two sections describe attempts to model the dynamics of multispecies metacommunities.

7.1 Competing species

Interspecific competition can influence the rates of local extinction and colonization. Levin (1974) analysed a model in which species cannot invade a population of the competitor, but if established at an unoccupied site could in turn resist invasion by other species. In such a situation with multiple stable states in local dynamics, patchy distributions of species might evolve in the absence of any environmental heterogeneity, enhancing regional species richness. At the local level, species richness is maximized at an intermediate migration rate, with many species occurring at low densities outside their strongholds owing to recurrent immigration, effectively as sink populations because of interspecific competition. In contrast, high migration rate might eliminate the priority effect (founder advantage) and reduce species richness. Priority effects in local dynamics are not all that common, but a few examples are suggestive, such as the rock lobsters and whelks on two small islands off the coast of South Africa, separated by only 4 km. On Malgas island the benthic communities are dominated by seaweeds and rock lobsters, on Marcus island by whelks and mussels. Barkai and McQuaid (1988) report observations and experiments indicating that rock lobsters at high density prevent colonization by whelks, and vice versa, giving rise to two completely different communities on otherwise comparable islands. At the scale of individual plants and other sessile organisms, competition for space is pre-emptive and might be considered to represent a priority effect (below).

Interspecific competition can be incorporated into classical metapopulation models with population turnover. To keep models tractable it has been customary to ignore local dynamics and to focus on the effect of competition in lowering colonization rate, in elevating extinction rate, or both. Such models are straightforward extensions of the single-species Levins model. But before turning to the models, let us consider a solid empirical example, the assemblage of *Daphnia* water fleas in rock pools (Ranta 1979; Hanski and Ranta 1983; Pajunen 1986; Bengtsson 1989, 1991). On islands in the northern Baltic, rock pools are small water-filled cavities in the bed rock, typically around 1 m^2 in surface area and some tens of cm deep. A diverse community of plants and animals lives in these pools (Järvinen and Ranta 1987), perturbed by summer droughts and winter freezing. The community includes three species of water fleas, *Daphnia magna*, *D. pulex* and *D. longispina*. The water fleas have the capacity to produce drought-resistant resting eggs (ephippia), but careful surveys conducted over many years have conclusively demonstrated that local populations also frequently go extinct and new ones are established (Hanski and Ranta 1983; Pajunen 1986; Bengtsson 1989). Local extinctions occur at a rate of 10% per year per population in pools occupied by single species, whereas in the presence of another species the extinction rate is increased to more than 15% (Table 7.1; Bengtsson 1988). In contrast, Pajunen (1986) and Bengtsson (1988) found no evidence for a reduced rate of colonization by these species of occupied pools compared with empty pools.

Table 7.1 Extinction rate in three species of *Daphnia* water fleas in rock pools on islands off the coasts of Finland and Sweden in the Baltic

Locality	Single-species pools		Two-species pools	
Flatholmen	0.13 ± 0.037	(82)	0.15 ± 0.046	(58)
Mönster	0.12 ± 0.038	(74)	0.42 ± 0.140	(12)
Ängskär	0.10 ± 0.025	(143)	0.17 ± 0.051	(54)
Tvärminne	0.11 ± 0.028	(123)	0.16 ± 0.052	(50)

The figures are extinction probabilities per population per year (SD; number of possible extinction events in brackets). Data from Bengtsson (1988, 1989) and Pajunen (1986).

A two-species patch model

To investigate interspecific competition in a network of many small habitat patches with ephemeral local populations we expand the Levins model to two species (Levins and Culver 1971; Horn and MacArthur 1972; Slatkin 1974; Hanski 1983a; Nee and May 1992). Instead of keeping track of the fraction of patches occupied by individual species, we have to keep track of the fraction of patches occupied by species 1 only (P_1), patches occupied by species 2 only (P_2), patches occupied by both species (P_{12}), and the fraction of empty patches (P_0). The following four equations give the rates of change of the fractions of different types of patch:

$$\frac{dP_0}{dt} = -(c_1A_1 + c_2A_2)P_0 + e_1P_1 + e_2P_2$$

$$\frac{dP_1}{dt} = c_1A_1P_0 - [e_1 + (c_2 - \gamma_2)A_2]P_1 + (e_2 + \varepsilon_2)P_{12}$$

$$\frac{dP_2}{dt} = c_2A_2P_0 - [e_2 + (c_1 - \gamma_1)A_1]P_2 + (e_1 + \varepsilon_1)P_{12}$$

$$\frac{dP_{12}}{dt} = (c_1 - \gamma_1)A_1P_2 + (c_2 - \gamma_2)A_2P_1 - (e_1 + \varepsilon_1 + e_2 + \varepsilon_2)P_{12}, \qquad (7.1)$$

where $A_1 = P_1 + P_{12}$ and $A_2 = P_2 + P_{12}$ are the fractions of patches in which species 1 and 2, respectively, are present. Assuming that the fraction of suitable patches is fixed, $P_0 + P_1 + P_2 + P_{12} = h$, one of the four equations can be eliminated (h allows us to vary the density of suitable patches in the landscape as in Section 4.4; alternatively, we could model the numbers rather than the fractions of different kinds of patch, with the colonization rate parameter scaled by the total number of patches). Even then, it is clear that moving from one to two species complicates the model greatly. The single-species extinction and colonization parameters e and c have been indexed to indicate possibly different values in the two species. Additionally, in the two-species patches the extinction and colonization rates are modified to account for interspecific effects, the extinction parameter is increased to $e + \varepsilon$ and the colonization parameter is reduced to $c - \gamma$.

Following Slatkin (1974), let us consider the two special cases of competition affecting only the colonization rate or the extinction rate. In the former case, and assuming that species 1 occurs in the network at equilibrium, species 2 cannot invade the network if:

$$\frac{\hat{P}_2}{\hat{P}_1} < \frac{\gamma_2}{c_2}, \qquad (7.2)$$

where the hat indicates the equilibrium value in the absence of competition. In other words, species 2 cannot invade unless its equilibrium distribution (number of local populations) when present in the network alone ($\hat{P}_2$) is large and the effect of the resident species (γ_2) is small. Reversing the subscripts shows that it is impossible for both species to prevent invasion by the competitor (because γ cannot be greater than c), and hence, in this model, either the two species coexist at the metapopulation level or one of them is excluded regardless of which species was originally present. In the extreme case, called 'lottery competition' (Sale 1977; Chesson and Warner 1981), a resident species in a patch entirely prevents establishment of the later-arriving species, as in Levin's (1974) model with multiple stable states in local dynamics. The lottery model is most applicable to competition for space, with one 'patch' being occupied by just one individual. In lottery competition, $\gamma = c$, and stable coexistence of the two species is impossible, because both $\hat{P}_2/\hat{P}_1 < 1$ and $\hat{P}_1/\hat{P}_2 < 1$ cannot be true at the same time. Coexistence might be attained in lottery competition if there is, e.g., environmentally-caused variation in birth rates (Chesson and Warner 1981; Ågren and Fagerström 1984) or some resource partitioning, which mechanisms would, however, promote coexistence also without the lottery mechanism.

In pure extinction competition, exemplified by the water fleas in rock pools, the presence of the competitor does not obstruct colonization but increases the risk of local extinction. Now it is possible to select parameter values (large values of ε) such that both species are able to prevent the competitor from invading the network, hence there is the possibility of a metapopulation-level priority effect (Hanski 1983a). In the special case of two identical species, which have the same values for all model parameters, a species is unable to invade a patch network occupied by its competitor at equilibrium if:

$$(c + e)c - (2c - \gamma\hat{P})\gamma < 0. \tag{7.3}$$

This is never true for feasible parameter values, hence it has been concluded that two identical competitors can always coexist in a patch network (Levins and Culver 1971; Slatkin 1974; Christiansen and Fenchel 1977; Hanski 1983a; Shorrocks 1990), contrary to a key result of the classical competition theory (Volterra 1926), the principle of competitive exclusion (Hardin 1960; Hutchinson 1978). According to this principle, two identical species cannot coexist, because an individual's competitive action harms equally both conspecifics and heterospecifics, and the relative abundances of the two species are expected to follow a random walk until one of them goes extinct (Chesson 1991). Does habitat patchiness, then, promote coexistence so much that a fundamental result of the competition theory is overturned? Such a conclusion would be unwarranted, for the following reason. The metapopulation result rests critically on the assumptions of two distinct time-scales and instantaneous growth to patch carrying capacity after population establishment, inherited from the single-species Levins model (Section 4.1). These assumptions give an unfair advantage to the species which happens to be regionally rarer. If the two species are really identical, it is difficult to see why the species to arrive first at a habitat patch would not have a better chance of winning the patch than the later-arriving species, because the first species enters into local competition with a higher initial local abundance. Therefore, as the regionally more common species is more likely to arrive first at a given patch, it should win more patches in local competition. This is a telling example of how expanding a model might convert a previously acceptable simplifying assumption into a deep pitfall. Ignoring local dynamics is a sensible first approximation in the single-species model, but may lead to a very misleading conclusion in the case of two species. Nonetheless, this shortcoming of the model does not invalidate all the general conclusions.

Fugitive species

Local competition often involves competitive asymmetries that prevent local coexistence (Connell 1983; Schoener 1983). A competitive asymmetry makes co-existence less likely also at the metapopulation level, though a higher level of asymmetry is compatible with coexistence in metapopulations than in local populations. But let us assume that the asymmetry in competitive ability is inversely correlated with an asymmetry in colonization ability, inferior competitors being superior colonizers and vice versa (though I refer to colonization ability only, it should be recalled that what really matters in metapopulation persistence is the ratio of colonization rate to extinction rate, c/e in eqn 4.2). In this case, metapopulation-level coexistence is not hard to achieve—one species is superior at exploiting previously empty patches (large c and large ε), the second species does well because it is little affected by competition (small c and small ε). Such a metapopulation-level mechanism of coexistence was termed fugitive coexistence by Hutchinson (1951) and Skellam (1951) in two pioneering studies, and it has been later studied by Hanski and Ranta (1983), Nee and May (1992), Moilanen and Hanski (1995), and others.

Nee and May (1992), May and Nowak (1994), Tilman *et al.* (1994, 1997) and Stone (1995) have extended the Levins model reparameterized to study the consequences of habitat destruction (Section 4.4) for two or more competing species. Assuming the competition–colonization trade-off, Tilman *et al.* (1994, 1997) and Stone (1995) find the at first surprising result that habitat destruction is most harmful for the persistence of competitively dominant species. The explanation is the assumed low colonization rate (small c/e) of the superior competitors. In the model of Tilman *et al.* (1994) the competitive dominants are originally most abundant species, whereas in Stone's (1995) model, inspired by coral-reef ecology, the competitive dominants are originally uncommon, making them especially vulnerable to habitat destruction. In all these models, habitat destruction may increase the relative and even the absolute abundance of the inferior competitor but superior colonizer (large c/e), because it will benefit of the decline in the abundance of the superior competitor (but see McCarthy *et al.* 1997 for a thoughtful discussion).

Figure 7.1 gives an empirical example of the effect of the number of habitat patches on metapopulation size in the three rock pool-living *Daphnia* species with apparent differences in colonization and competitive abilities (Hanski and Ranta 1983; although see Bengtsson 1991). The different patch networks occur on 14 islands, which can here be considered as replicate landscapes with different numbers of habitat patches available. Note that the numbers of rock pools occupied by

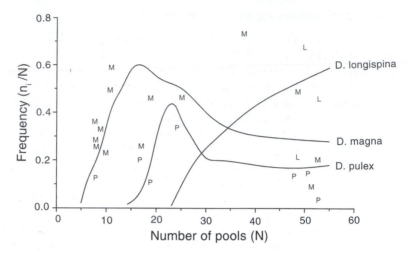

Fig. 7.1   Fractions of rock pools occupied by three *Daphnia* species as a function of the number of pools on 14 islands off the coast of Finland. The letter symbols show empirical results ('M' is *D. magna*, 'P' is *D. pulex*, and 'L' is *D. longispina*), the lines are model-predicted results (from Hanski and Ranta 1983). It was not possible to parameterize the three-species model rigorously, and the prediction represents just a choice of plausible parameter values that seemed to fit the empirical data reasonably well. Nonetheless, these results demonstrate that metapopulation-level competition is a plausible explanation of the observed interspecific differences in the distribution of the species.

D. magna and *D. pulex*, the presumed inferior competitors (Hanski and Ranta 1983), first increase with *decreasing* patch number as the superior competitor but worse colonizer (*D. longispina*) fades out from the community. Ultimately, of course, all species are adversely affected by reduced patch number.

There are several other putative examples of fugitive coexistence, including competition among fungi in the laboratory (Armstrong 1976), and mosses (Marino 1991), plants (Tilman 1982; Grubb 1986; Tilman *et al.* 1994) and parasitoids (Lei and Hanski 1998; Section 11.2) in the field. Another good rock pool example involves two species of corixid water bug, *Arctocorisa carinata* and *Callicorixa producta*. A replacement series experiment revealed a complete overlap of resource use in the two species (Pajunen 1982). Strong interference competition (cannibalism) by the larger species (*A. carinata*) gives it a decisive edge in local competition, summarized by the competition coefficients of 2.6 (effect of one *A. carinata* on one *C. producta*) and 0.4 (effect of one *C. producta* on one *A. carinata*; Pajunen 1979). Locally, the larger *A. carinata* soon replaces the smaller *C. producta*, but the smaller species has about twice the migration rate of the superior competitor, and hence the inferior competitor is typically more frequent in newly-filled rock pools (Pajunen 1979). This example provides very strong evidence for the fugitive mechanism—long-term coexistence is almost certainly based on inversely correlated asymmetries in competitive and colonization abilities.

The voter model and the segregation mechanism

The models discussed so far assume that all patches and populations are equally connected to each other—the models make the mean-field assumption. Like in single-species models, this is a sensible simplifying assumption for many purposes, but especially in the case of two interacting species many interesting phenomena emerge when this assumption is relaxed. Empirically, competition is often strongly localized in natural populations, most notably in plants and other sessile organisms competing for space, which is a good reason to consider spatially explicit approaches.

Durrett and Levin (1994) discuss the following 'voter' model that represents, in a simple manner, interactions among similar plant species competing for space (for related and more elaborated models see Inghe 1989; Pacala 1997; Pacala and Levin 1997). The name of the model relates to the notion of voters (here lattice cells) expressing their preference among several candidates (here the competing species) in an election. To take a simple example, assume that the cells can be in one of two states, representing the presence of one of the two competing species, and assume that one species at a time can occupy a particular cell. Let us further assume that there is a probability Δ that cell x is in the same state at time $t + 1$ as it was at time t, and that there is a probability $1 - \Delta$ that it is in the state of a randomly selected neighbour out of eight neighbours. This assumption reflects local competition and migration. When the model is iterated, the striking feature to emerge is clustering of the states among the cells (Fig. 7.2)—nearby cells are affected by each other and their common neighbours, and they tend to acquire the same state by amplifying (local interaction) and 'memorizing' (local migration) the small-scale fluctuations in cell occupancy due to the stochastic rules of state change (Pacala and Levin 1997). Pacala (1997) discusses in detail how this segregation mechanism, which is a general feature of models of this type, might explain the coexistence of many similar plant species (see also Section 1.2). In this model, coexistence is enhanced rather than deterred by similarity among the competing species (Pacala and Levin 1997), though it should be noted that coexistence among identical competitors is not stable in the sense that a rare species would have an advantage (Chesson 1991). Spatial segregation does not prevent one of the species gradually drifting to extinction, even if time to extinction might be long. Notice also that the segregation mechanism facilitates coexistence when migration distances are short, in contrast to the fugitive mechanism, where the inferior competitor but superior colonizer survives due to long migration distances.

7.2 Predator–prey metapopulations

The asymmetric interaction between a predator and its prey has the potential to wreck local stability and to create oscillations (Lotka 1925; Volterra 1926; Nicholson and Bailey 1935). A voluminous literature explores the various mechanisms that tend either to inhibit or to amplify the oscillatory tendency in predator–prey interactions (May 1976; Hassell 1978; Crawley 1993). For instance, it

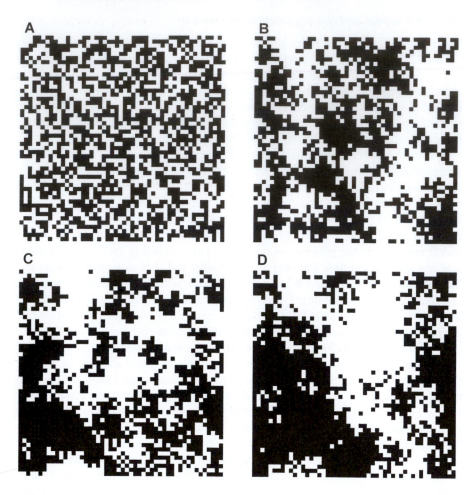

Fig. 7.2   Patterns of spatial distribution in two competing species generated by a two-dimensional voter model after 1, 100, 400 and 1000 time units (panels from [a] to [d]). $\Delta = 0.5$.

is well known that increasing the strength of self-regulation in prey or predator increases stability (May 1976). Variance among the prey individuals in the risk of being predated (or parasitized) increases stability (Hassell 1978), while time delays due to, e.g., predator numerical response decrease stability (May 1976). All this happens in coupled local populations. When the model is extended to space two additional factors influence stability: the degree of asynchrony in local dynamics and the strength of coupling via migration of the prey and predator populations. In this section I discuss a model that is akin to the above-described competition model and assumes completely asynchronous (independent) local dynamics. I then briefly review the predator–prey metapopulation theory more generally, including, in the next section, dynamics that ensue when migration is restricted to nearby local populations.

In the spirit of other patch models, a predator–prey metapopulation model can be constructed by writing down equations for the rates of change in the fractions of patches that are empty (E), occupied by prey only (P), and occupied by both prey and predator (X). The predator is not assumed to survive without prey, hence predator-only patches do not occur. Let us make the mean-field assumption (all patches are equally strongly connected), whereby colonization rate is proportional to the product of occupied and empty patches, and let us assume that all conspecific populations have an equal risk of extinction. For simplicity, the prey–predator patches are not assumed to contribute to the colonization of empty patches by the prey (for models relaxing this assumption see Holt 1997). Denoting by $h = E + P + X$ the fraction of suitable patches, the rates of change in the prey-only and in the prey–predator patches are given by (May 1994):

$$\frac{dP}{dt} = c_P P(h - P - X) - c_X PX - e_P P$$

$$\frac{dX}{dt} = c_X PX - e_X X. \tag{7.4}$$

The globally stable equilibrium point of this model is given by:

$$\hat{P} = \frac{e_x}{c_x}$$

$$\hat{X} = (h - h_X)\frac{c_P}{c_P + c_X}, \tag{7.5}$$

where:

$$h_X = \frac{e_X}{c_X} + \frac{e_P}{c_P}.$$

Thus the predator population persists ($\hat{X} > 0$) if the density of suitable patches (h) exceeds the threshold value h_X, which is given by the sum of the extinction/colonization ratios for the prey and the predator. Prey persists alone in patch networks with $h_P < h < h_X$, where h_P is the single-species threshold patch density for persistence, e_P/c_P (Section 4.4). Figure 7.3 shows the fractions of prey-only and prey–predator patches as functions of the fraction of suitable patches h; compare this with Fig. 4.5.

The model illustrates the possibility of predator–prey coexistence in a patch network when the local interaction is unstable, but also the greater vulnerability of the predator than the prey to habitat destruction (decreased h). Results on checkerspot butterflies and their specific larval parasitoids support both predictions (Table 7.2). The parasitoids significantly increase the risk of local extinction of their host (Lei and Hanski 1997; Section 11.2), and hence the local interaction is unstable, but in large patch networks both species survive by virtue of recurrent colonizations.

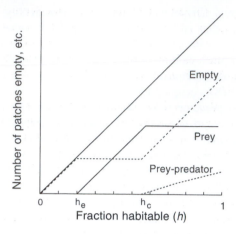

Fig. 7.3 The predicted changes in the fractions of empty patches, prey-only patches, and prey–predator patches when the fraction of suitable patches (h) changes (eqn 7.4; May 1994).

In smaller patch networks, the parasitoid is typically absent and only the host butterfly remains (Table 7.2), reminiscent of the frequent absence of infections like measles from small human communities (Anderson and May 1991; Keeling and Grenfell 1997). Finally, even the host is absent from the smallest patch networks (Fig. 9.3).

Patch occupancy models of the type of eqn 7.4 assume entirely independent dynamics in local populations and a possibly high rate of local extinctions; the predator–prey metapopulations persist due to asynchronous local dynamics and sufficiently high rate of colonization. Eliminating asynchrony would give the predator access to all prey populations simultaneously, which would lead to a reduction in prey numbers at least in the short term. For this reason, Levins (1969) suggested in his very first metapopulation study that pest control measures should be

Table 7.2 Number of habitat patches (H), host-only patches (H_P), and host–parasitoid patches (H_X) in six butterfly–parasitoid metapopulations in Finland

Host	Parasitoid	Area	H	H_P	H_X
M. cinxia	*C. melitaearum*	Main Åland	1176	353	72
		Kumlinge	96	23	–
		Föglö	25	15	2
		Sottunga	14	6	–
M. diamina	*C. melitaearum**	Tampere	94	35	–
E. aurinia	*C. bignellii*	Joutseno	113	42	Present

The host species are checkerspot butterflies, the parasitoids are their specific larval parasitoids (*Cotesia*; J. Pöyry *et al.*, unpublished; see also Section 11.1).
*Parasitizes *M. diamina* in Sweden (Eliasson 1995), where the host is more common than in Finland.

applied synchronously in all pest populations (Ives and Settle 1997 have recently suggested that this might not be the optimal strategy if the pest has natural enemies with dynamics coupled with prey dynamics). In predator–prey metapopulation models without any mechanisms generating local asynchrony, the overall metapopulation dynamics are unstable if local dynamics are unstable (Reeve 1988; Rohani *et al.* 1996).

A critical question, then, is how much asynchrony there exists in natural populations and which factors promote asynchrony (A.D. Taylor 1988). Asynchrony may be generated by several mechanisms, including spatial differences in habitat patches (Maynard Smith 1974), non-uniform migration among local populations (Hilborn 1975), demographic stochasticity in small populations (Nachman 1987, 1991; Wilson *et al.* 1998) and environmental stochasticity (Crowley 1981; Reeve 1988). Asynchrony is more likely to be achieved in large patch networks (A.D. Taylor 1988 and references therein). The role of migration is a complex issue, and no general conclusions have emerged, other than that very low and very high rates of migration do not favour persistence (A.D. Taylor 1988). A very low colonization rate would not suffice to compensate for extinctions, whereas very high migration rates would syncronize dynamics in the entire metapopulation. If migration rates in the prey and the predator are very different, migration may even destabilize otherwise stable local dynamics, by effectively decoupling the interaction locally and eliminating the effect of stabilizing local mechanisms (Rohani *et al.* 1996).

Empirical evidence for predator–prey metapopulation dynamics remains limited, especially from field populations (A.D. Taylor 1988; Harrison and Taylor 1997). Some of the best examples include laboratory (Huffaker 1958), greenhouse (Nachman 1987, 1991) and orchard studies on mites (Walde 1994). A recent study on protists in an experimental laboratory system constructed of connected bottles supported metapopulation-level persistence in a predator–prey system (Holyoak and Lawler 1996). Limited field evidence may reflect scarcity of appropriate studies and enormous data requirements for conclusive tests of model predictions (Steinberg and Kareiva 1997); predator–prey metapopulation dynamics might not be unusual in systems that accord with the main model assumptions, specific predators or parasitoids attacking prey living in small discrete habitat patches. Two insect examples are the tephritid fly *Urophora cardui* and its two *Eurytoma* parasitoids (Zwölfer 1982; Eber and Brandl 1994) and the butterfly–parasitoid system described in Chapter 11 (see also Table 7.2 and Section 12.5). The interaction between the prickly-pear cactus (*Opuntia* spp.) and the moth *Cactoblastis cactorum*, which was introduced to Australia to control the cactus, has been considered to represent a classical example of predator–prey metapopulation dynamics (Dodd 1959), but more recent observations suggest alternative interpretations (Monro 1967; Caughley 1976; Harrison and Taylor 1997). A genuine example of herbivore–plant metapopulation dynamics with herbivore-caused plant extinctions is represented by the interaction between *Hadramphus spinipennis*, an endangered monophagous weevil on Mangere Island, New Zealand, and its host plant *Aciphylla dieffenbachii* (Schöps *et al.* 1998).

7.3 Predation and complex spatial dynamics

Pioneering work by Turing (1952) employed the diffusion model approach (Section 1.2) to study predator–prey dynamics with migration in continuous space. In such models, both reproduction and migration are continuous processes. Consider the following scenario. Increasing density of prey facilitates the growth of the predator population, but increasing density of the predator inhibits the growth of the prey. Additionally, the predator population inhibits its own growth, whereas the prey population is assumed to enhance its own growth (Kareiva 1990). Thus the prey is an 'activator' and the predator is an 'inhibitor' of population growth. These two tendencies might bring about a stable equilibrium point in the absence of migration. When migration (the diffusion term in the model) is added, and if the predator moves faster than the prey, the equilibrium can become unstable, and a spatial pattern in population densities may evolve in an entirely homogeneous environment. Intuitively, this can be explained by much of the inhibiting power of the predator being located in a 'wrong' place by its faster movement rate than that of the prey. Relatively high predation rate outside the current patches of high prey density inhibits the expansion of the prey, though the spatial pattern may be temporally dynamic.

Diffusive instability in the reaction–diffusion models has been said to occur under fairly restrictive conditions (Levin and Segel 1976; Mimura and Murray 1978). For instance, pattern formation requires inverse density dependence in the prey dynamics, which would indeed be an unlikely condition, where it not that type II functional response of the predator yields the same effect (Turchin *et al.* 1998). In any case, another class of model, discrete-time spatial contact models, produces similar complex spatial dynamics with less restrictive assumptions. These models assume discrete population dynamics, such as occur in species with non-overlapping generations (Neubert *et al.* 1995). The models are constructed as sets of integro-difference equations (Kot and Schaffer 1986; Kot 1989; Hastings and Higgins 1994; Neubert *et al.* 1995). Much of the variety in model predictions is generated by different forms of the redistribution kernel (Neubert *et al.* 1995), in particular how 'fat' is the tail of the distribution of migration distances. The important message from these models is that predator–prey dynamics with migration can generate complex spatial patterns in entirely homogeneous landscapes. Empirical evidence is so far practically lacking, partly perhaps because there are not many predator–prey pairs which are dynamically closely coupled and in which there is a substantial difference in the migration rates of the two species.

Perhaps the best terrestrial example so far of spatial pattern formation due to predator–prey interaction is the dynamics in the tussock moth *Orgyia vetusta* on the Californian coast, where it has persistent small outbreaks (Harrison 1997), not explicable by spatial variation in environmental conditions (Harrison 1994, 1997). The tussock moth has flightless females, and hence a low rate of migration, much lower than in a set of specialist parasitoids (Harrison 1994). Both observational and experimental studies have shown that the rate of parasitism is especially high just outside the outbreak area, apparently due to migration of parasitoids from the outbreak population, which tends to prevent the outbreak from expanding, while experimental populations farther away from the outbreak suffered a low rate of parasitism and began to grow (Harrison and Wilcox 1995; Brodmann *et al.* 1997; Harrison 1997).

Dynamics in discrete space with local migration

Let us return to fragmented landscapes. The patch model described in Section 7.2 demonstrated how a predator and its prey might persist regionally despite a high extinction risk locally. To explore predator–prey metapopulation dynamics further, let us now relax two of the more restrictive assumptions of the patch models by making migration local and by modelling changes in local population sizes explicitly. The models discussed below assume that habitat patches are located as cells in a regular lattice, hence the name coupled-map lattice model. The following host–pathogen model is due to White *et al.* (1996). It extends previous work by Hassell *et al.* (1991) and Comins *et al.* (1992).

Assume that within each cell a host population grows exponentially on its own, the population size becoming multiplied by a factor R in each generation. Fraction $1 - \exp(-vY)$ of the host becomes infected by the pathogen, where Y is the number of free-living infective pathogen particles and v is a parameter. Each infected host individual releases S pathogen particles, of which a fraction q_2 survives until the next host generation. Finally, assume that fraction q_1 of the currently existing pathogen particles survives one host generation as free-living particles (Briggs and Godfray 1996; White *et al.* 1996). With these assumptions, local dynamics are given by the following pair of equations:

$$X(t + 1) = RX(t)\exp(-vY[t])$$
$$Y(t + 1) = q_1 q_2 Y(t) + q_2 SX(t)(1 - \exp(-vY[t])). \tag{7.6}$$

The model reduces to the Nicholson–Bailey model, familiar from host–parasitoid dynamics (Hassell 1978; Hassell *et al.* 1991), by setting q_1 equal to 0, in other words by removing the possibility of long-term pathogen survival. Like the Nicholson–Bailey model, eqn 7.6 has an unstable equilibrium point (White *et al.* 1996), and some extra features would be needed to have the pair of species coexist locally. White *et al.* (1996) extended the model to the metapopulation level by assuming a set of local populations on a lattice. Emigration rate was assumed to be constant and the migrants were divided equally among the eight neighbouring cells. Figure 7.4 portrays the predicted dynamics on a 30 by 30 lattice as a function of two parameters, the host emigration rate and the survival rate of free-living pathogen particles. Depending on the parameter values, five distinct types of dynamics emerge (White *et al.* 1996):

- When pathogen survival is high the host goes extinct. This is especially likely for high host emigration rate, which increases the amplitude of fluctuations and leads to occasional release of very large numbers of pathogen particles.

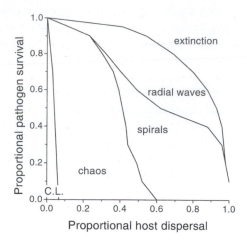

Fig. 7.4   Five types of spatial dynamics generated by iterating eqn 7.6 on a 30 by 30 lattice. Parameter values: $r = 2$, $q_2 = 1$, $S = 1$, $v = 1$ and $m_Y = 0.89$ (pathogen migration rate). The type of dynamics is plotted against the pathogen survival rate q_1 and host migration rate m_X. (Based on White *et al.* 1996.)

- With somewhat lower pathogen survival and/or host emigration rate, the interaction is persistent at the metapopulation level, with host and pathogen populations producing 'radial waves' propagating from a central focus. These waves vary in period, intensity and shape, unlike the apparently similar waves generated by reaction–diffusion models.
- Low pathogen survival and high host migration generate 'spirals' and more stable metapopulations. The spiral structures are characterized by population densities forming spiral waves, which rotate in either direction around a relatively immobile focal point.
- With even lower host emigration rate, the pattern becomes chaotic, without obvious spatial patterns, but population densities are relatively stable at the metapopulation level.
- Finally, a 'crystal lattice' pattern is observed for very low host migration rate; now the dynamics freeze into an entirely static pattern.

In general, metapopulation persistence in this class of predator–prey model is increased by the size of the arena and by relatively low migration rate (Hassell *et al.* 1991; Wilson *et al.* 1998), which both help maintain asynchrony in the metapopulation. For parameter values that lead to low mean local abundances, such as small values of v in eqn 7.6, demographic stochasticity may increase average abundances by increasing the numbers of host populations that increase in size after predator extinction. Eventually, though, these patches become recolonized by the predator, enabling rapid predator increase (Wilson *et al.* 1998).

The range of spatial dynamics generated by eqn 7.6 and related models is astonishing. This is not the full story, however. Numerical simulations have revealed that the type of spatial pattern and metapopulation dynamics can switch between two distinct kinds in a seemingly erratic fashion. Thus, for certain values of parameters, the dynamics might represent radial waves for thousands of generations, then turn to distinct spiral waves, which after some time switch back again to radial waves (White *et al.* 1996). Such long-term 'transient' behaviour, which might never lead to any more stationary dynamic behaviour, has been detected in other spatially explicit metapopulation models (Hastings and Higgins 1994; Hendry and McGlade 1995). These observations have profound implications on the way we analyse models and draw conclusions about the behaviour of real interacting metapopulations. At the same time, unfortunately, it becomes increasingly difficult to envision an empirical system that could be used to critically test the complex spatio-temporal dynamics predicted by the models.

7.4 Metacommunity dynamics

There are several ways of extending the above models to assemblages of more than two species, or to metacommunities. Holt (1997) expands the predator–prey model in Section 7.2 to a chain of three species, with the conclusion that metapopulation dynamics may constrain food chain length—the higher-level predators require increasingly large patch networks to persist (Table 7.2). More complex dynamics with multiple stable states might occur if the top predator makes the interaction between the basal species and the intermediate predator more stable (Holt 1997). Holt (1995, 1997) has also developed metapopulation models with two prey species, two types of habitat patch, and a predator. These models demonstrate a kind of landscape-level apparent competition (Holt 1977) effect, with the prey species that is the worse colonizer suffering more from predation, just as in the case of habitat destruction without predation (Section 7.1). With two types of patch each with a specialist prey but the predator using both patch types and prey species, the predator might cause prey extinction in one patch type and thereby itself become confined to the alternative type of habitat. Empirically, this would generate a pattern hard to explain, absence of a prey species from suitable habitat with no hint of the causal mechanism (predation)!

Caswell and Cohen (1991) have developed related patch models for many species. Using the metapopulation framework, they have studied models of species succession (Connell and Slatyer 1977), in which colonization and extinction rates depend on the kinds of species already present. The same models clarify the intermediate disturbance hypothesis (Grubb 1977; Connell 1978; Huston 1979) and the related idea of predator-mediated coexistence (Paine 1966; Harper 1969; Connell 1971), which posit that an intermediate disturbance or predation rate often increases the likelihood of regional coexistence of fugitive species and superior competitors. A major conclusion emerging from the work of Caswell and Cohen (1991) is that the outcome of metacommunity dynamics is critically affected by processes operating at three different scales: the within-patch scale, the scale of migration and the scale of disturbance/predation. The competition and predator–prey models in this chapter provide good examples.

Community assembly has been studied theoretically by assuming a finite or an infinite mainland species pool and Lotka–Volterra dynamics in the focal community with a given distribution of interaction coefficients. Communities may be assembled by assuming simultaneous invasions, accepting those combinations of species in which species have positive densities (May 1974), or somewhat more realistically by assuming sequential invasions, in which the ability of single species to invade the current community is tested. The latter can be accomplished by numerically integrating the system of equations to determine the outcome of invasion. The following general conclusions emerged from Case's (1991) study (the approximate methods used by Post and Pimm (1983) and Drake (1988) in their pioneering studies turned out to be faulty—Morton *et al.* 1996). The probability of successful colonization decreases with the size of the resident community and the strength and variance of interspecific interactions. Thus a superior competitor might fail to invade a species-rich community because of the web of direct and indirect interactions. If invasion is successful, it may lead to the extinction of one or more resident species; invasion-caused extinctions are increasingly likely with increasing size of the resident community. Different invasion sequences may lead to alternative stable species assemblages. The example in Fig. 7.5 on the success rate of avian introductions around the world seems to support the conclusion that communities with more species are more difficult to invade. On the other hand, there is no conclusive evidence that these introductions have led to extinctions of resident

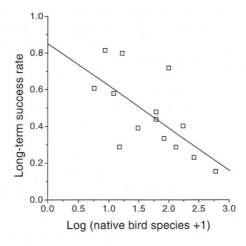

Fig. 7.5   The success rate of avian introductions to islands and mainland sites plotted against the size of the resident bird community. (Based on Case 1991.)

species (Case 1991), implying rather weak interactions. The Hawaiian bird fauna provides one counter-example, with later-introduced species having caused problems for species introduced earlier (Moulton and Pimm 1983, 1986).

These community build-up studies could be extended to classical metapopulations with colonizations occurring from other habitat patches rather than from the mainland, to expand the two-species models of competition (Section 7.1) and predation (Section 7.2) to metacommunities. Holt's (1997) work is a start in this direction. Another line of inquiry could be directed at communities of species with no or very weak interspecific interactions, typically consisting of species at the same trophic level. Community assembly then consists of just summing up the independent occurrences of species. This approach might seem trivial and uninteresting, but it is not. It is possible that many communities consist of species that interact so weakly that no great damage is done by ignoring interactions completely. In the following section I describe a model of this type that generates species–area curves and distribution–abundance relationships for large assemblages of species.

7.5 Species-richness of communities and distribution of species

Two multispecies distributional patterns have become firmly established in ecology. The first pattern is the species–area (SA) curve, the increasing number of species with increasing patch (island) area (Williams 1943; Preston 1948; MacArthur and Wilson 1967; Connor and McCoy 1979; Rosenzweig 1995). The second pattern is the positive relationship between the distribution and average local abundance of species, which I here call the distribution–abundance (DA) curve (Hanski 1982a; Brown 1984; Lawton 1993; Gaston 1994). Amazingly, these two patterns have been studied entirely independently, probably because no explanatory variable has been employed that would explicitly connect the two dependent variables, species number on islands and the extent of distribution of individual species.

The missing link is provided by the key variable of metapopulation models, the incidence J_{ij}, defined as the long-term probability of species i being present in habitat patch j (Section 5.3). Considering a set of R patches and a community of Q species, there are two obvious ways of summing up the incidences. The sum of the J_{ij} values across the species gives the expected number of species in patch j, S_j, while the sum of the J_{ij} values across the patches gives the expected distribution of species i, D_i. In the metapopulation literature, regional distribution is typically measured by the fraction of occupied patches, $P_i = D_i/R$. Note that S_j and P_i are the key variables of the SA and DA curves, respectively. Below I describe a model (Hanski and Gyllenberg 1997) which is built upon the incidence concept and which provides a unified framework for the study of species–area and distribution–abundance relationships.

The model

Let us denote by $K_{ij} = w_i A_j$ the 'carrying capacity' of species i in patch j, where w_i is the constant density of species i and A_j is the area of patch j. m_A and σ_A^2 are the mean and the variance of patch areas, and m_w and σ_w^2 the mean and the variance of species' densities. Without loss of generality, I assume that m_w equals unity.

In the spirit of the Levins model, the rate of change in the probability p_{ij} of species i being present in patch j, in the absence of any interspecific interactions, is given by (Section 5.2):

$$\frac{dp_{ij}}{dt} = C_i(t)[1 - p_{ij}] - e_{ij} p_{ij}, \tag{7.7}$$

where $C_i(t)$ sets the colonization rate of empty patches and e_{ij} sets the extinction rate of extant populations. For simplicity, let us assume that e_{ij} is given by K_{ij}^{-1}, which is supported by many empirical studies (Gilpin and Diamond 1976; Hanski 1994a; corresponding to $x = 1$ in eqn 5.9; the rate parameter e in eqn 5.9 has been absorbed in the unit of patch area; Hanski and Gyllenberg 1997). The incidences are then given by:

$$\hat{p}_{ij} = J_{ij} = \frac{\hat{C}_i w_i A_i}{\hat{C}_i w_i A_j + 1}, \tag{7.8}$$

where $\hat{C}_i$ is the equilibrium value of $C_i(t)$. There are two distinct cases to consider, which require different assumptions about $C_i(t)$. In mainland–island metapopulations, colonization takes place from a permanent mainland community, as in the dynamic theory of island biogeography (MacArthur and Wilson 1967). In classical metapopulations, species are distributed in a network of habitat patches without an external mainland, and colonization occurs from the presently occupied habitat patches. Box 7.1 derives explicit formulas for species number on islands and the distribution of species in mainland–island metapopulations and gives a recipe for calculating these quantities numerically for classical metapopulations.

Predictions and ecological implications

The most widely used model of SA curves is the power function model (Arrhenius 1921), $S = kA^z$, which is generally employed in the $\log_e$-transformed form (Preston 1960; MacArthur and Wilson 1967; Connor and McCoy 1979):

$$\log S = \log k + z \log A.$$

This model has the obvious drawback of being unbounded, contrary to common sense and empirical results (Schoener 1976; Connor and McCoy 1979; Buys *et al.* 1994; Williams 1995). However, the power function model is widely used because ecologists have found that $\log S$ generally increases roughly linearly with $\log A$ for a large range of island areas. Field studies have typically focused on the measurement

Box 7.1 Slopes of the species–area (SA) and distribution–abundance (DA) curves

In mainland-island metapopulations, all colonists arrive from the permanent mainland community, where the density of species i is given by w_i. $\hat{C}_i$ is given by cw_i, where the value of c decreases with increasing distance to the mainland. With this choice of $C_i(t)$, eqn 7.8 becomes:

$$J_{ij} = \frac{cA_j w_i^2}{cA_j w_i^2 + 1}.$$

Assuming that Q (species number on mainland) is large and $\log w$ is uniformly distributed with zero mean, we obtain, after some calculation, the expected number of species on island j as:

$$S_j = \Sigma_i J_{ij} = \frac{Q}{4\sigma_w\sqrt{3}}\log \Gamma,$$

where (dropping the subscript j):

$$\Gamma = \frac{1 + cAe^{2\sigma_w\sqrt{3}}}{1 + cAe^{-2\sigma_w\sqrt{3}}}.$$

The slope of the SA curve is then given by:

$$\frac{\partial \log S}{\partial \log A} = \frac{1 - \Gamma^{-1}}{\Gamma \log \Gamma}.$$

Very similar results are obtained for back-to-back exponential and lognormal distributions of w, although the expressions are more complicated in these cases (Hanski and Gyllenberg 1997).

The distribution of species i is given by:

$$P_i = \frac{1}{R}\Sigma_j J_{ij} = 1 - \frac{1}{2\sigma_A\sqrt{3}}\log \frac{cw_i^2 + q_1}{cw_i^2 + q_2},$$

and the slope of the DA curve is given by (dropping the subscripts):

$$\frac{\partial \log (P/[1 - P])}{\partial \log w} = \frac{2cw^2 q}{(cw^2 + q_1)(cw^2 + q_2)P(1 - P)}.$$

where:

$$q_1 = e^{-m_A + \sigma_A\sqrt{3}}, q_2 = e^{-m_A - \sigma_A\sqrt{3}}, q = \frac{1}{2\sigma_A\sqrt{3}}[q_1 - q_2].$$

In classical metapopulations, the colonization rate of empty patches is proportional to the pooled abundance of the species in the landscape, $C_i(t) = cw_i\Sigma p_{ij}(t)A_j$. Substituting eqn 7.8 into the expression $\hat{C}_i = cw_i\Sigma J_{ij}A_j$ for the equilibrium of $C_i(t)$ gives:

$$1 = cw_i^2 \sum_{j=1}^{R} \frac{A_j^2}{\hat{C}_i w_i A_j + 1},$$

from which $\hat{C}_i$ can be solved provided that $cw_i^2\Sigma A_j^2 > 1$, which is a necessary and sufficient condition for species i to persist in the patch network. The slopes of the SA and DA curves can then be calculated numerically using eqn 7.8.

of the slope of the linear regression of log S against log A. The present model predicts approximately linear SA curves (Box 7.1), which are comparable with many empirical results (Fig. 7.6). The model makes several testable predictions:

- The slope increases with increasing isolation (decreasing c), which has been observed for many archipelagos (MacArthur and Wilson 1967; Rosenzweig 1995). The model even explains the exception, small slopes in the most isolated oceanic archipelagos (Schoener 1976; Connor and McCoy 1979), on the realistic assumption that in very isolated archipelagos colonization occurs largely among the islands rather than from the mainland (see below).
- Denoting the per-year colonization and extinction probabilities by λ and μ, one obtains (Hanski and Gyllenberg 1997) $cA = (\lambda/\mu)E[w^2]$, where E denotes the expected value. The value of the slope of the SA curve is a function of cA in this model (Hanski and Gyllenberg 1997), and hence the slope is a function of the ratio of colonization to extinction probabilities as has been previously suggested (Schoener 1976).
- The slope increases with decreasing variance of the species-abundance distribution, σ_w; this prediction has not been tested.
- The slope is greater for mainland–island metapopulations (archipelagos) than for classical metapopulations without an external mainland (Fig. 7.7), which is one of the key empirical findings in the SA literature (MacArthur and Wilson 1967; Rosenzweig 1995). The difference between the slope values is usually attributed to 'transient' species inflating the species number in small study plots on the mainland (Connor and McCoy 1979), but the present model shows that shallow slopes are generally expected for metapopulations without an external mainland.
- There is a good quantitative agreement between the predicted and observed values of the SA slope (Fig. 7.7).

In the literature on the distribution–abundance relationship (Hanski *et al.* 1993c; Lawton 1993; Gaston 1994), the emphasis has been in the demonstration that a positive relationship exists, whereas there is no widely accepted model for the DA curve. Typically, the rarest species on the mainland do not occur on any island ($P = 0$), whereas the commonest species are found on all ($P = 1$) or most islands. This observation suggests a logistic model for the DA curve, with P increasing from zero to one with increasing density of the species. Hanski and Gyllenberg (1997) assumed that P is a function of the logarithm of density, log w. This relationship can be linearized with the logit-transformation, $\log[P/(1 - P)]$. Box 7.1 gives the derivative of $\log[P/(1 - P)]$ with respect to log w. The model predicts that this slope (of the DA curve) ranges from 1 to 2 (Fig. 7.6). There are no extensive data for quantitative testing of this prediction, because field studies have seldom recorded the densities of species on the mainland. Figure 7.6 gives one example. In the model, the slope of the DA curve depends on three parameters, the mean (m_A) and the variance (σ_A) of island areas, and the parameter combination cw_i^2. It is worth stressing that this explanation of the DA curve is a version of the metapopulation hypothesis

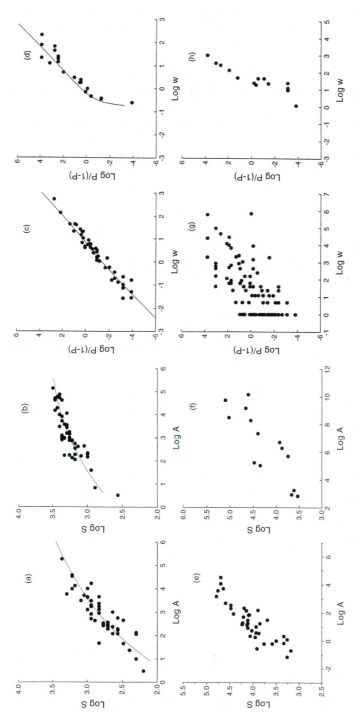

Fig. 7.6 Theoretical and empirical examples of approximately linear SA and DA curves. (a) and (b) show the logarithm of species number against the logarithm of island area in the mainland–island and classical metapopulations, respectively, predicted by the model. (c) and (d) show corresponding examples of predicted DA curves, with the logarithm of $P/(1 - P)$ plotted against the logarithm of w. The continuous lines show the expected values, the dots give stochastic realizations obtained by assigning species to a set of 50 islands using their predicted incidences on these islands. (e) to (h) give corresponding empirical examples, for moths on islands (Nieminen 1996a,b) (e and g) and for birds on mainland (Witkowski and Pŕonka 1984; Harms and Opdam 1990) (f and h). The following parameter values were used in the model examples: $Q = 50$, $m_A = 3$, $\sigma_A = 1$, $\sigma_w = 1.5$ and $c = 0.01$ (mainland–island model) or $c = 0.00005$ (classical metapopulation).

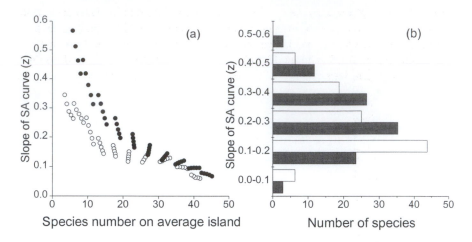

Fig. 7.7 (a) The predicted slope of the SA curve plotted against the species number on an average island ($m_A = 3$) in the mainland–island (filled dots) and classical metapopulations (open circles), respectively. Results were obtained for the following parameter values: $R = 50$, $Q = 50$, $m_A = 3$, $\sigma_A = 1$, $\sigma_w = 1.5..2$ (five equally spaced values) and $\log c = -5..5$ (10 equally spaced values) in the mainland–island model and $\log c = -15..-5$ in the classical metapopulation model. (b) Distribution of empirical slope values, shown separately for data sets from archipelagos (mainland–island metapopulations, filled bars, $n = 35$) and mainlands (classical metapopulations, open bars, $n = 16$). The difference between the two sets is significant at the 5% level (data from Connor and McCoy 1979, omitting data sets in which areas covered <3 orders of magnitude).

(Hanski 1982a; Gyllenberg and Hanski 1992; Hanski *et al.* 1993c; Section 9.4), but an explanation which explicitly recognizes differences in species' carrying capacities (Nee *et al.* 1991; Venier and Fahrig 1996). The model does not include the effect of migration on local dynamics (the rescue effect), which might steepen the DA curve in real metapopulations (Gyllenberg and Hanski 1992; Section 9.4).

There is a large literature on the species–area relationship (Rosenzweig 1995), and many models have been described that purport to explain the observed patterns. Some models are entirely phenomenological, based on the canonical lognormal species abundance distribution (Preston 1948, 1960; May 1975), or random (Coleman 1981) or non-random (Williams 1995) but fixed placement of individuals in space. Other models are mechanistic and dynamic, but less general than the present one in ignoring interspecific abundance differences (Schoener 1976; Wissel and Maier 1992; Caswell and Cohen 1993). The latter models fail to predict realistic SA curves without making the implausible assumption of complete density compensation among competing species. On the other hand, hypotheses ignoring extinction–colonization dynamics fail to predict the effect of isolation on the slope of the SA curve. It thus seems that a satisfactory model of the SA curve involves ecological differences among the species, such as are reflected in the DA curve. Both the SA and DA curves have been explained in the literature by the same 'competing' hypotheses of ecological specialization (habitat heterogeneity) and extinction–colonization dynamics (Connor and McCoy 1979; Lawton 1993). It is satisfying that these hypotheses are here merged into the same model, which also unites the two empirical patterns.

Part II

FIELD STUDIES

Metapopulation ecology, like population ecology in general, has two facets, a set of concepts and models that constitute the metapopulation theory, and a body of empirical knowledge that has been accumulated over the years by field ecologists. The mingling of the two has not been as thorough as one would hope, partly because the theoretical and empirical researchers tend to read their own literature and to follow their own research traditions. The two traditions are necessary, but there has not been enough emphasis on modelling of generally important issues with clear-cut testable predictions, and not enough emphasis on empirical work addressing significant issues for model construction. Some comfort may be taken from the observation that, in this respect, metapopulation ecology is no worse, and might be better, than some other fields of ecology.

The following three chapters discuss empirical studies and some conceptual issues that are pertinent to the brand of metapopulation ecology covered in this book. Many other questions could have been included, but my choice of topics was guided by the conceptual frame established in Part I. The following chapters include several references to the Glanville fritillary butterfly, which is the focal species in Part III, because these examples could not be replaced with anything else. More comprehensive reviews of many topics could be presented, but I have often opted for describing especially informative examples rather than summarizing limited and heterogeneous data, which might lead to the false impression that sound resolution of a particular empirical question has already been reached. I also feel that it is not productive to dismiss some ecological scenarios as uninteresting if some alternative scenarios are more frequent by somebody's count. Any real-world situation is interesting, and a better understanding of a multitude of examples adds to our comprehension of ecological interactions in general.

I start with the fundamental question about the spatial structure of populations. Metapopulation ecology makes the simplifying assumption that landscapes consist of discrete habitat patches occupied by more or less discrete local populations. How common is this situation? How good is the evidence for source–sink population structures? Classical metapopulation ecology is especially concerned with population turnover; what do we know about population turnover empirically? Does increasing fragmentation of populations and high turnover rate towards range margins set the limit of species' geographical ranges? Our empirical understanding of these and many other similar issues about metapopulation dynamics remains fragmentary, partly because the classical approach to population ecology has for a long time been focused on individual populations, but also because of the practical difficulties of investigating population ecology on large spatial scales.

Metapopulation models typically assume that we can identify the suitable habitat of a species independently of the presence of the species. Chapter 9 reviews the types of evidence that can be mustered in empirical studies for the presence of empty but suitable habitat. Some of this evidence is indirect and based on the influence of habitat patch area and isolation on the presence of local populations. Though not used in the very simplest models, patch area and isolation effects on local extinction and colonization represent some of the most useful information that can be gleaned in empirical studies and used in models; the incidence function model (Section 5.3) in particular is constructed with the help of area and isolation effects. The remaining sections in Chapter 9 deal with the distribution and abundance of species, a theme that is much broader than can be meaningfully covered in this context, but to which metapopulation ecology contributes a distinct perspective. The final section summarizes the core–satellite species hypothesis, which I proposed in 1982 and which has inspired many empirical studies.

Chapter 10 addresses the application of metapopulation concepts and models to conservation. For other recent reviews on this topic see Doak and Mills (1994), Fahrig and Merriam (1994), Harrison (1994), McCullogh (1996) and Hanski and Simberloff (1997). Metapopulation ecology has largely replaced the dynamic theory of island biogeography as the basic population ecological model for conservation. This replacement involves paradoxes that are discussed in Chapter 10. A key conservation worry is habitat destruction, involving habitat loss and fragmentation as well as habitat deterioration; metapopulation ecology is rightly expected to make predictions about the population-biological consequences of habitat destruction. The reverse question might also be raised—how should we attempt to introduce a species into an empty network of suitable habitat patches? The rest of Chapter 10 deals with selected case studies in which the metapopulation approach has been employed in one way or another.

8

Spatial structure of populations

8.1 The nature of populations

Some species have practically continuous populations over large areas of suitable habitat; one example is the common shrew (*Sorex araneus*) in coniferous forests in Fennoscandia. Even in these cases, the environment is not quite as continuous as it might seem at first, and population density is likely to vary from one place to another. For most species, most landscapes are patchy, because the species have evolved specific habitat requirements and because the landscapes are complex habitat mosaics. It would be interesting to have a detailed description of spatial population structure for some large assemblage of species in some large area. Denoting by h_i the percentage of the landscape consisting of suitable habitat for species i, one wonders what would be the distribution of the h_i values in large communities? And what would be the distribution of p_i, the fraction of the suitable habitat that is actually occupied at one point in time? It is frustrating that we have such basic data for only a handful of individual species, not for communities.

By systematically sampling a large heterogeneous landscape one might estimate the product $h_i p_i$, given by the fraction of sampling points in which the species was observed. Figure 8.1 gives two examples on birds and ground beetles in areas of 500 and 30 km^2, respectively, in southern Finland. In both cases roughly half of the species occurred in less than 10% of the total area, that is, half of the species had strikingly patchy distributions at the regional scale. Unfortunately, it is not possible to infer from these data whether species have restricted distributions because of small h_i, because of small p_i, or both.

It is commonly assumed that many species are structured into few large populations surrounded by many small ones (Schoener and Spiller 1987; Harrison 1991, 1994; Schoener 1991; Simberloff 1994, 1995). Given that the sizes of the habitat patches suitable for a particular species are affected by many largely independent processes, one might expect a lognormal distribution of patch areas. The *ca.* 1600 habitat patches (dry meadows) in which the Glanville fritillary butterfly breeds in southern Finland (Part III) have indeed a lognormal size distribution, with an average area of 1400 m^2, minimum of 4 m^2, and maximum area of 3 ha. Assuming that the extinction rate scales as the inverse of the patch area, the patch-specific extinction rate also is lognormally distributed. There might be factors that reduce the variance in extinction rates, but clearly we have reason to assume much variation from population to population. Whether a metapopulation is likely to persist for a long time just because of low risk of extinction of the largest local

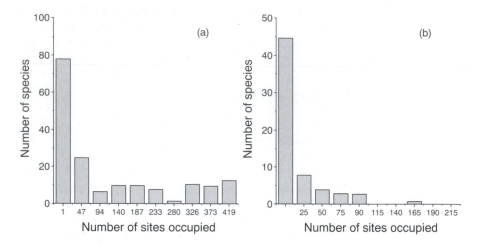

Fig. 8.1 Distributions of occupancy frequencies of birds (a) and ground beetles (b) on large sampling grids located in heterogeneous landscapes. In (a), the study area consists of 465 contiguous sampling quadrants of 1 km² each (Lammi, southern Finland, 1994–1996; T. Pakkala and J. Tiainen, unpublished). In (b), there are 240 sampling sites of 250 by 250 m² placed systematically within a continuous area of 30 km² (H. Kinnunen, unpublished).

populations becomes a question of how large the largest habitat patches are. Large variance in patch sizes by itself does not mean that long-term persistence is not based on extinction–colonization dynamics. In the Glanville fritillary metapopulation, the largest patches are 10 000 times bigger than the smallest patches, but in this case even the largest populations are small enough, and fluctuate so much, that they have a substantial risk of extinction (Section 11.2).

Large communities typically show a lognormal distribution of species abundances (Preston 1948, 1962; MacArthur 1957; May 1975). It is not always clear what is meant by 'abundance' in this context, but most often it probably refers to an estimate of local density. For many species in most communities local density is so low that one would expect stochasticity to play a significant role in local dynamics and to lead to local extinctions, thus reducing the area of occupied suitable habitat. Naturally rare species may have evolved efficient means of reaching suitable habitat, but nonetheless the predominance of rare species in local communities suggests that extinction–colonization dynamics are widespread in nature.

Changing views about the structure of butterfly populations

Some butterfly species, including the monarch butterfly in North America (Scott 1986) and the red admiral in Europe (Ford 1945), perform spectacular long-distance migrations. Other species are thought to be very sedentary, for example the silver-studded blue (*Plebejus argus*) in North Wales, where in 40 years it failed to colonize habitat patches that were only a few kilometres from large populations (Thomas and Harrison 1992; Lewis *et al.* 1997). Differences in the mobility of butterfly species are so striking that an idea developed of their falling into two types, those with 'open' or 'closed' population structures. The notion of closed populations can be traced back to the early studies by E. B. Ford and his colleagues, who observed that the vast majority of individuals in many species remain in the natal habitat patch (Ford and Ford 1930; Ford 1945). Paul Ehrlich and his colleagues produced additional support for the sedentary behaviour of butterflies with their studies on the checkerspot *Euphydryas editha* and other species in California (Ehrlich *et al.* 1975; Singer and Ehrlich 1979; Ehrlich 1984). Thomas (1984) eventually declared the distinction in plain words.

The evidence on which the concept of a closed population was established was not very strong, however. Mark–recapture studies had suggested that most individuals stay put, but nobody had conducted extensive mark–recapture work simultaneously in many habitat patches (populations) to really establish the extent of migration. There are severe logistical problems in doing so as well as unresolved statistical problems in estimating the rate of between-population migration (Ims and Yoccoz 1997; but see Box 2.1). Nonetheless, when mark–recapture studies were extended to networks of habitat patches in the 1990s, the conclusion soon emerged that the closed butterfly populations are not nearly as closed as previously thought (Hanski *et al.* 1994; Hanski and Kuussaari 1995; Nève *et al.* 1996; Thomas and Hanski 1997). Lifetime emigration rates of specialist butterflies from typical habitat patches (0.01 to 1 ha) are often 10–30% or more (Thomas and Hanski 1997), and although most migrants travel only some hundreds of meters, some individuals reach patches several kilometres away from the natal patch (Table 8.1). Records of butterflies in places clearly not suitable for breeding, such as inner cities, light vessels, etc., support the view that individuals of even the most sedentary species occasionally move relatively long distances (Dennis and Shreeve 1996).

The changing views about the spatial structure of butterfly populations are worthy of reflection for several reasons. Butterflies represent one of the best known taxa. If our understanding of their spatial population structures was misleadingly simplified only a few years ago, our understanding of the population structures of many other taxa is probably so even today—and I do not intend to say that the final word has yet been said on butterflies! There is probably nothing very special about butterflies, and similar population structures might be expected in many other taxa, especially those which are specialized in their feeding habits or microhabitat selection. It is also instructive to observe that while the idea of closed butterfly populations was germane to the classical metapopulation concept, with little migration among local populations, the present view with more migration calls for the use of models including the effect of migration on local dynamics (Sections 4.3 and 5.5). Often there is no clear distinction between local and metapopulation dynamics, no two distinct time-scales, rather the two blend to each other gradually, but there is still a definite need to account for the spatial structure of populations.

Table 8.1 Results of mark–release–recapture studies on butterflies in which several local populations were studied simultaneously

Species	Number of habitat patches studied	Longest observed migration distance (km)	Longest observed colonization distance (km)
Hesperia comma	2	0.3	8.6
Colias alexandra	11	8.0	–
Colias meadii	3	1.3	–
Lysandra bellargus	3	0.3	–
Pseudophilotes baton	2	1.5	–
Maculinea arion	2	5.7	–
Plebejus argus	4	0.05	*ca.* 1
Proclossiana eunomia	13	4.7	>5
Melitaea cinxia	50	3.0	–
Mellicta athalia	3	1.5	2.5
Eurodryas aurinia	–	–	15–20
Euphydryas editha	9	5.6	4.4
Euphydryas anicia	4	3.0	–
Melanargia galathea	44	7.3	–

From Hanski and Kuussaari (1995), who give the original references.

8.2 Population turnover

Much evidence for population turnover (Diamond 1969, 1971, 1984; Terborgh and Faaborg 1973; Lynch and Johnson 1974; Jones and Diamond 1976; Diamond and May 1977; Reed 1980; Schoener and Spiller 1987; and many others) was accumulated in the wake of the dynamic theory of island biogeography (MacArthur and Wilson 1967), which is closely related to metapopulation models (Section 10.1). Some of the evidence has been subsequently discredited on the grounds that it did not come from 'real' populations but rather from more or less arbitrary subdivisions of 'true' populations (Smith 1975; Simberloff 1976), or because it did not represent stochastic extinctions as assumed by the theory (Simberloff 1974). The question about the actual causes of population extinction is indeed an important one and continues to evoke dispute (next section), and there are problems with the estimation of extinction rates from the kind of data that ecologists typically have available (Box 8.1). Nonetheless, there is no doubt that small populations face a high risk of extinction and that with increasing population size the risk of extinction practically always decreases (Diamond 1984; Schoener and Spiller 1987; Schoener 1991; Fig. 9.2). The widely observed nested subset pattern of island and habitat patch occupancy in communities that are entirely or primarily structured by extinctions testifies to the significance and predictability of area-related and hence population size-related risk of extinction (Patterson 1987; Cutler 1991; Wright *et al.* 1997). Similarly, there is little argument that the rate of colonization would not very generally decrease with increasing isolation (Fig. 9.2), though different taxa have, of course, different powers of migration, and hence a given level of physical isolation might mean very different levels of effective isolation for different taxa.

Given the definite relationships between population size and extinction risk, and between habitat patch size and extinction risk, it is not very informative to tabulate isolated rates of population extinction without reference to patch or population sizes, but at least such figures reflect the magnitude of extinction rates in the kinds of populations that ecologists have studied. Fahrig and Merriam (1994) summarized the results of 19 studies on plants and animals, reporting annual extinction rates mostly ranging from 5 to 30%. In another review of 21 studies, Schoener (1983) found that annual extinction rate increased from 1–10% in terrestrial vertebrates and plants to 10–100% in terrestrial arthropods (Schoener actually analysed annual turnover rate, including extinctions and colonizations, but because he divided the number of turnover events by the sum of the species number in the beginning and at the end of the study period, the measured rate is also an approximate estimate of the extinction rate; Box 8.1). Schoener (1983) observed that the annual extinction rate declined approximately linearly with the generation length of the species. In a plant study on seven species of Asteraceae on small islands off Vancouver Island, British Columbia, Cody and Overton (1996) found high annual turnover rates ranging from 0.09 to 0.57. Fischer and Stöcklin (1997) studied 26 remnant sites of nutrient-poor calcareous grasslands in the Swiss Jura mountains. These sites had a total of 1181 local populations of 185 species in 1950, but only 719 populations in 1985. Local populations that had gone extinct tended to be small in 1950, and they tended to represent habitat specialist species with a short life cycle.

8.3 Mechanisms of population extinction

The most significant *cause* of population extinction is undoubtedly habitat loss. Metapopulation studies attempt to improve our understanding of how habitat loss at the landscape level, involving the fragmentation of the remaining habitat, affects metapopulation persistence. The most significant *mechanism* of population extinction in more or less stable habitats is environmental stochasticity, and the most significant *correlate* of the actual risk of population extinction is small population size. So much is widely agreed. Disagreement remains about the magnitude and relative significance of these processes. Rather than review all possible mechanisms of extinction, I comment here on the environmental causes and mechanisms and the role of migration in extinctions. The comprehensive example on the Glanville fritillary butterfly in Section 11.2 illustrates how, in a particular metapopulation, a wide range of extinction mechanisms typically operates.

Classical metapopulation models assume that local extinctions are stochastic, in other words that extinctions represent the disappearance of populations from habitat patches in which the environmental conditions remain generally favourable for the species. By definition, the environment is not to be blamed if an extinction is caused by demographic or genetic stochasticity, or by interaction with a competitor or a

Box 8.1 Estimates of population turnover rate

The basic measure of population turnover on a true or a habitat island is defined as:

$$T_\Delta = \frac{E_\Delta + C_\Delta}{S_t + S_{t+\Delta}},$$

where E_Δ and C_Δ are the observed numbers of extinction (disappearance) and colonization (appearance) events over the time period Δ, and S_t and $S_{t+\Delta}$ are the observed numbers of species at the beginning and at the end of the time period. Diamond and May (1977) observed that this formula is likely to underestimate the true turnover rate per unit time (usually one year) if Δ is large; an extinction may be followed by colonization during the same time interval, hence no turnover is recorded. Russell *et al.* (1995) present an equation to estimate the actual probabilities of extinction (μ) and colonization (λ) in unit time,

$$T_\Delta = \frac{\mu[1 - (1 - \mu - \lambda)^\Delta]}{\mu + \lambda},$$

which can be fitted to empirical data on turnover events as a function of Δ, the census interval. When applied to extensive data on breeding birds on island off the coast of Britain, Russell *et al.* (1995) found that the model gave a systematic misfit due to long-term changes in species numbers on islands. They then went on to include a trend in species number in their model.

Clark and Rosenzweig (1994) and Rosenzweig and Clark (1994) estimated turnover rate by directly examining a Markov chain model of island occupancy, with λ denoting the probability of colonization in unit time, μ denoting the probability of extinction, and δ the probability that a species present at the beginning of the time interval is absent at the end. δ thus equals $\mu(1 - \lambda)$, which is the probability that an existing population disappears and is not replaced in unit time. The stationary probability of species being present is given by $\lambda/(\lambda + \delta)$. Clark and Rosenzweig (1994) show how the above probabilities can be estimated with maximum likelihood from regular or sporadic census data. Parameters can be estimated for individual species if long series of census data are available, or they can be estimated for islands using pooled data for many species, in which case only a few censuses are needed. In the latter case, one has to assume that all species have the same extinction and colonization probabilities, which might, of course, not be the case. A nice feature of the maximum likelihood estimates is that the result can be visualized as a graph showing the likelihood of different combinations of parameter values having produced the data. Figure 8.3 gives two examples.

predator. More problematic is the operation of environmental stochasticity, which might involve changes in the quality of resources. For instance, host plants for herbivores might wither during a drought causing elevated mortality. Is this an instance of 'stochastic' extinction, or 'deterministic' extinction due to the habitat having turned unsuitable? Caughley (1994), Harrison (1994), Simberloff (1994) and Thomas (1994a) among others have suggested that most extinctions are due to temporary or more permanent environmental changes, not to populations fluctuating to extinction in stable environments. No doubt examples of all kinds of extinction can be found, but there is so much empirical evidence to demonstrate the extinction-proneness of small populations in apparently suitable habitat patches (Section 9.2)

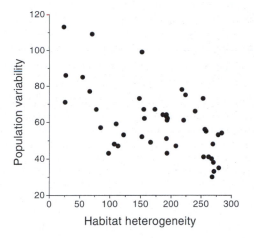

Fig. 8.2 Relationship between a measure of spatial heterogeneity of habitat patches and population variability in the bush cricket *Metrioptera bicolor*. Each symbol represents one population. (Based on Kindvall 1996b.)

that it would be unreasonable to attribute the bulk of extinctions to permanent habitat change.

One of the mechanisms by which increasing patch area might reduce the probability of population extinction is the changing-environment scenario (Section 2.2). Environmental conditions change in time and any particular habitat type, or a combination of environmental conditions, might disappear temporarily. Populations in large habitat patches might have low extinction rates because large patches are mosaics of many habitat types, all of which are unlikely to become unfavourable at the same time. Figure 8.2 gives a nice empirical example showing how the level of temporal variability in bush cricket populations decreases with increasing habitat heterogeneity (Kindvall 1996b). Increasing population variability is expected to increase the risk of extinction, hence here it is environmental heterogeneity which most likely reduces the risk of population extinction.

Extinctions in successional habitats

Classical metapopulation models were constructed with stochastic extinctions in mind, but the models can be adapted to situations in which the environment is in flux. For instance, Stelter *et al.* (1997) have studied the dynamics of the grasshopper *Bryodema tuberculata* inhabiting open and dry habitat patches on gravel bars along rivers in Central Europe. The dynamics of the habitat patches are affected by succession and floods. The effect of catastrophic floods were deemed to have a twofold effect on metapopulation persistence. If floods are frequent many local populations become eliminated simultaneously, increasing the probability of metapopulation extinction. On the other hand, if floods are very infrequent, a metapopulation might go extinct because succession eliminates local populations at a high rate. Another example of metapopulation dynamics in a network of successional habitat patches is the marsh fritillary (*Eurodryas aurinia*) metapopulation in a network of small forest clearings in eastern Finland (Klemetti 1998). The host plant *Succisa pratensis* occurs commonly on clearings 2 to 10 years old. Long-term persistence of the metapopulation requires that new clearings appear at a sufficiently high rate to compensate for the inevitable local extinctions. Other examples of metapopulations living in successional habitats include the Furbish lousewort (*Pedicularis furbishiae*; Section 10.5), the checkerspot butterflies *Euphydryas gillettii* (Debinski 1994) and *Mellicta athalia* (Warren 1987) and the grasshopper *Trimerotropis saxatilis* (Gerber and Templeton 1996). Gopher mounds (Wu and Levin 1994) and prairie dog colonies (Reading *et al.* 1993) provide temporary patchy habitats for plants and animals, respectively, with distinct metapopulation structures at larger spatial scales.

Reduced risk of extinction: the rescue effect

A particular question about population extinction in the metapopulation context is to what extent extinction risk is reduced by migration from other populations enhancing the size of the focal population—the rescue effect (Brown and Kodric-Brown 1977). Section 4.2 described the concept and the population dynamic consequences of the rescue effect, including alternative stable equilibria. How strong is the empirical evidence for the rescue effect?

There is little doubt that, in principle, immigration can increase the sizes of local populations, which is the core idea in source–sink dynamics (next section). Given that extinction risk generally decreases with increasing population size (Fig. 9.2), immigration should logically reduce extinction risk. Unfortunately, demonstrating the rescue effect in practice is not easy; merely showing that there is substantial migration among populations is not enough (Stacey *et al.* 1997). I resort here to the large data base on the Glanville fritillary butterfly to present one good example. Table 8.2 summarizes the results on the annual extinction rate of populations that had 1, 2, 3–5 and >5 larval groups. Extinctions were modelled with logistic regression, including the following explanatory variables (Hanski *et al.* 1995b): patch area, regional trend in population sizes, and a measure of the pooled size of the neighbouring populations, denoted by S (Box 11.1). A significant rescue effect is manifested as a significant effect of S on extinctions, because when there are many populations close to the focal one (large S), there is, on average, much immigration to that population. The results demonstrate a significant rescue effect in the case of populations that had 1 or 2 larval groups, but not in the case of the larger populations (Table 8.2). This makes sense, because a given number of immigrants has a greater impact on the dynamics of small than large populations.

Table 8.2 The rescue effect reduces the risk of local extinction in small populations of the Glanville fritillary butterfly

Population size	Extinct	n	Average S	The rescue effect	
				t	P
1	Yes	150	2.55		
	No	76	2.84	−2.97	0.003
2	Yes	46	2.78		
	No	58	3.12	−2.24	0.025
3–5	Yes	46	2.88		
	No	202	2.75	−0.63	0.527
>5	Yes	14	3.31		
	No	204	2.83	1.42	0.155

The table gives the sizes of local populations in terms of the number of larval groups in autumn 1993; the numbers of populations that went extinct and survived; a measure (S) that is proportional to the numbers of immigrants arriving at the population; and t-test results for the rescue effect, which was measured by the effect of S on extinction (logistic regression model, which also included the effects of patch area and regional trend in population sizes on extinction).

Increased risk of extinction: emigration losses

Populations in small patches are often likely to receive large numbers of immigrants in relation to the size of the local population, but the other side of the coin is that the per-capita emigration rate is likely to be high in small patches with large perimeter-to-area ratio. Many empirical studies on insects have demonstrated that per-capita emigration rate increases with decreasing patch area (Kareiva 1985; Turchin 1986; Back 1988; Kindvall 1995; Hill *et al.* 1996, Kuussaari *et al.* 1996; Sutcliffe *et al.* 1997). Box 2.1 describes a method for quantifying the scaling of emigration rate with patch area in mark–recapture studies of metapopulations. This scaling imposes a lower limit on the area of occupied habitat patches. Furthermore, as immigration rate decreases with increasing isolation, the deterministic critical minimum patch size for occupancy increases with increasing isolation. Thus populations might be absent from small isolated habitat patches because the local growth rate plus the immigration rate is less than the emigration rate (Thomas and Hanski 1997). The extent to which migration rather than extinction–colonization dynamics accounts for the observed patch area and isolation effects on occupancy remains an empirical question which cannot be resolved without good data on population sizes and on the rate and scale of migration.

8.4 Sources and sinks

Michael Singer, Chris Thomas and their colleagues have studied Edith's checkerspot butterfly (*Euphydryas editha*) in the Sequoia National Forest in California for 20 years (Singer and Thomas 1996; Thomas *et al.* 1996). The butterfly used to live in rocky outcrop areas surrounded by coniferous forests, where females oviposit on two host plants, *Pedicularis semibarbata* and *Castilleja disticha* (both Scrophular-iaceae). In the late 1960s, the environment was changed by logging, and by 1978 the butterfly had colonized a new habitat, clear-cuts, where females oviposited on a novel host plant, *Collinsia torreyi* (Thomas *et al.* 1996). In the 1980s, the butterfly used the habitat mosaic of rocky outcrops and clear-cuts with different larval host plants. Survival from egg to adult was higher on *C. torreyi* in clear-cuts than on *P. semibarbata* on rocky outcrops (Moore 1989), giving an order of magnitude higher population growth rate in the former habitat (Thomas *et al.* 1996). Migration rate between the two habitats was asymmetric, twice as high from clear-cuts to outcrops than vice versa. In the outcrop patches, butterfly density decreased with increasing distance from the nearest clear-cut, suggesting that the clear-cut acted as a source and migration increased the density of butterflies on outcrops (Thomas *et al.* 1996).

A series of environmental perturbations occurred in 1989 to 1992, eventually killing virtually all host plants in clear-cuts but not on rocky outcrops. Following a sequence of three disasters, by 1993 the density of the butterfly in clear-cuts had been reduced to 10^{-5} of the density in 1986 (Thomas *et al.* 1996). The source populations had thereby been wiped out, and an exceptional experiment had been created—what would happen in the populations on rocky outcrops? Larval density on outcrops declined to one third, and the reduction was most severe in areas close to the former source populations (Thomas *et al.* 1996). After one more generation, the isolation effect disappeared and the outcrop populations did not show signs of being reduced to extinction. Extinction was not expected, because the butterfly had survived on the outcrops in the absence of source populations 30 years earlier, before the clear-cut areas existed. The outcrop populations were not true sinks, but they were pseudo-sinks in the sense of Watkinson and Sutherland (1995)— immigration had pushed the equilibrium population density beyond the local carrying capacity.

The results of Thomas *et al.* (1996) provide an exceptionally convincing example of source–sink dynamics. In this case, the sink was not a true sink, because it did not go extinct in the absence of immigration, but it could have been with much the same results before the extinction of the source. Excepting extreme cases with no reproduction in the sink, distinguishing between true sinks and pseudo-sinks is difficult in practice without experimentally reducing population density and thereby observing whether the growth rate becomes positive at low densities (Keddy 1981; Watkinson 1985). The butterfly example is instructive also in demonstrating that source populations might be more vulnerable to extinction due to environmental vagaries than (pseudo-)sink populations. In such cases, the long-term persistence of a source–sink metapopulation might depend on the presence of sink populations as envisioned in the model of Gyllenberg *et al.* (1997; Section 3.3). Another scenario in which marginal habitat might be critical for long-term persistence is exemplified by the forest landscape in Finland, where the remaining small patches of old-growth forest are scattered in the midst of successional managed forests. Some bird species, such as the three-toed woodpecker (*Picoides tridactylus*) and the pygmy owl (*Glaucidium passerinum*) have high densities in the remaining patches of old-growth, but they also breed in small numbers in managed forests, especially close to old-growth fragments (R. Virkkala, T. Pakkala, personal communication). The managed forests, though probably true sink habitats, might nonetheless facilitate connectivity among the isolated fragments of old-growth and thereby promote long-term persistence. Similar results have been reported for mammalian populations in tropical forests in Queensland, Australia (Laurance 1991).

Although there are not many empirical studies on source–sink metapopulations as instructive as that by Thomas *et al.* (1996) on the checkerspot butterfly, there is little doubt that many species have local populations living under environmental conditions so different that there are substantial differences in birth and death rates. Pulliam and Danielson (1991), Paradis (1995), Pulliam (1996) and Stacey *et al.* (1997) have reviewed a range of examples, including studies on plants, birds and mammals, with different amounts of information available. Ehrlich and Murphy (1987) and Tscharntke (1992) have highlighted the importance of 'reservoir' populations, or long-term predictable source populations, in insect metapopulation dynamics. Geographically marginal populations can often be true sinks. For instance, Järvinen and Väisänen (1984) found in a 7-year study that a northern population of the pied flycatcher (*Ficedula hypoleuca*) would have gone extinct without immigration from more southern populations. Detailed ecological studies on the plants *Cakile edentula* (Keddy 1981; Watkinson 1985) and *Sorghum intrans* (Watkinson 1985) enabled Watkinson *et al.* (1989) to parameterize a simulation model and to conclude that the study areas had both source and true sink populations. Watkinson *et al.* (1989) found that *C. edentula* was most abundant in a sink population; it received a high level of seed migration from a nearby source population.

Evolution in source–sink metapopulations

Let us return to the butterfly example. Singer and Thomas (1996) studied the oviposition preference of *E. editha* females in the two types of habitat, rocky outcrops and clear-cuts. The host plant which was used in clear-cuts, *Collinsia torreyi*, occurred also on the outcrops, but it was a catastrophically bad host on the outcrops, as it regularly dried out before the larvae had time to reach the stage required for winter diapause (Rausher *et al.* 1981; Singer and Thomas 1996). Quite sensibly, most females on outcrops completely avoided laying eggs on *C. torreyi*; instead they laid on *Pedicularis semibarbata*. In clear-cuts, where *C. torreyi* was the only host available, females oviposited on it, although they still preferred *P. semibarbata* or were neutral when tested in an experiment (Singer and Thomas 1996). Female preference influenced their movement behaviour between the two habitats (Thomas and Singer 1987; Singer and Thomas 1996)—those individuals most strongly preferring *P. semibarbata* were most likely to move from clear-cuts to outcrops, while females most likely to leave the outcrop were those that either

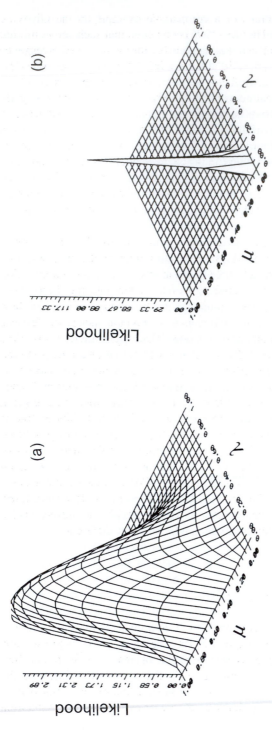

Fig. 8.3 Two examples of surfaces giving the likelihoods of different combinations of colonization (λ) and extinction (μ) probabilities explaining the observed patterns of presences (P) and absences (A) of species on islands. In (a), the maximum likelihood estimates of λ and μ were calculated for the hypothetical series *PPAAAPPA*, giving $\hat{\lambda} = 0.5$ and $\hat{\mu} = 0.75$. The likelihood surface is not sharply peaked, which indicates considerable uncertainty in the estimated values of λ and μ. In (b), the surface is for real data on many species of plants on 93 islands (from the study of Abbott and Black 1980). The maximum likelihood estimates $\hat{\lambda} = 0.1$ and $\hat{\mu} = 0.05$ are the pooled estimates for all the species. (Reproduced from Clark and Rosenzweig 1994.)

preferred *C. torreyi* or were neutral (for a comparable example on the Glanville fritillary see Fig. 6.2). Singer and Thomas (1996) concluded that such active mixing of butterflies among the habitat patches was responsible for the observed divergence in the host preferences of butterflies in the two habitat types. More recently, Boughton (1998) has observed another factor affecting gene flow between the clear-cut and outcrop populations. Clear-cuts are relatively sunny and have a slightly earlier butterfly flight season than outcrops. It turns out that *C. torreyi* remains edible long enough for the clear-cut-born butterflies to reproduce, but the plant senesces before most immigrants from the outcrops have arrived. As a result, butterflies moving from outcrops to clearings have much lower reproductive success than those moving in the opposite direction, and this asymmetry, ultimately due to the phenologies of the two habitats, drives a gene flow out of clearings into outcrops irrespective of which is the source habitat in the sense of higher population growth rate.

In heterogeneous environments, local populations might be permanently maladapted because of migration from other habitats with contrasting selection pressures (Felsenstein 1976; Hedrick 1986). Generally, one could expect sink populations to be more maladapted, because they are the net importers of individuals and genes (Dias 1996; Pulliam 1996). In the butterfly example, however, host preference was more maladapted in the source population, as most females continued to prefer the non-existing host (*P. semibarbata*) and thereby lost time (Parmesan *et al.* 1995; females eventually accept a less preferred host if the preferred one is not found). The cost of maladaptation is here slight in comparison with another well-studied species, the blue tit (*Parus caeruleus*) in southern France (Blondel 1993; Dias *et al.* 1996). The birds breed in both deciduous and evergreen forests, at the time that coincides with high insect availability in deciduous forests but three weeks too early in coniferous forests. Breeding success is thereby reduced in coniferous forests, which are sink habitats, surrounded by the more extensive source habitats (deciduous forests). Interestingly, in Corsica the area of evergreen forests is much greater than on mainland France, and in Corsica blue tits are adapted to evergreen forests and maladapted to breed in the small patches of deciduous forest (Lambrechts and Dias 1993; Dias and Blondel 1996). Dias (1996) has termed such a change a source–sink inversion, including a switch in habitat preference.

8.5 Small-bodied habitat specialists

Which kinds of species are most likely to exhibit classical metapopulation structures and dynamics in fragmented landscapes? Murphy *et al.* (1990) suggested the following suite of attributes: small body size, high rate of population increase, short generation time and high habitat specificity. This list practically eliminates the species that have traditionally received the greatest attention in conservation biology, large vertebrates, with large body size, low growth rate, long generation time and relatively low habitat specificity. But the list of Murphy *et al.* (1990) is not really restrictive, as it comfortably accommodates most insects and hence the vast majority of species on earth.

There are good reasons to agree with Murphy *et al.* (1990). To start with habitat specificity, many habitat types have a fragmented distribution, and therefore species specializing in such habitats have necessarily fragmented populations. Small body size means that the number of individuals in an individual habitat fragment can nonetheless be relatively large, so large that they constitute a breeding population, especially because small species tend to have lower migration rate than large species. High growth rate implies that, after population establishment, local populations either go quickly extinct or grow to the local carrying capacity, in agreement with the assumption of two time-scales in many metapopulation models. Short generation time implies, among other things, that population oscillations are not buffered by great longevity of individuals—the population is vulnerable to environmental vagaries, increasing its risk of extinction. A final point not directly following from the list of Murphy *et al.* (1990) is interaction between local environmental conditions and large-scale weather perturbations. When there is such an interaction, which is most likely in habitat-specialist species with short generation time, local dynamics become to some extent asynchronous, facilitating metapopulation-level persistence (Murphy *et al.* 1990). The group of species which Murphy *et al.* (1990) had particularly in mind is butterflies, but there is no reason to suspect that similar conclusions would not apply to many other insects and other taxa as well.

Beetles living in dead aspen trees

The majority of forest-dwelling insect species lives in patchily distributed microhabitats, such as decaying wood, mushrooms, rare host plants, animal remains, etc. A good example from boreal forests is beetles living in large dying and dead aspen trees. In Finland, *ca.* 500 beetle species occur in this microhabitat, including some 40 specialist species (J. Siitonen, personal communication). Siitonen and Martikainen (1994) sampled 240 such tree trunks, half of which were located in eastern Finland, the other half in Russian Karelia, at the same latitude and 100–300 km east from the Finnish study sites. The significance of this sampling design is based on the new border that was established after the Second World War, when a large part of south-eastern Finland had to be surrendered to the USSR. The new border divided a previously homogeneous and largely forested landscape. After the war, the forests on the Finnish side became very intensively managed and almost no old-growth forest is presently left. In Russia, forestry has been extensive but not intensive until recently, and much decaying wood remained even in the exploited forests, with plenty of large aspen trees.

The Finnish samples included five species considered to be nationally scarce but no species listed as endangered or vulnerable in Finland. In striking contrast, the Russian samples contained 18 rare species, including two listed as extinct in

Finland, four species listed as endangered and one listed as vulnerable in Finland. As similar aspen trees were sampled on both sides of the border, it is difficult to attribute the difference in these results to any other factors than those involved in forest management, and especially to a difference in the density of large decaying aspen trees. Long considered something of a pest by Finnish foresters, aspen trees on the Finnish side have been extensively removed or killed by herbicides. On the Russian side such trees remain common. The individual tree trunks are habitat patches for the specialist species that might live for many generations in a single trunk. Eventually the trunk decays away and the local population necessarily goes extinct. In Finland, the density of suitable host trees seems to be so low that no positive extinction–colonization equilibrium exists for many species, whereas on the Russian side the same species can persist. The results of Siitonen and Martikainen (1994), like numerous anecdotal observations concerning specialized forest beetles elsewhere in Europe (Hanski and Hammond 1995), are consistent with the predicted threshold density of habitat patches for metapopulation persistence.

Patchy populations

Beetle species living in dead aspen trees might have several generations in a single trunk, and hence the individuals in a single trunk might well be considered to represent a local population, even if a relatively ephemeral one. The fungus beetles *Bolitophagus reticulatus* in Europe (Nilsson 1997) and *Bolitotherus cornutus* in North America (Whitlock 1992) represent two other well-studied examples of this type. In other cases, just a single generation develops in a particular microhabitat, while adult insects disperse widely and mate within a larger area. This sort of 'patchy population' structure (Harrison 1991, 1994) without distinct local breeding populations is best analysed by means other than metapopulation models (Section 1.2), even if, in reality, the metapopulation structure with multigeneration local populations merges gradually to the patchy population structure. Examples of patchy populations include many herbivorous insects colonizing clumps of host plant (Solbreck 1991; Harrison *et al.* 1995; Dempster *et al.* 1995) and dung and carrion insects (Hanski and Cambefort 1991). Wilson's (1980) trait-group scenario illustrates possible evolutionary consequences of such patchy population structure. In contrast, Maynard Smith's (1964) haystack model, a generalization of the local mate competition model (Hamilton 1967), assumes multiple generations in local populations and is related to genuine metapopulation models.

8.6 The end of the species

It is often assumed, and occasionally demonstrated, that the density of species is highest in the centre of its geographic range and declines to zero towards range margins (Hengeveld and Haeck 1982; Brown 1984, 1995; Hengeveld 1990; Lawton 1993; Maurer and Villard 1994). Obvious exceptions are ranges of terrestrial species abruptly ending at the end of the land. The conventional ecological argument for the ideally bell-shaped distribution of population density (Brown 1984, 1995) assumes that density is determined by a number of environmental factors that change continuously in space; that there is a unique combination of these factors that leads to the maximum density; and that density gradually decreases when the value of one or more of the environmental factors changes from the optimal value. If the environmental gradient is steep, gene flow from the large populations in the centre of the range might prevent local adaptation at range boundary and prevent the range from expanding outward (Kirkpatrick and Barton 1997).

One of the better-studied examples of density variation across geographical ranges is grassland sparrows in North America (Curnutt *et al.* 1996). Apart from the high density in the centre of the range declining towards range margins, Curnutt *et al.* (1996) detected an increase in temporal variance towards the edge of the range; Thomas *et al.* (1994) have reported the same result for butterflies. Thus average density is low but variability is high in the marginal populations. Curnutt *et al.* (1996) suggest that the observed pattern is due to net movement of individuals from the more productive central areas, which function as sources, to the less productive marginal areas with sink populations. The marginal populations might often be exterminated by unfavourable conditions, but the populations will be re-established by migration from the centre.

Assuming that the suitable habitat becomes increasingly fragmented, or that the risk of local extinction increases (Lennon *et al.* 1997), away from the centre of the distribution, the dynamics become increasingly affected by local extinctions and colonizations. Thus Rapoport (1982) found that the distribution of the palm *Coperinicia alba* became increasingly patchy towards the range margin, and Mehlman (1997) has shown empirically for sparrows in North America that population turnover rate increases towards the edge of the range. A diminishing fraction of the suitable habitat is expected to be occupied with decreasing average patch size and density, and ultimately a 'hard' range limit may be set by the general threshold condition for metapopulation persistence (Section 4.1; Prince and Carter 1985, Lennon *et al.* 1997). In contrast, genetic models in which gene flow from the centre of the range prevents local adaptation and thereby sets the range limit along an environmental gradient predict a 'soft' boundary, with population density declining asymptotically to zero (Kirkpatrick and Barton 1997).

Svensson's (1992) work on the whirling beetles in Sweden gives one of the more detailed descriptions of range boundaries. These beetles live in various kinds of small water body and often have a classical metapopulation structure (Nürnberger and Harrison 1995). Two species reach their northern range limit, whereas one species has its southern range limit, in central Sweden (Svensson 1992). In all three species, the fraction of occupied habitat patches increases for some 300–500 km away from the range limit (Fig. 8.4), suggesting that the actual limit is set by some processes increasing extinction rate, or reducing colonization rate, towards the range

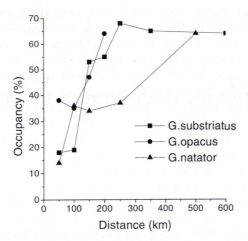

Fig. 8.4 The fraction of occupied habitat patches for three species of whirling beetles (*Gyrinus*) as a function of the distance from the geographical range limit. (Based on Svensson 1992.)

boundary. If the range boundary is set by metapopulation dynamics, it might be expected to oscillate in response to extinction–colonization dynamics (Lennon *et al.* 1997). A small-scale model of such oscillations in a sparse habitat patch network is given by the metapopulation dynamics of the American pika described in Section 12.4.

9

Mapping species occurrence on habitat availability

9.1 Empty habitat

Experienced field biologists are often amazingly successful in predicting the occurrence, or the absence, of particular species at particular sites. They know the species' ecological requirements, and they base their predictions on the assumption that where the requirements are met, within the geographical range of the species, the species is to be found—abundance and distribution reflect the response of local populations to local conditions (Brown 1995, p. 51). Although there is no reason to doubt that this conclusion would not capture a big element of truth, it does not capture the entire truth. Metapopulation dynamics with population turnover imply that a significant fraction of suitable habitat is unoccupied, even if it is not the same patches of habitat that are unoccupied all the time. Apart from being contrary to the expectations of many biologists, the question about empty habitat has profound implications for conservation (Hanski and Simberloff 1997).

How much empty habitat there is, is also a matter of scale. On very small scales, most habitat is 'empty' most of the time, in the sense that no individual occupies a particular spot of space (Gaston 1994). On very large continental scales, much habitat remains empty, as evidenced by successful establishment of alien species after introductions across natural and practically absolute migration barriers (Elton 1958; Drake *et al.* 1989). The bone of contention is how much empty habitat there is at the intermediate metapopulation scale—habitat within the ordinary migration range of the species but which nonetheless remains unoccupied for several generations.

Species and landscapes are expected to differ in the extent of empty but suitable habitat. Common sense and models (Chapter 4) suggest that species that are locally abundant, species which occur in relatively unfragmented environments, and species with great powers of migration, have little empty habitat. The opposite is true of low-density species in environments that are highly fragmented in relation to the movement capacity of the species. These latter species—species which have a wide geographical distribution but are scarce everywhere—represent the 7th type of rare species in the classification of Rabinowitz *et al.* (1986). It would be surprising indeed if such species, with small local populations, prone to local extinction and lacking in power to send out large numbers of migrants, would manage to occupy all of the suitable habitat all the time.

Types of evidence

Five types of evidence support the widespread occurrence of empty habitat. The weakest evidence consists of an expert's failure to find the species where all its ecological requirements seem to be met. Such instances are commonplace for biologists conducting surveys of endangered species, which are typically rare species in fragmented landscapes. No ecologist familiar with the species–area curves is very surprised to find that scores of species are absent from small fragments of apparently suitable habitat, although it is not usually clear exactly why they are absent (Section 7.5). The catch here is, of course, that one can never be absolutely sure by observation only that a particular site has all the resources required by the species, although in practice the best experts seldom make large mistakes. A more objective approach to identifying empty but suitable habitat involves a statistical discrimination between occupied and unoccupied habitat based on a set of environmental attributes. Any habitat patch that is deemed 'suitable' by such a classification but is actually found to be empty is considered 'empty but suitable' (Lawton and Woodroffe 1991; Doncaster *et al.* 1996).

Another type of evidence for empty habitat consists of observations of natural colonizations: a site was empty for some time but became subsequently occupied. This is the sort of evidence that many field studies on population turnover (Section 8.2) have accumulated. A critical person might retort by questioning whether the habitat really was suitable before it became occupied. In fact, colonizations are often believed to happen, though not necessarily instantly, after improvement in habitat quality (Thomas and Jones 1993; Thomas 1994a,b; Thomas and Hanski 1997). In many other cases, however, there is no hint of habitat change before colonization. In any case, a stronger test of empty habitat consists of experimental introductions of species to empty sites, which has been done repeatedly with many taxa, following Boycott's (1930) pioneering studies on freshwater snails dwelling in small ponds. Figure 9.1 summarizes results for butterflies in Britain and Ireland. Admittedly, most established butterfly populations again went extinct in less than a decade (Fig. 9.1), but this is not an adequate demonstration of intrinsically unsuitable habitat, because a large fraction of 'natural' small populations will not last for any longer, even on protected reserves (Warren 1992: table 11.4). Primack and Miao (1992) describe an informative plant example.

Yet another line of evidence for the occurrence of empty habitat consists of the observation that the frequency of patch occupancy often decreases with increasing isolation and decreasing patch area. These patterns could be explained by some element of habitat quality changing with patch isolation and area, but this is an unlikely explanation especially in the case of isolation, because the relevant metapopulation studies are usually conducted in such small areas that climatic and other major environmental variation among the patches is simply not present. Very small patches might be intrinsically unsuitable, either because the amount of space becomes limiting (territorial species) or because habitat quality is different in the very smallest patches. The interpretation that the declining incidence of patch

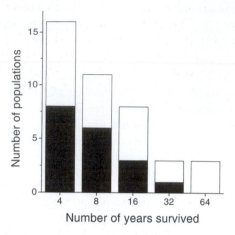

Fig. 9.1   Distribution of the number of years that successful introductions of butterfly populations in Britain and Ireland have survived. There were altogether 171 introductions of 32 native species, of which 41 were successful, in the sense that the population survived for at least 3 years. The shaded part of the histogram shows populations that are known to have gone extinct. (Based on data in Warren 1992: table 11.8.)

occupancy with increasing isolation is not due to low habitat quality is often backed up by knowledge about the migration distances of the organism. Thus we know from tens of field studies on butterflies that, in many species, individuals usually move only up to 1–2 km (Table 8.1). Therefore, if suitable habitat for such species occurs in small patches isolated by several kilometres, one could expect much habitat to be empty at any given time.

9.2 Patch area and isolation

The spatial structure of landscapes influences the abundance and distribution of species in several ways, many of which are catalogued in texts on landscape ecology (Forman and Godron 1986; Urban *et al.* 1987; Turner 1989; Wiens 1997). Meta-population ecology deals very effectively with two effects, the effect of patch area on extinction and the effect of patch isolation on colonization. These effects are so often so fundamental that it is appropriate to designate them as the first-order landscape effects on population biology. It has become customary, and quite rightly so, in current field studies on metapopulation dynamics to document the effects of habitat patch area and isolation on patch occupancy, while the area and isolation effects on extinction and colonization are essential ingredients of many spatially realistic metapopulation models (Chapter 5). Well-documented area and isolation effects thus establish an important link between theoretical and empirical studies,

akin but even stronger than the link between the species–area curve and the dynamic
theory of island biogeography. Though we are not, any more, surprised to find the
effects of patch area and isolation in the regional distribution of species, we should
not fail to appreciate their significance. The patch area and isolation effects on
habitat occupancy, when detected at the appropriate spatial scale for the focal
species, strongly suggest that the regional distribution of the species is dynamic and
hence that a metapopulation approach of some kind is worth considering in
subsequent studies.

Patch-level evidence

To recapitulate the main reasons why we expect habitat patch occupancy to decrease
with decreasing patch size and with increasing isolation, recall that small patches are
expected to have small local populations with high risk of extinction, whereas
isolated patches seldom receive immigrants and hence have a small probability of
colonization when empty (Fig. 9.2). Assuming that the species occurs in a stochastic
balance between local extinctions and colonizations, these effects on the processes
of extinction and colonization become translated into effects on the pattern of patch
occupancy at the steady state (Section 5.3).

The mapping of the processes to distributional patterns is especially straightforward in mainland–island metapopulations, in which the source of colonization
remains unchanged and isolation is simply the distance from the mainland to the
habitat patch. In classical metapopulations without an external mainland, the
measurement of isolation is complicated by three factors. First, isolation refers to
isolation from the existing local populations, not just from any habitat patches, and
the set of extant populations varies as populations go extinct and new ones are being
established. Second, while estimating the level of isolation, one should take into
account the influence of all populations that might function as sources of colonists.
This is not always done in empirical studies, which often measure isolation by the
distance to the nearest local population, or by the distance to the nearest local
population exceeding some particular size. There is really no excuse for using such
obviously deficient measures instead of the sensible measure described in Box 11.1
or some comparable measure. Third, measurement of isolation is usually based on
straight-line distances between pairs of habitat patches, ignoring the type of the
intervening landscape. One example which clearly shows that the latter can make a
big difference is due to Åberg *et al.* (1995), who studied the occupancy of small
forest fragments suitable for the hazel grouse (*Bonasia bonasia*). The influence of
straight-line isolation on occupancy was much greater for habitat patches surrounded
by unsuitable coniferous forests than for patches surrounded by farmland.

Field studies have accumulated much evidence for the effects of patch area and
isolation on occupancy. Table 9.1 summarizes the results of a sample of empirical
studies covering a wide range of taxa. Most studies have focused on single species,
which is unfortunate, because interspecific comparisons would be especially
informative. To take two examples of the latter, the area effect was clearly different

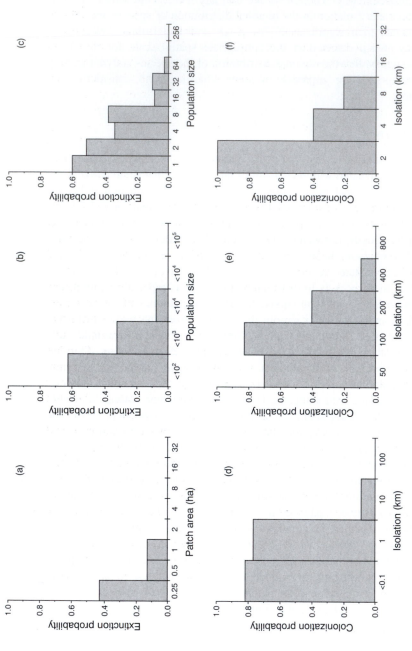

Fig. 9.2 Effects of patch size or population size on extinction risk and of patch isolation on colonization probability. Extinction of local populations: (a) the bush cricket *Metrioptera bicolor* in discrete habitat patches (*n* = 90; extinction in 1 year; Kindvall and Ahlén 1992); (b) pooled data for four species of spider on small islands in the Bahamas (*n* = 222 island-populations, extinctions in 4 years; Schoener and Spiller 1987); (c) the silver-studded blue butterfly (*Plebejus argus*) in discrete habitat patches (*n* = 55; extinctions in 7 years; Thomas 1994b). Colonization of empty habitat patches: (d) the silver-spotted skipper (*Hesperia comma*) on dry meadows in England (*n* > 100; colonization over 9 years; Thomas and Jones 1993); (e) the bush cricket *Metrioptera bicolor* in discrete habitat patches (*n* = 43; colonization in 1 year; Kindvall and Ahlén 1992); (f) the checkerspot butterfly *Euphydryas editha* in discrete habitat patches in California (*n* = 59; colonization over 9 years; Harrison *et al.* 1988).

for three species of shrew on islands in lakes (Peltonen and Hanski 1991), in a manner that was consistent with the known biology of the species (Hanski 1992, 1993). In an experimental study using habitat patches of red clover in an agricultural landscape, Kruess and Tscharntke (1994) found that herbivorous insects were less affected by isolation than parasitoids, with implications for biological control and the persistence of natural enemies in fragmented landscapes. In the case of relatively immobile species with little migration among patches in recently fragmented landscapes, patch area and time since isolation can be expected to explain much of the variance in patch occupancy, in the same manner as in other non-equilibrium metapopulations (Brown 1971; Diamond 1984; Harrison 1991, 1994), including tropical and temperate national parks (Newmark 1995, 1996). Soulé *et al.* (1988) have described a convincing example on chaparral-requiring bird species in southern California, with 80-year-old small (<10 ha) fragments having lost practically all the species.

In special circumstances, the area and isolation effects can be reversed. Thus the pollinator-transmitted smut *Ustilago violacea* spread faster to more isolated host populations of the plant *Silene alba*, apparently because isolated populations attracted disproportionate numbers of pollinator visits per plant (Thrall and Antonovics 1995). Nieminen and Hanski (1998) studied the occurrence of noctuid moths on islands in northern Baltic, off the south coast of Finland, where immigrants to islands could arrive both from other islands and from the mainland (as described by eqn 4.4). We compared a set of five small islands located in a compact group ('group islands') with a set of five comparable islands scattered across a large expanse of sea ('scattered islands'). We found that moths that are strong fliers but uncommon on the islands occurred more frequently on the scattered than group islands, apparently because these moths can detect and fly to an island from a long distance, and hence the group islands 'compete' for migrants coming from the mainland. In contrast, weakly flying species with many large populations on the islands occurred more frequently on the group islands, where between-island migration substantially increased their colonization rate (Nieminen 1996b).

Patch area and isolation might be related to population density in habitat patches, either because of patch area- and isolation-dependent emigration and immigration rates or because of some confounding environmental factors (the latter especially in the case of patch area). Empirical studies have reported both positive and negative relationships between patch area and density (Kareiva 1983; Stamps *et al.* 1987; Bowers and Matter 1997). It is apparent that population density is influenced by so many factors, which themselves might be correlated in a complex manner with patch area, that no general pattern is likely to emerge, in contrast to the effects of patch area and isolation on patch occupancy.

Network-level evidence

In mainland–island metapopulations, the predicted effects of patch area and isolation on habitat patch occupancy are strictly effects at the level of individual habitat

Table 9.1 Sample of field studies demonstrating the effects of habitat patch size and isolation on patch occupancy

Taxon	Habitat patches	Study area	Area effect	Isolation effect	Test	Reference
Rust pathogen *T. ulmariae*	Host plant populations	20 km²	Yes	2 of 4 cases	Logistic regression	Burdon et al. (1995)
Daphnia water fleas	Experimental tanks, 4–300 L	NA	Yes	NA	Sign test	Bengtsson (1993)
Fly *Urophora cardui*	ca. 300 patches of host plant	35 km²	Yes	NA	Logistic regression	Eber and Brandl (1994)
Bush cricket *M. bicolor*	ca. 100 vegetation patches	NA	Yes	Yes	Correlation	Kindvall and Ahlén (1992)
Grasshopper *O. caerulescens*	312 dry meadows	20 km²	Yes	Yes	U test	Appelt and Poethke (1997)
Butterfly *Plebejus argus*	157 vegetation patches in nine networks	NA	Yes	Yes	Contingency tables	Thomas and Harrison (1992)
Butterfly *Hesperia comma*	Grazed meadows	10 km²	Yes	Yes	Logistic regression	Thomas and Jones (1993)
Butterfly *Melitaea cinxia*	50 small dry meadows	15 km²	Yes	Effect on density	Logistic regression	Hanski et al. (1994)
Butterfly *Melitaea cinxia*	1502 meadows	3500 km²	Yes	Yes	Logistic regression	Hanski et al. (1995a)
Butterfly *E. editha*	ca. 50 meadows	450 km²	NA	Yes	Logistic regression	Harrison et al. (1988)
Butterflies many spp.	Vegetation patches	NA	2 of 5 species	5 of 5 species	Logistic regression	Thomas et al. (1992)
Spiders	108 small islands	NA	Yes	NA	–	Schoener and Spiller (1987)
Bull trout *S. confluentus*	High-elevation watersheds	NA	Yes	NA	Logistic regression	Rieman and McIntyre (1995)
Lizards	Introduced to 30 islands	1000 km²	Yes	NA	–	Schoener and Schoener (1983)
Frog *Rana lessonae*	ca. 200 ponds	NA	No	Yes	Logistic regression	Sjögren Gulve (1994)

Table 9.1 continued

Taxon	Habitat patches	Study area	Area effect	Isolation effect	Test	Reference
Wart-biter *D. verrucivorus*	70 habitat patches	NA	Yes	Yes	Logistic regression	Hjermann and Ims (1996)
Bird *Sitta europaea*	*ca.* 50 woodland patches	–	Yes	Yes	Logistic regression	Verboom *et al.* (1991)
Bird *T. urogallus*	Forest fragments	250 km²	Yes	Yes	–	Rolstad and Wegge (1987)
Bird *Bonasia bonasia*	74 isolated woodlands	5250 km²	Yes	Weak	Logistic regression	Åberg *et al.* (1995)
Birds	Habitat fragments	450 km²	Yes	No*	Regression	Soulé *et al.* (1988)
Shrews	108 islands in lakes	NA	Yes	1 of 3 cases	Analysis of variance	Peltonen and Hanski (1991)
American pika *O. princeps*	77 piles of rock	5 km²	Yes	Yes	–	Smith (1980)
Dormouse *M. avellanarius*	238 woodlands	NA	Yes	Yes	Logistic regression	Bright *et al.* (1994)

*Time since isolation had a significant effect on patch occupancy.

Table 9.2 Effects of average patch area and patch number in 4 km² squares on the fraction of occupied patches (*P*) in replicate Glanville fritillary metapopulations (Part III)

Patch area			Patch number / 4 km²		
Average area (ha)	Occupancy		Number of patches	Occupancy	
	n	*P*		*n*	*P*
<0.01	23	0.24	1	61	0.21
0.01–0.1	138	0.24	2–3	70	0.32
0.10–1.0	88	0.40	4–7	58	0.25
>1.0	6	0.56	>7	66	0.41

The replicate patch networks in the 4 km² squares were divided into four classes based on average patch area or patch number (from Hanski *et al.* 1995a).

The effects of average patch area and density on occupancy (*P*) were tested with analysis of variance on ranks, using the four patch area and density classes shown in the table. Both effects were highly significant (area: $F_{3,251} = 5.69$, $P = 0.001$; density: $F_{3,251} = 4.21$, $P = 0.006$; no significant interaction).

patches, as there is no direct nor indirect interaction between the populations on different patches. In classical metapopulations the situation is different because colonization is due to migration from the existing local populations. One consequence of these interactions among the local populations in classical metapopulations is that patch area and isolation effects are expected to occur at the level of patch networks as well as at the level of individual patches. Thus the proportion of habitat patches that are occupied at one point in time is predicted to increase with increasing average patch area, and with decreasing average isolation, of the patches in the network.

The only quantitative test of the network-level effects of average patch area and isolation which I know of comes from the study of the Glanville fritillary butterfly (Part III). Having surveyed the entire distribution of the species in southern Finland (Section 11.1), we divided the entire study area into 2 by 2 km² squares and treated the patch networks in these squares as replicates (Hanski *et al.* 1995a). The average habitat patch area within a square ranged from less than 0.01 to more than 1 ha, and the number of patches per square ranged from 1 to more than 20. The predicted effects were observed (Table 9.2)—the fraction of occupied patches increased from 0.21 in the squares with average patch size <0.01 ha to 0.56 in the squares with the average patch size >1 ha, and the fraction of occupied patches increased with increasing number of patches in the squares. In view of the pivotal role of the network-level effects of patch area and isolation in classical metapopulation theory, it would be very useful to have several other empirical studies examining these effects. In the epidemiological literature it is well established that the long-term persistence of many infectious diseases in host populations is possible only if the host population density is greater than a threshold value (Anderson and May 1991); this is directly comparable to the isolation effect in metapopulations of free-living organisms. A comparative study on mammalian nematodes showed that the proportion of host individuals infected (prevalence) increases with host population density and host body size (Arneberg 1996). Making the seemingly reasonable assumption that transmission rate increases with host population density and body size (larger bodies produce more parasites), these results nicely parallel patch area and isolation effects on the fraction of occupied patches in metapopulations.

9.3 Network size

The Levins model and related deterministic models (Chapter 4) describe the dynamics of metapopulations in large patch networks. These models may fail badly when applied to networks with only a small number of patches, as they ignore the extinction–colonization stochasticity that is always associated with local extinctions and colonizations. Section 4.5 reviewed modelling results on the risk of metapopulation extinction in finite patch networks, and the conclusion was reached that there should be at least 15–20 patches in a well-connected network of small patches with extinction-prone local populations to ensure long-term metapopulation persistence. It is worth stressing the similarity between the threat that extinction–colonization stochasticity poses to small metapopulations and the threat that demographic stochasticity poses to small local populations (MacArthur and Wilson 1967; Goodman 1987a,b; Ebenhard 1991; Lande 1993). These predictions assume independent fates of local populations and individuals, respectively. In the presence of correlated stochasticity, that is regional stochasticity in the case of metapopulations and environmental stochasticity in the case of single populations, the critical patch and individual numbers for long-term persistence may be much greater, and indeed any 'threshold' sizes become increasingly blurred.

I turn again to our butterfly study to test the predicted effect of network size on metapopulation persistence. Here it is not sensible to use patch networks in the 2 by 2 km² squares as replicates, as was done above, because these networks are not dynamically independent units. For the present purpose, I use as replicates the 127 semi-independent patch networks (SINs) into which Hanski *et al.* (1996c) divided the entire Glanville fritillary network with *ca.* 1600 habitat patches. The SINs are separated from each other by a sufficient distance, usually >1.5 km, to make interactions between the respective metapopulations weak but not so weak as to completely exclude recolonization from another SIN. Figure 9.3 shows that roughly one third of the SINs with <10 patches were occupied, whereas occupancy reached practically 100% when the network had 15 or more patches, in good agreement with the theoretical prediction (Section 4.5). The observed level of patch occupancy in networks with <15 patches is significantly and substantially lower than predicted by random placement of local populations among the habitat patches (Fig. 9.3). Other butterfly examples discussed by Thomas and Hanski (1997) support the conclusion that a viable network of habitat patches should have more than 15–20 patches. Results on the specialist parasitoids of checkerspot butterflies suggest that more than 20–40 host populations might be needed for the persistence of the parasitoid

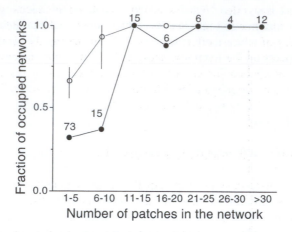

Fig. 9.3 Fraction of patch networks that were occupied by a metapopulation of the Glanville fritillary in 1993 as a function of the number of patches in the network. The networks have been arranged in groups with 1–5, 6–10, etc., patches. The number of networks in each class is given by the small number. The open dots give the null hypothesis obtained by randomizing the positions of the existing local populations among the habitat patches (average, minimum and maximum values in 10 randomizations; based on material described in Hanski *et al.* 1995a, 1996c).

metapopulation (Table 7.2). This figure is higher than the corresponding figure for the host butterflies, possibly because of inevitable fluctuations in the number of host populations or because not all patches in the parasitoid examples were well connected.

Less well-documented examples of the effect of network size on metapopulation persistence include *Daphnia* water fleas in rock pools on islands in the Baltic. Interspecific competition increases the rate of local extinctions in these species (Table 7.1). Different islands constitute almost entirely independent replicates, with different numbers of rock pools. The results from Finland and Sweden show that species number increases with the number of rock pools in the network (Fig. 7.1). Interspecific competition is expected to reduce the fraction of pools occupied by each species at equilibrium; this also is observed (Fig. 7.1). Although it is likely that processes other than extinction–colonization dynamics, including slight niche differences (Ranta 1979) and variation in the stability of rock pools among the islands (Pajunen 1986), play some role in this system, the long-term persistence of these species seems to hinge significantly on network size. In the epidemiological literature the persistence of many diseases depends on host population size (Anderson and May 1991); this is analogous to the network size effect in metapopulations of free-living organisms.

With increasing number of patches in the network, the pooled area of suitable habitat in the network increases, and one could argue that the network size effect is just the familiar area effect. This is exactly so, but for species with a distinct metapopulation structure the network size effect might, in fact, be fundamental for the understanding of the area effect (Section 7.5). Increasing the sizes of individual patches in a network will not increase network size, but it will increase the expected life-time of the metapopulation via reduced local extinction risk. Note that the effect of network size on metapopulation life-time is best assessed in relation to the expected life-time of local populations (eqn 4.19).

9.4 Distribution–abundance relationship

Two widely recognized patterns in ecology deal with the distribution of species—the species–area curve (MacArthur and Wilson 1967) and the distribution–abundance relationship (Hanski 1982a; Hanski *et al.* 1993c; Lawton 1993). I refer to the latter as the DA curve for short. 'Distribution' in this context can mean different things depending on the spatial scale of interest (Gaston 1994); in this chapter I refer to distribution in the sense of the fraction of suitable patches occupied in a patch network; this fraction is denoted by P. The DA curve describes a positive relationship between the average size of existing local populations and the extent of distribution. In other words, widely distributed species are generally found to be locally more abundant than species with a more restricted distribution, a pattern which was known already to Darwin (1859) but which has been quantified only since the early 1980s (Hanski 1982a; Brown 1984). Gaston and Lawton (1990) have summarized the overwhelming empirical support for interspecific DA curves. There is no need to repeat that evidence here, instead I discuss the mechanisms that might lead to interspecific DA curves. I also discuss intraspecific DA curves, with data points representing the abundance and distribution of one species at different times or in different patch networks.

Unlike with the species–area curve, for which several statistical models have been proposed (Connor and McCoy 1979), including the well-known power function model, $S = kA^z$ (Arrhenius 1921), the DA curve lacks a widely recognized statistical model. In the metapopulation context, when distribution is measured by the fraction of occupied patches, P, the logistic model is a sensible choice, with P as the dependent variable and log-transformed local abundance as the explanatory variable (Section 7.5). Ideally, abundance should be measured as the average population density in all existing local populations, but in practice this is usually impossible and local abundance (density) is estimated from a sample of local populations.

A common feature of both the species–area and distribution–abundance curves is that they have been detected on many spatial scales (Gaston and Lawton 1990; Hanski *et al.* 1993c; Lawton 1993; Gaston 1994; Rosenzweig 1995) and that they have been given several explanations. There is little hope of discovering a single mechanism that could explain these relationships at all spatial scales and in all taxa. Nonetheless, much interesting biology must be involved in these patterns. Here I review the mechanisms that seem most relevant for the DA curve at the metapopulation scale. These mechanisms can be divided into two kinds, those that involve metapopulation dynamics in a broad sense and those that do not.

Non-dynamic mechanisms

Before considering any biological mechanisms, let us observe that the DA curve can simply be an artefact of sampling (Brown 1984; Wright 1991; Hanski *et al.* 1993c; Gaston 1994)—the presence of rare species is easily overlooked, hence their distribution is easily underestimated. Taken to its extreme, the sampling hypothesis implies that all species actually occur everywhere; this, however, is an implausible proposition and certainly incorrect for many taxa that are well known or are easy to observe (Brown 1984; Gaston *et al.* 1996). Nonetheless, sampling most likely makes a contribution to the DA curve in most studies (Hanski *et al.* 1993c).

The simplest biological mechanism that could explain the DA curve is spatially aggregated resource availability. If the occurrence of species closely maps the availability of their resources, and if locally abundant resources are also more widely distributed, which is often plausible, the positive DA curve is expected for the consumer. Though this hypothesis only relocates the ultimate explanation one trophic level downwards, there is little doubt that this is exactly what often happens, especially when the resources are other living organisms. For instance, Dixon and Kindlmann (1990) found a positive relationship between the abundances of 12 aphid species and the abundances of their host plants, and Gilbert (1991) found the same for *Heliconius* butterflies. Gaston *et al.* (1997) discuss other examples and evidence for this mechanism.

The most interesting non-dynamic explanation of DA curves is Brown's (1984) niche breadth hypothesis, according to which generalists with broad environmental tolerances have large local populations and are widely distributed, whereas specialists have small local populations and have more restricted distributions. This is a seemingly testable hypothesis, assuming that niche breadth can be measured independently of abundance and distribution, though as we see below, Brown's hypothesis can also be seen as an element in a more comprehensive explanation of the DA curves. Gaston and Lawton (1990), Hanski *et al.* (1993c) and Gaston *et al.* (1997) found little support for a positive correlation between niche breadth and local abundance, and they found several counter-examples. Theoretically, one could expect that a specialist does well under the conditions into which it has adapted, and hence no positive DA curve would be expected if samples reflect species' abundances on their specific resources. In practice, samples are often less specific, and one could expect that specialists seem uncommon because of their small 'carrying capacities' in the habitat patches sampled. In this case, Brown's hypothesis becomes compounded with the resource availability hypothesis. The evidence for the niche breadth–distribution relationship is also ambiguous, when studies suffering from sampling artefacts are excluded (Gaston *et al.* 1997).

Despite these negative empirical results, it still seems likely that there is more than a grain of truth in the combination of the resource availability and Brown's hypotheses: many highly specialized species are rare, whereas many of the abundant and widespread species are generalists. But even if such relationships were widely observed, these hypotheses remain an incomplete explanation of the DA curves in metapopulations, because these hypotheses are completely silent about the dynamics.

Metapopulation dynamic mechanisms

There are two different reasons a positive DA curve could be expected in metapopulations, dubbed the carrying capacity hypothesis and the rescue effect hypothesis (Hanski 1991b). The former mechanism assumes a set of species which use the same habitat patches but which have different carrying capacities, and hence have different expected population sizes, in the habitat patches. Assuming realistically that extinction risk increases with decreasing population size and that colonization rate of empty patches increases with numbers of individuals in the occupied patches, the equilibrium fraction of occupied patches increases with the local carrying capacity (Section 4.1). Therefore, in an assemblage of species with different carrying capacities, one would expect the locally more abundant species to have a wider distribution (Hanski 1991b; Nee *et al.* 1991).

The carrying capacity hypothesis assumes that local abundance affects distribution but not vice versa. Under the rescue effect hypothesis there is causality in both directions—immigration reduces the risk of extinction of small populations, and immigration rate increases with increasing number of existing local populations (Section 4.2). Only the rescue effect hypothesis predicts an intraspecific DA curve in the course of changing distribution of species. However, examining an intraspecific distribution–abundance relationship is not necessarily a critical test of the rescue effect hypothesis, because it is possible that model assumptions are violated. For instance, contrary to the assumption in the models, there might be no distinction between local and metapopulation time scales but instead distribution and abundance might change concurrently simply because they reflect general changes in population size. Presumably, this is what happens in samples taken from a single local population.

Assuming that the assumptions of the metapopulation model are met, one can make some more specific predictions. Using the model in Section 4.3, eqn 4.12, let us construct an assemblage of species by varying just one parameter, which would correspond to having a guild of species sharing largely but not entirely the same biology. In this model, if species differ only in their intrinsic growth rate (r), extinction rate (e), colonization ability (β) or mortality during migration (α'), the assemblage is always expected to show a positive DA curve. Only when the species differ in their emigration rates (m) could a negative relationship emerge, as high emigration rate might reduce local abundance but increase distribution. Based on these predictions, one could expect that the exceptional species with high local

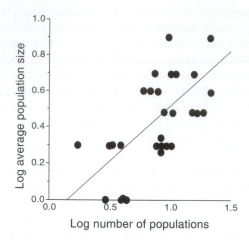

Fig. 9.4 Intraspecific distribution–abundance relationship in the Glanville fritillary butterfly. The data points represent patch networks in replicate 4 km² squares, omitting squares with <12 patches. Distribution is measured by the number of local populations in the patch network. (Based on Hanski *et al.* 1995a.)

abundance but restricted distribution have a large *r/m* ratio. Söderström's (1989) study of forest mosses seems to support this prediction. Species with high emigration rate and hence small *r/m* should be exceptions in the opposite direction: low local abundance but relatively large distribution. Data on British butterflies (Hanski *et al.* 1993c) and the contrast between British mammals and birds (Gaston *et al.* 1997) support this prediction, but other data sets failed to support it (Hanski *et al.* 1993c).

Brown (1995) has suggested that a positive DA curve in a community on islands refutes the metapopulation mechanism. This only applies to the rescue effect hypothesis—a positive DA curve is very much expected under the carrying capacity hypothesis even in mainland–island situations, as was formally demonstrated in Section 7.5 (for examples see Fig. 7.6). It is more difficult to come up with good evidence for the rescue effect hypothesis, because interspecific differences are likely to blur patterns potentially reflecting the rescue effect. Figure 9.4 gives a putative example in which metapopulation dynamics with the rescue effect seem to be the primary explanation of the DA curve. The data points come from different metapopulations of the Glanville fritillary butterfly, occupying the semi-independent patch networks described in Section 9.3 above. Here we can practically exclude the non-dynamic mechanisms, as well as the carrying capacity hypothesis, because the amount of suitable habitat in the network did not explain variation in average population sizes (Hanski *et al.* 1995a). This example is especially strong because there is independent evidence for the rescue effect in these metapopulations (Table 8.2; Hanski *et al.* 1995b).

Apart from metapopulation dynamics in the usual sense, a positive DA curve might be generated by density-dependent habitat selection, with the range of habitat types used by the species expanding with increasing population density (O'Connor 1987; Wiens 1989). This mechanism is likely to be important only on relatively small spatial scales, because density fluctuations are not likely to be well correlated over large areas (Hanski and Woiwod 1993b). Density-dependent habitat selection might explain both intraspecific and interspecific DA curves. An interspecific pattern generated by this mechanism would easily be mixed with Brown's niche breadth hypothesis, with the more widely distributed species showing greater realized habitat niches.

In conclusion, there is some evidence for many of the mechanisms that have been proposed to explain the positive distribution–abundance relationship (Hanski *et al.* 1993c; Gaston *et al.* 1997). I suggest that the most likely explanation to account for the DA curves at regional scales is the carrying capacity version of the metapopulation mechanism, with interspecific differences in population densities being due either to differences in species' niche breadths as in Brown's hypothesis, differences in resource availabilities to different species, or even to interactions with natural enemies. In this scenario, interspecific differences in species' average abundances influence distribution via the effects of local abundance on extinction and colonization rates. The model in Section 7.5 gives a formal description of this hypothesis.

9.5 Core and satellite species

In species-rich communities, the distribution of species' abundances is typically lognormal (Preston 1948, 1960; May 1975; Rosenzweig 1995). In contrast, the sizes of individual metapopulations in a metacommunity, as measured by the fraction of occupied habitat patches, P ('distribution'), are constrained to between zero and one. Therefore, the distribution of P values in a metacommunity is truncated at zero and unity. But what is the shape of this distribution? According to the core–satellite species hypothesis (Hanski 1982a), the distribution is bimodal, most species being either widely distributed (P large) or rare (P small), or unimodal but with the mode close to zero or unity.

The notion that the frequency distribution of species' distributions is bimodal has a long history. In a remarkable early paper on sampling statistics in ecology, Raunkiaer (1910) divided plant species into five equally large frequency classes based on the number of quadrants they occupied (50 quadrants were usually sampled). He observed that the numbers of species in successive frequency classes first declined but typically increased again in the highest class, with 81 to 100% of quadrants occupied. This pattern became known as Raunkiaer's law. Subsequent authors cast doubt on the biological significance of the law by observing that the pattern depended on quadrant size (Gleason 1922; Goodall 1952; Greig-Smith 1957). Furthermore, Raunkiaer (1934) himself, Preston (1948), Williams (1950) and

Nee *et al.* (1991) have suggested that bimodality is an artefact of sampling, as P is constrained to be equal to or less than unity; very abundant species simply cannot have a distribution greater than $P = 1$. The quantitative arguments by Raunkiaer (1934) and Williams (1950) to demonstrate that the shape of the distribution depends on the spatial scale of sampling assume that individuals are randomly distributed in space and that their mean abundances are lognormally distributed. Sampling and the spatial scale are indeed paramount considerations when one attempts to explain the distribution of species' frequencies in quadrant samples taken from homogeneous vegetation (Hanski 1982b), but it is more difficult to see the relevance of sampling in the case of classical metapopulations consisting of local populations with relatively independent dynamics. Brown (1984) suggested that bimodality is due to some species being generalists and therefore present in most habitat patches, while others are specialists and are therefore confined to a small number of patches, the other patches being unsuitable for them. However, the core–satellite species hypothesis attempts to explain the distribution of species in a set of patches that are all suitable, hence habitat specialization is not a relevant issue. Furthermore, even if one were to examine the distribution of species in a set of dissimilar patches, it is not obvious why interspecific differences of the type conceived by Brown (1984) should necessarily yield a bimodal distribution of P values.

Metapopulation dynamic reasons to expect bimodality

Two different metapopulation dynamic scenarios yield a bimodal distribution of P values (Hanski 1982a; Hanski and Gyllenberg 1993). The original core–satellite species model (Hanski 1982a) assumes that the per-population extinction rate decreases linearly to zero with increasing P due to strong rescue effect. In this case, the Levins model (Section 4.1) is reduced to the logistic model for P

$$\frac{dP}{dt} = c'P(1 - P),$$ (9.1)

where $c' = c - e$ in terms of the parameters of the Levins model. This model has a stable equilibrium at $P = 1$ for positive c'. Assuming however that c' is a random variable with variance much greater than the mean, the distribution of P values in the long course of time is bimodal, P being most of the time close to 1 or close to 0. This result implies that in an assemblage of species occupying independently the same set of habitat patches, most species are either common (P close to 1) or rare (P close to 0) at one point in time, even if all species were identical (the same distribution of c' values). For some parameter values all species would be rare, or common, but in no case should the majority of species have an intermediate value of P, assuming that all patches are suitable for colonization by all the species. To prevent permanent global extinction, this model assumes some migration from outside the metapopulation (Hanski 1982b; Hanski and Gyllenberg 1993). The model also predicts shifts in the status of the species from the core to the satellite class and vice versa.

The essential element in this model is the rescue effect, which can be modelled in a more realistic manner (Hanski and Gyllenberg 1993) than in the original model (Hanski 1982a). Hanski and Gyllenberg (1993) show that bimodality might arise not only because of temporal variance in parameter values but also because of interspecific variation in parameter values. To see this recall that with strong rescue effect the fraction of occupied patches increases rapidly from a small value to a large value with a change in parameter value(s), in the extreme case creating alternative stable equilibria (Section 4.2). The reason for *P* tending to be either small or large is the positive feedback generated by the rescue effect in metapopulation dynamics. Given a distribution of parameter values for different species, the distribution of the respective *P* values may easily become bimodal, as most parameter values lead to either a low or a high value of *P* (Hanski and Gyllenberg 1993). Naturally, if all species have a very low, or a very high, value of the key parameter(s), there might be no bimodality, but then all species have a small or a large *P*. If there are large differences in patch sizes, and hence in the expected life-times of the respective local populations, the metapopulation might survive periods of rarity (small *P*) due to low risk of extinction in the very largest habitat patches (Hanski and Gyllenberg 1993). When bimodality is generated by interspecific differences in parameter values, rather than by temporal variance as in the original model, core–satellite switching is expected only in the case of alternative stable equilibria (Hanski and Gyllenberg 1993).

Empirical results

One problem in testing the core–satellite species hypothesis is heterogeneity in patch sizes and qualities. When the patch occupancy distribution is constructed for a community of many species, the assumption that all the patches are equally suitable for all the species may be violated. Patches generally differ greatly in size. A large number of very small patches is liable to generate a small fraction of occupied patches, because very small patches with high extinction risk have in any case a small probability of being occupied. If there is much variation in patch size, one should either exclude entirely the smallest patches with minimal contribution to the overall metapopulation dynamics, or calculate metapopulation size as the fraction of the pooled patch area that is occupied, thereby giving more weight to the larger patches.

The core–satellite species hypothesis is best tested with data for a single species from many independent networks, for which the predicted distribution of *P* values is the same as for one metapopulation in the long course of time. This sort of data are difficult to obtain, and the only data set which I know of comes from our study on the Glanville fritillary butterfly (Part III). There is huge variance in patch areas in this material, from 12 m^2 to 3 ha, and as the numerous very small patches can hardly support local populations, there is a need to take variance in patch sizes somehow into account while calculating the *P* value. Figure 9.5 gives the results of two complementary analyses in which the smallest patches were either excluded or *P*

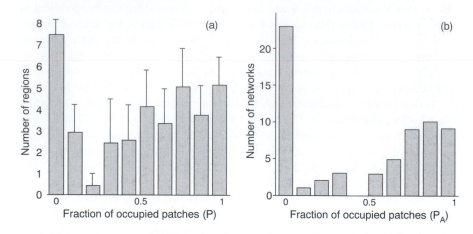

Fig. 9.5 The frequency distribution of occupied habitat (*P*) in replicate patch network in the Glanville fritillary butterfly. The two panels give the results of two complementary analyses. In (a) the replicate networks come from 4 km² squares into which the study area was divided, small patches (<500 m²) were omitted, and squares with <5 patches were omitted because the respective *P* values are unreliable and limited to a few possible values. The lines give standard deviations, obtained from 100 systematically different divisions of the study area into 4 km² squares (from Hanski *et al.* 1995a). In (b) the *P* values denote the occupied fraction of the pooled patch area in semi-independent patch networks, again omitting networks with <5 patches. (Based on Hanski *et al.* 1995b.)

was calculated as the fraction of the pooled patch area that was occupied. In both analyses, the distribution is clearly bimodal (Fig. 9.5). This is an especially strong test of the core–satellite hypothesis, because there are independent data to demonstrate the operation of the rescue effect in these metapopulations (Table 8.2; Hanski *et al.* 1995b).

Figure 9.6 shows four other examples of bimodal distributions of *P* values in four assemblages of plants and animals. Other studies finding (usually qualified) support for bimodality are listed in Table 9.3, along with studies finding no trace of bimodality. On balance, the evidence is not compelling one way or another. Though the following might seem a misguided attempt to save the hypothesis, a number of potential reasons for non-bimodal distributions, typically with a predominance of very small *P* values, may be listed:

- The spatial scale of the study might be too small or too large; the hypothesis applies to the metapopulation scale. It is well known that the distribution of geographical ranges on continental scales are strongly right-skewed (Gaston 1994 and references therein).

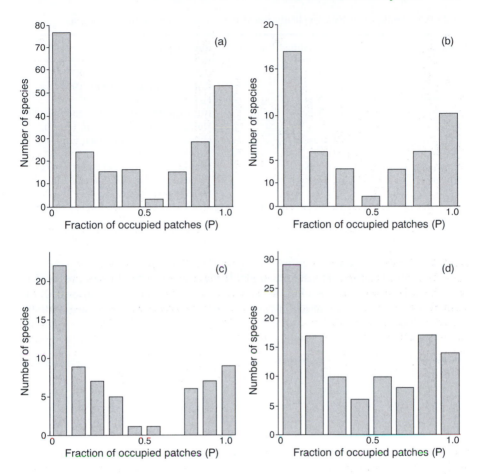

Fig. 9.6 Examples of bimodal core–satellite distributions. (a) Anthropochorous plants inhabiting small villages surrounded by forest (Linkola 1916; analysis in Hanski 1982b); (b) British butterflies (data from Pollard *et al.* 1986, analysis in Hanski *et al.* 1993c); (c) intestinal helminths in three species of grebe (Stock and Holmes 1988); and (d) cynipine gall wasps on oaks (Cornell 1985).

- When using data for many species sampled in the same patch network, as is customary, not all patches might be suitable for all the species, thereby leading to underestimation of P for these species.
- If there are large differences in patch sizes or qualities, and many very small or low-quality patches are included in the data set, calculating P as the fraction of occupied patches might not give a meaningful measure of distribution (see above for recommendations).

Table 9.3 Studies that have explicitly tested the core-satellite species hypothesis

Taxa	Qualified support	References
Prairie plants	Yes	Gotelli and Simberloff (1987), Collins and Glenn (1990, 1991)
Anthropochorous plants	Yes	Hanski (1982b)
Macroparasites	Yes	Bush and Holmes (1986), Stock and Holmes (1988), Esch *et al.* (1990), Haukisalmi and Henttonen (1994), Arneberg (1996)
Insects	Yes	Hanski (1982a,c), Menendez (1993)
Insects	No	Williams (1988), Gaston and Lawton (1989), Obeso (1992)

One criticism of the core–satellite model is that species are not often observed to switch between the core (*P* close to 1) and the satellite status (*P* close to 0), as predicted by the original model (Lawton and May 1983; Gaston and Lawton 1989). Such switching is, however, not predicted to occur if bimodality is caused by interspecific differences in parameter values (Hanski and Gyllenberg 1993). Nonetheless, core–satellite switching would constitute particularly compelling evidence for the hypothesis.

10

Metapopulation dynamics and conservation biology

10.1 Paradigm shift

Conservation biology evolved into a distinct field of biology in the 1970s, with two major contributions from population biology, the dynamic theory of island biogeography (MacArthur and Wilson 1963, 1967) and population genetics applied to endangered species, particularly in the study of drift and inbreeding in small populations (Simberloff 1988). The island biogeographic theory quickly attracted much attention (Fig. 10.1). In this theory, species number on true and habitat islands settles to a dynamic equilibrium between ongoing local extinctions and recolonizations. The island theory clearly shares key underpinnings with metapopulation models—the division of nature into discrete fragments of habitat, with movement of individuals among relatively unstable local populations. There is also an apparent difference, as the island biogeographic theory deals with communities, not individual species, and as its key statistic is species richness. This difference is however more apparent than real, as the island biogeographic model can be seen as a composite of models for independent species, with the community-wide immigration and extinction rates being simply sums of the respective species-specific rates (Simberloff 1969, 1983; Gilpin and Diamond 1981; Section 7.5). The underlying concept of population dynamics is the mainland–island metapopulation model (Section 4.1).

Within a decade, the dynamic theory of island biogeography came to dominate much of conservation biology, with a series of nearly simultaneous papers (Terborgh 1974, 1975; Diamond 1975; Wilson and Willis 1975) advocating a set of 'rules' of refuge design ostensibly derived from the island theory (Hanski and Simberloff 1997). Although some of the rules were not, in fact, based on this theory (Simberloff 1988), they became popular in the conservation community and were reproduced in textbooks and published in newspapers. The theory of island biogeography and the rules of refuge design provided a paradigm for conservation biology. The dominance of the island biogeographic paradigm was so strong that empirical studies which today would clearly be seen as belonging to metapopulation research were published as island biogeographic studies, with no mention of the term 'metapopulation' (Fritz 1979 and Weddel 1991 are two examples).

In the middle of the 1970s critical views began to appear. It was noted that most ecological publications citing the theory simply interpreted the species–area curve in

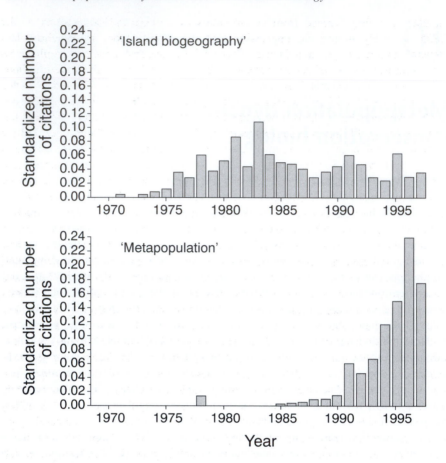

Fig. 10.1 Numbers of citations of the key words 'island biogeography' and 'metapopulation' in the BIOSIS data base in 1970–1997, standardized by the relevant total number of papers.

terms of the theory, largely ignoring alternative explanations (Simberloff 1974), and that there was little empirical evidence for continuing local extinctions of the sort envisioned in the theory (Lynch and Johnson 1974; Simberloff 1974). Nevertheless, ecologists tended to view the theory favourably until around 1980, when doubt about widespread local extinctions became pervasive (Gilbert 1980; Schoener and Spiller 1987; Williamson 1989). From this point onwards, interest in the dynamic theory of island biogeography has been waning, while interest in metapopulation ecology has been increasing (Fig. 10.1). This pattern of changing attention is surprising in view of the conceptual and even formal similarities between the two theories.

There are at least four plausible reasons for the shift in interest (Hanski 1996; Hanski and Simberloff 1997). First, conservation biology along with much other population biology turned from communities to species and populations in the 1980s, possibly giving the impression that the island theory is somehow less relevant than metapopulation theories. Second, it became fashionable to emphasize a non-equilibrium view of nature instead of the traditional equilibrium view (Wiens 1977, 1984; Chesson and Case 1986). Though there is really no difference in this respect between the island model and metapopulation models, an illusory difference might have been created by distinct emphases. In the island theory, the emphasis has been on the equilibrium number of species, to the extent that the theory received the nickname 'equilibrium' theory. In contrast, in metapopulation theories the emphasis has been on population turnover, possibly creating a sense of a non-equilibrium theory (Hanski and Simberloff 1997). In reality, of course, both theories are similar in assuming unstable local populations and allowing for equilibrium at the regional scale. Third, the island theory assumes a permanent mainland community, which is sufficient to prevent global extinction, whereas there is no such refuge in Levins's metapopulation model—the classical metapopulation might well go globally extinct. The metapopulation model is more appropriate than the island model for endangered species with a perilous existence in fragmented landscapes. Fourth, an important reason for the early success of the island theory and the failure of Levins's metapopulation concept, which were published nearly simultaneously in the late 1960s, is in the application of the models. MacArthur and Wilson were able to tell field ecologists what they should measure in practice, whereas Levins was less successful in this respect. Only in the 1990s, with the development of spatially realistic metapopulation models (Chapter 5) has the contact between metapopulation theory and empirical studies become strong. Incidentally, the same point is made forcefully clear by Wilson's (1994) interpretation of why Hamilton's (1963, 1964) theory of kin selection became so firmly established once the idea was picked up by a few influential individuals (including Wilson himself). Hamilton succeeded dramatically not only because kin selection is such a great idea but because, on top of that, he was able to tell us something new about the real world in concrete, measurable terms. He provided the tools for real, empirical advances in sociobiology (Wilson 1994, p. 317). How else could it be, but so real has been the gap between theory and empirical studies in much of ecology that such an obvious point seems like a revelation.

The shift from the dynamic theory of island biogeography to metapopulation theories (Fig. 10.1) is an excellent example of Kuhn's (1970) paradigm shift (Hanski 1989, 1996; Merriam 1991; Hanski and Simberloff 1997). Much of the criticism of the island biogeographic theory could be directed, and has been directed (Harrison 1994; Simberloff 1994), against metapopulation theories, hence it is hard to imagine objective scientific reasons for accepting one while rejecting the other. There are, however, certain differences in the applications of these theories that are probably significant. One apparently important issue is the spatial scale. The dynamic theory of island biogeography was originally developed to explain patterns on large spatial scales, whereas the metapopulation concept is associated with fragmentation of our ordinary landscapes. Not many of us are familiar with oceanic islands, which initially supplied many key examples for the theory of island biogeography, but all of us are familiar with the loss and fragmentation of some natural habitats. Species' distributions become fragmented with the fragmentation of their habitat, and metapopulation models of more or less isolated local populations connected by some migration become the natural choice. It is not a coincidence that whereas birds provided the bulk of early examples for the island theory, butterflies have gained a somewhat similar status in current metapopulation studies especially in Europe (Thomas 1994b; Hanski and Kuussaari 1995; Thomas and Hanski 1997). The main ecological data interpreted in terms of the island theory were simply species–area relationships, showing that, other things being equal, large sites tend to have more species than small ones. One of the rules of refuge design (Simberloff 1988) expresses this relationship as a mandate for conservation planners. Conservationists recognized that astute opponents could turn this emphasis on non-viable small populations against conservation, and though there are reasons to value small sites also in the context of the island theory, the main salvation of small sites was the shift by conservationists to the metapopulation paradigm (Hanski and Simberloff 1997). Finally, extinction rates on the often large islands considered in the island biogeographic studies are low and difficult to measure, whereas extinctions of small populations in fragmented landscapes are commonplace and relatively easy to document, providing a strong empirical basis for the meta-population models.

At present, the metapopulation concept is used widely in conservation (Soulé 1987; Soulé and Kohm 1989; Western and Pearl 1989; Falk and Holsinger 1991; Fiedler and Jain 1992; Harrison 1994). Enthusiasm has reached such a level that Doak and Mills (1994), Harrison (1994), Thomas (1994a), and Hanski and Simberloff (1997) see a potential danger in the widespread application of the metapopulation approach to species that might not be spatially structured in the way assumed by the models. Harrison (1994) goes as far as suggesting that even the general message for conservation drawn from classical metapopulation models is false—few species persist regionally in a balance between local extinctions and recolonization of unoccupied habitat. Harrison (1994) is targeting her criticism primarily and rather literally against the Levins model. She might have failed to note that metapopulation-level persistence is also predicted by more realistic models incorporating, for instance, the effects of patch area and isolation on local dynamics. There is an opportunity for misapplication, but this is not unique to metapopulation models. The metapopulation paradigm has served and continues to serve a useful function by stimulating conservation biologists to gather data that are critical for the development of effective conservation strategies for many species: movement rates between sites, reproduction and mortality rates that might vary from site to site, population size-dependent extinction risk, and the like (Hanski and Simberloff 1997). Additionally, there is already abundant support for the use metapopulation models of one kind or another for a better understanding of population dynamics of many species, though it would be unwise to place unrealistic hopes for what can be achieved in conservation by metapopulation models alone.

In the following sections, I first discuss metapopulation persistence in fragmented landscapes and the establishment of metapopulations in currently empty patch networks. The remaining sections review the application of metapopulation concepts in the conservation of the northern spotted owl and some other taxa as well as the spatially extended population viability analysis in general.

10.2 Habitat fragmentation and metapopulation persistence

Habitat destruction poses the single greatest threat to the long-term survival of species on earth (Barbault and Sastrapradja 1995). Habitat destruction has three major components, straightforward loss of habitat, increasing fragmentation of the remaining habitat, and deterioration of habitat quality. By fragmentation I mean that the remaining habitat of fixed total area is located in ever smaller and more isolated discrete fragments (patches). Habitat loss and fragmentation usually occur together, but it is useful to maintain a conceptual distinction between the two processes. Apart from the effects of patch area and connectivity of suitable habitat, which are the focus of metapopulation studies, habitat fragmentation has many other consequences (Saunders *et al.* 1991). For instance, many empirical studies have investigated the physical and biotic consequences of increasing the (relative) amount of habitat edge that necessarily accompany fragmentation (Harris 1988; Temple and Cary 1988; Yahner 1988; Mills 1995). Such 'edge effects' and other consequences of habitat fragmentation on habitat quality are beyond the scope of this volume. Edge effects and habitat quality may nonetheless influence metapopulation dynamics via effects on effective patch areas and hence on extinction rate, on colonization rate, and by creating source–sink dynamics.

Habitat loss, or habitat fragmentation?

The theory described in Section 5.1 suggests that in the early stages of habitat destruction, while the remaining suitable habitat is still well connected, populations are primarily affected by habitat loss. This means that the expected population sizes are reduced in proportion to the remaining total area of the habitat. With increasing habitat destruction, the degree of connectivity between the remaining fragments is reduced, and habitat fragmentation begins to amplify the impact of habitat loss, leading to a highly non-linear response of species to the diminishing amount of suitable habitat (Fig. 10.2a).

Andrén (1994, 1996, 1997) has attempted to assess to what extent habitat fragmentation as opposed to habitat loss has affected populations of birds and mammals in fragmented forest landscapes. He reasoned that population responses can be explained by habitat loss alone when the so-called random placement hypothesis cannot be rejected—the occurrence of the species in habitat fragments accords with what would be expected for equally large samples from an un-fragmented area. In particular, the random placement hypothesis does not predict

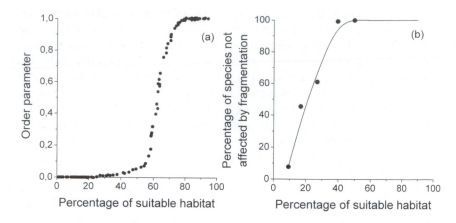

Fig. 10.2   Habitat destruction and fragmentation. (a) A theoretical result depicting the ratio between the largest connected patch and the total amount of suitable habitat (called order parameter) against the fraction of suitable habitat remaining (Bascompte and Solé 1996; Section 5.1). (b) The fraction of empirical studies of forest birds and mammals that failed to detect an effect of habitat fragmentation (as opposed to simple habitat loss) against the percentage of the landscape covered by suitable habitat. (Based on Andrén 1994.)

any isolation effect, nor any effect of patch area on population density. The results in Fig. 10.2b show that the random placement hypothesis could not be rejected in studies in which more than half of the landscape consisted of suitable habitat. The rapid increase in the number of studies demonstrating fragmentation with increasing habitat destruction accords with the theoretical expectation, although populations of birds and mammals seemed to tolerate more habitat loss before the consequences of fragmentation began to appear than was predicted by the simple landscape model (Fig. 10.2a). Detailed studies on individual forest-living bird species support the conclusion that fragmentation effects become increasingly severe when less than 30% of the landscape remains suitable for the species, and that many species are likely to go regionally extinct when the coverage of the habitat is less than 5–10% (Rolstad and Wegge 1989; Åberg *et al.* 1995; Enoksson *et al.* 1995; Angelstam 1997). These figures are based on the percolation theory-based landscape models, which are relevant for, e.g., many vertebrates living in originally extensive habitats. There are, of course, vast numbers of other taxa that have adapted to naturally fragmented habitats covering only a small fraction of total landscape area. In the case of such species, the percolation theory and related landscape models do not predict that species' persistence is not threatened if at least 30% of the original habitat area is retained.

   The reason for the quantitative difference between model predictions and empirical observations in Fig. 10.2 is most likely just in the simplicity of the model, not allowing for any migration across unsuitable habitat and assuming random loss of habitat. Andrén (1994) found in his simulations that until 60–80% of the habitat had been lost, isolation was minimal and the suitable habitat patches were usually separated from each other by just one spatial unit. Habitat loss might be aggregated, or the distribution of suitable habitat might be fractal, which will generally increase connectivity for a given level of habitat loss and thereby facilitate persistence (Lavorel *et al.* 1993; With and King 1997; With *et al.* 1997). Clumping of suitable habitat patches is generally advantageous for metapopulation persistence (Adler and Nürnberger 1994), essentially because clumping reduces the cost of migration for an average individual. Naturally, introducing some further complications, such as enhanced spread of a disease among closely connected populations (Hess 1996; Lei and Hanski 1997) might change this conclusion.

The neighbourhood habitat area of a patch network, H_n, defined in Section 5.2, is a practical measure of the amount of habitat accessible to individuals in a fragmented landscape. H_n decreases with habitat loss and fragmentation. A simple metapopulation model including the effects of isolation and patch area on colonization and extinction (Section 5.2) predicts that the fraction of occupied patches (P) decreases roughly linearly with decreasing logarithm of H_n until a threshold value is reached and the metapopulation goes extinct (Fig. 10.3a). Figure 10.3b gives the empirical results for 127 patch networks available to the Glanville fritillary butterfly in SW Finland (Section 11.1). Networks with small H_n were not occupied, as expected, but the predicted linear increase of P with increasing log H_n was not observed, possibly because the simple model (Section 5.2) includes no variation in habitat quality and assumes that the metapopulations occur at stochastic steady state. Both assumptions are to some extent violated in the Glanville fritillary metapopulations (Sections 12.2 and 12.3).

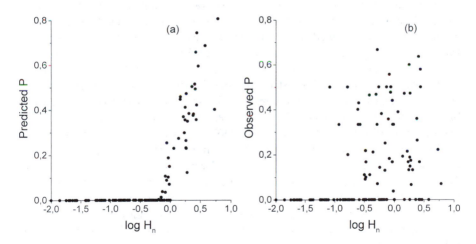

Fig. 10.3 (a) Model-predicted equilibrium fraction of occupied patches ($\hat{P}$) against the logarithm of the neighbourhood habitat area (H_n) in 127 patch networks of the Glanville fritillary butterfly (eqn 5.3). Parameter values: $\alpha = 2$ and $c = 10$. (b) Empirical results for the same material, P now giving the fraction of patches in the networks occupied by the butterfly in 1996.

The message from this combination of theoretical and empirical results is clear-cut. With increasing habitat destruction, the impact of habitat loss becomes amplified by the additional fragmentation effects, and in highly fragmented landscapes the effects of patch size and isolation begin to dominate. Although habitat requirements vary among the species and types of landscape, the values suggested by Fig. 10.2 for birds and mammals are so large that most modern landscapes are well within the range where fragmentation effects can be expected to be significant.

The metapopulation approach is primarily concerned with population dynamics in landscapes where there is little suitable habitat left and the fragmentation effects dominate. In the early stages of habitat destruction, where habitat loss dominates, processes other than classical metapopulation dynamics are likely to be more important; examples are the various landscape mosaic effects described in landscape ecology (Wiens 1997 and references therein). The use of the species–area relationship to predict species loss (Simberloff 1986; Whitmore and Sayer 1992; May *et al.* 1995) is most legitimate for landscapes with no substantial fragmentation effects, although there are, unfortunately, no good examples to demonstrate what difference habitat fragmentation actually makes for predictions based on the species–area curve. For an apparently successful recent application of the use of the species–area curve to predict extinctions and imminent extinctions see Brooks *et al.* (1997).

Habitat subdivision

One of the issues raised by the dynamic theory of island biogeography in conservation was the SLOSS problem (Diamond 1975; Simberloff and Abele 1976; Gilpin and Diamond 1980; Higgs and Usher 1980; Diamond and May 1981; and many others)—should we aim at establishing a *single large or several small* reserves with the same total area to maximize the number of species preserved? In the metapopulation context, the same can be asked about the likelihood of long-term persistence of metapopulations (Quinn and Hastings 1987). Island biogeographic studies typically tackled the SLOSS question simply by enumerating the numbers of species found in areas of different sizes. In the metapopulation context, it is natural to consider the dynamics by allowing stochastic local extinctions and recolonizations, whereby the question changes to one about persistence in different kinds of patch network. Can habitat subdivision increase, or does it reduce, the probability of long-term persistence?

Theory does not give an unequivocal answer to this question. Increasing subdivision leads to ever smaller populations with increased risk of local extinction, but this might be counteracted by increasing independence (asynchrony) in the dynamics of subdivided local populations enhancing recolonization. Whether habitat subdivision is beneficial or not thus depends on how fast extinction risk increases with decreasing population size (Quinn and Hastings 1987) and on how much there is asynchrony in the dynamics of subdivided populations (Hanski 1989). We might expect subdivision to be especially beneficial in multispecies communities in which much of variability in population sizes is generated by interspecific interactions. Holyoak and Lawler (1996) found subdivision to increase persistence in an experimental protist predator–prey system, but Forney and Gilpin (1989) and Burkey (1997) have reported the opposite result. Forney and Gilpin (1989) reared *Drosophila* either in one large or two small (connected or disconnected) plastic bottles with the same amount of larval medium; the extinction rate was lower in large bottles. Burkey (1995) analysed the effect of subdivision by first empirically deriving the relationship between extinction rate and fragment area for mammals, birds and lizards, then comparing the predicted loss of species from either a single large area or five independent and disconnected small areas of the same pooled size. He found that the cumulative extinction probabilities were ultimately higher for the subdivided systems than for single large areas, although in the short-term the probability of extinction might be lower for the subdivided system. Burkey's (1995) study did not consider migration among the patches.

It is evident that there is no universal answer to the SLOSS problem, nor do we need one, as with the spatially realistic metapopulation models (Chapter 5) we can contrast the probability of metapopulation survival in particular patch networks case by case. What is needed, however, are multispecies algorithms of site selection for conservation incorporating spatial dynamics of the species. From the conservation viewpoint, the key variable is generally the total amount of suitable habitat, not the degree of fragmentation, though clearly this book would not have been written if I believed that the spatial configuration of the habitat makes no difference at all.

The Levins rule

In highly fragmented landscapes suitable habitat occurs in small discrete patches and it is appropriate to apply the classical metapopulation approach. Modelling habitat destruction as loss of patches in the Levins model has led to the seemingly very useful prediction that the number of empty patches at steady state remains constant for all levels of habitat destruction down to the level where the metapopulation goes extinct—the Levins rule (Section 4.4). Here I recount one anecdotal observation that seems to support the Levins rule and one quantitative test that fails to support it.

The anecdote is about the occurrence of the white-backed woodpecker (*Dendrocopos leucotos*) in Finland and Sweden (Andrén 1997). Using data from the extensive virgin forest area in the Białowieza National Park along the border between Poland and Belorussia, the amount of suitable habitat was estimated at 0.64 (64% of the landscape), of which 88% was occupied. Thus the fraction $0.64 \times 0.88 = 0.56$ of the total area was occupied and, according to the Levins rule, the threshold amount of suitable habitat necessary for long-term survival is given by the amount of suitable habitat minus the amount of occupied habitat at equilibrium, or $0.64 - 0.56 = 0.08$. In both Finland and Sweden, the white-backed woodpecker has become threatened because of habitat loss, with less than 100 breeding pairs remaining in both countries. The estimated amount of suitable habitat in those regions in

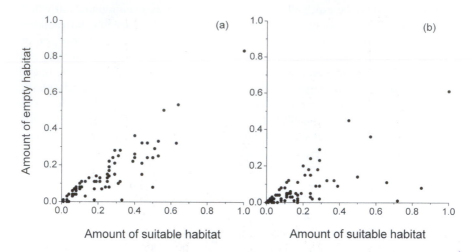

Fig. 10.4 A test of the Levins rule using data on the Glanville fritillary butterfly. Panel (a) gives the fraction of empty but suitable habitat patches plotted against the number of suitable patches in replicate networks in 4 by 4 km² regions (the number of suitable patches was scaled by dividing it by the maximum number in the data set, 72 patches). Panel (b) gives the same information but now the patches have been weighed by their areas.

Finland and Sweden where the white-backed woodpecker has barely survived, 9 and 7%, respectively, agrees remarkably well with the predicted threshold, 8% (Andrén 1997). In two areas in Norway, suitable habitat covers 50% of the landscape and the white-backed woodpecker has remained relatively common (Stenberg and Hogstad 1992).

To test the Levins rule more rigorously, I take advantage of the extensive data base on the Glanville fritillary butterfly (Part III). In the model, one compares landscapes with a smaller or greater density of suitable habitat. I therefore divided our entire study area of 50 by 70 km² into 4 by 4 km² regions, and compared the respective patch networks in these equal-sized areas. A problem with this approach is that the squares dissect natural patch networks, which adds extra variance to the results. Figure 10.4 shows the fraction of empty patches plotted against the fraction of suitable habitat, which was measured by the number of patches within the square. There is much variability in the results, but the trend which is evident is increasing amount of unoccupied habitat with increasing density of suitable habitat, not the predicted constant amount of unoccupied habitat (Fig. 4.5). A likely explanation of the increasing trend is large variance in patch areas, which is predicted to generate a positive relationship (Gyllenberg and Hanski 1997; Section 4.4). Weighing the patches with their areas in the calculation of the fraction of occupied patches (P) leads to the opposite result, with the amount of empty habitat tending to decrease with increasing area of suitable habitat, though once again with large variance. The inevitable conclusion is that this quantitative test of the Levins rule fails to produce

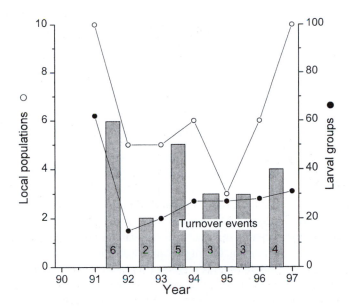

Fig. 10.5 The number of local populations, the pooled number of larval groups, and the number of turnover events in an introduced metapopulation of the Glanville fritillary butterfly on the island of Kumlinge in SW Finland, with 17 suitable habitat patches within an area of 2 by 4 km^2. The network was empty in 1991 before the introduction. Only two local populations have persisted throughout the period 1991–1997 (M. Kuussaari and I. Hanski, unpublished).

any useful insight to the dynamics of the butterfly metapopulation. As there is no reason to assume that the Glanville fritillary would represent an unusual example, at least as far as insect metapopulations go, I conclude that the Levins rule has no value for practical applications, though it has heuristic value in highlighting the significance of patch density in metapopulation dynamics and, more generally, the adverse consequences of habitat loss on long-term survival of species.

Corridors and metapopulation persistence

The value of ecological corridors is a much-debated issue in conservation biology (Simberloff and Cox 1987; Noss 1987; Saunders and Hobbs 1991; Simberloff *et al.* 1992). From the metapopulation perspective, the key function of corridors is to increase connectivity among habitat patches and thereby to increase the rate of migration and recolonization. Disconnected small stepping-stone patches might have the same function (for an illuminating example on the American pika see Section 12.4). To what extent, and which kind of, corridors actually function in the intended manner is an empirical question for which no general answers are likely to emerge. The empirical literature has produced a wide range of conflicting results on the significance of corridors (Bennett 1990; Merriam and Lanoue 1990; Dunning *et al.* 1995; Haas 1995; Hill 1995; Machtans *et al.* 1996; and many others). What is a beneficial movement corridor for some species, for instance narrow forest-dividing corridors of cleared land (Sutcliffe and Thomas 1996), can be harmful for other species, for instance by facilitating the spread of edge-preferring predators to forest interior (Rich *et al.* 1994).

Translocations to enhance colonization rate

Metapopulation persistence hinges on the balance between extinctions and colonizations. In this perspective, there is little difference between reducing extinction rate, for instance by improving habitat quality, and increasing colonization rate by translocations. Translocations might, of course, lead to undesirable gene flow if distant populations are mixed, which should generally be avoided. If inbreeding increases the risk of local extinction, mixing of populations might, however, be beneficial in leading to increased fitness (Oostermeijer *et al.* 1995). Notwithstanding possible genetic effects, and assuming that no real harm is done to the source populations by removing the individuals for translocation, I cannot see any strong arguments against managed colonizations at the scale of metapopulations in situations where severe habitat fragmentation has practically eliminated the possibility of natural recolonization. One example in which translocations might be the only option is *Centaurea corymbosa* (Asteraceae), an endemic plant in southern France with an extremely small range (3 km^2). The species is such a poor colonizer that there is little chance of natural colonization although suitable habitat is available (Colas *et al.* 1997). Carefully designed translocations (Lubow 1996) could also serve as invaluable experimental tests of population biological models (Armstrong and McLean 1995). In the next section, I discuss translocations to empty habitat patch networks with the intention of establishing a new metapopulation.

10.3 Metapopulation establishment

Conservationists are increasingly concerned with habitat restoration (Buckley 1989; Jordan *et al.* 1987; Berger 1989) and re-introductions of species (Griffith *et al.* 1989; Kleiman 1989; Serena 1995). If long-term persistence of species is expected to depend on metapopulation dynamics, restoration and re-introductions should be planned at the regional rather than at the local scale. Re-introductions of British butterflies to isolated sites have generally produced only temporary success at best (Fig. 9.1; Thomas and Hanski 1997), as could be expected from the known extinction-proneness of small butterfly populations (Warren 1992; Hanski *et al.* 1995b; Thomas and Hanski 1997). Butterfly introductions into patch networks have produced better results (Fig. 10.5; Nève *et al.* 1996; Thomas and Hanski 1997), although there are so few well-documented metapopulation-level introductions of any taxa that the evidence remains anecdotal. Many successful introductions might nonetheless hinge on metapopulation dynamics without our knowledge. The moth *Cactoblastis cactorum* was introduced to Australia to control the prickly-pear cactus (*Opuntia*). The control was successful, and the interaction between the cactus and the moth was attributed to predator–prey metapopulation dynamics by Nicholson (Monro 1967; subsequent studies have, however, questioned this interpretation: Caughley 1976; Harrison and Taylor 1997).

It is self-evident that a landscape with much suitable habitat, whether in one or many pieces, is more favourable for (meta)population establishment than a landscape with little habitat. If a species is introduced into a fragmented landscape, a question that necessarily arises is how the introduction should be attempted, given a finite number of individuals to introduce. Which patches should be selected for introduction? Would it be better to introduce all individuals into one patch, giving it a large initial population size, or would it be preferable to divide the individuals among several patches, to spread the risk of immediate failure?

Spatially realistic metapopulation models (Chapter 5) can be used to compare the likelihood of successful introduction from different habitat patches. Figure 10.6 gives an example based on the incidence function model parameterized for the Glanville fritillary butterfly (Section 12.1). The result is not surprising—the most favourable patches for introduction are those that are large and located in the center of the network. Nonetheless, it is useful to have an objective method to assess the effects of patch size and spatial location on the likely outcome of the introduction. A relatively simple model is sufficient for this purpose, because it is likely that alternative reasonable models will predict the same ranking of patches in terms of their value for introduction.

To gain a better understanding of the second question, into how many similar patches a given number of individuals should be introduced, let us make some experiments with the n-population simulation model described in Section 5.5. Local dynamics are given by the Ricker model:

$$N_i(t+1) = N_i(t)e^{r(1-N_i[t]/K)}, \tag{10.1}$$

where $N_i(t)$ is the size of population i at time t, r is the intrinsic growth rate, and K is the local carrying capacity. As in Section 5.5, stochasticity is added by assuming that each population goes extinct with probability:

$$E_i(t) = 1 - \frac{(1-e_l)N_i(t)^2}{e_s^2 + N_i(t)^2}, \tag{10.2}$$

where e_l and e_s are two parameters, defining a sigmoid relationship between population size and extinction risk (extinction risk increases rapidly with decreasing $N(t)$ when e_s is small, and the asymptotic extinction rate in large populations is given by e_l). To construct a patch network, let us assume that there are 20 habitat patches forming a chain with the two ends joined to each other; all patches are hence in identical positions in the network. The fraction of individuals in population i migrating to population j at time t is given by

$$m_{ij} = \frac{e^{-\alpha|i-j|}}{C}, \quad (10.3)$$

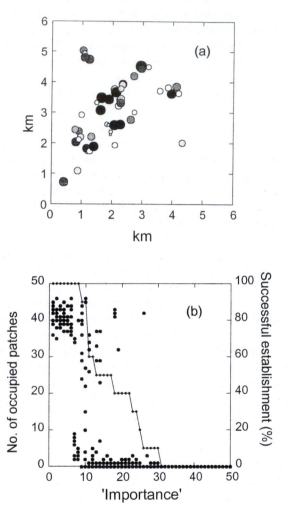

Fig. 10.6 An example of the use of a spatially realistic metapopulation model (the incidence function model; Section 5.3) to assess the significance of specific local populations for metapopulation dynamics. The map in (a) shows the locations and relative sizes of 50 habitat patches. The darkness of the shading indicates the probability of successful re-invasion of the network from the focal patch. In (b), the patches have been arranged in order of decreasing 'importance'. The vertical axis gives the percentage of independent simulations in which only the focal patch was initially occupied and which persisted for >20 years (line) and the number of patches occupied after 20 years (dots; based on Hanski 1994c).

where α is a parameter setting the scale of migration (much long-distance migration when α is small), and the scaling constant C equals the sum of the values in the numerator across the index j. $1/C$ gives the fraction of individuals that remain at the natal site (note that the value of C depends on α). A currently empty patch is colonized if the immigrants to the patch survive the extinction risk specified by eqn 10.2. In the simulations, I calculated first local growth, then migration, and lastly the risk of extinction in each time unit. Each initial condition of n individuals divided equally among p randomly selected patches was iterated for 100 time steps and replicated 100 times. The fraction of metapopulations that survived until the end of the simulation was scored for each parameter combination.

A general feature of the results is that the probability of successful metapopulation establishment increases with increasing number of individuals used for introduction (n), decreasing stochasticity (small e_l and e_s), increasing population growth rate (r), and increasing migration rate (decreasing α). Figure 10.7 shows selected results contrasting a sedentary species with a mobile species, and a species in which small populations are, or are not, sensitive to stochasticity. In the case of species with long-distance migration and rapidly decreasing risk of extinction with increasing population size, it is generally better to place the n individuals into one rather than in several patches (Fig. 10.7d), as spreading the individuals among many patches would create many small extinction-prone populations. Only when n becomes a large fraction of the local carrying capacity should the introduced individuals be spread among several patches, to avoid the initial population overshooting the carrying capacity. The other extreme is shown in Fig. 10.7a, a relatively sedentary species with relatively high risk of extinction even in large populations; the success rate is low unless there are many individuals to introduce, but now it is generally preferable to place the initial n individuals into several patches. In the example in Fig. 10.7b, the optimal number of starting populations is intermediate. This species is sedentary and even large local populations are vulnerable to stochastic extinction. Paradoxically, when $p = 1$ the highest success rate occurs with intermediate n, below the local carrying capacity. The reason is that a population started at an intermediate density will increase, in the first generation, beyond the carrying capacity, and export more individuals than a population started at the carrying capacity.

The main conclusion emerging from these simulations is that one might occasionally do better by dividing the n individuals available for metapopulation establishment among several habitat patches rather than placing them all in one patch. The relative advantage of several initial populations is greatest when:

- the number of individuals available for introduction (n) is large
- the species has a high rate of population increase (r)
- the species has a limited colonization distance (large α)
- and when environmental stochasticity is strong and hence large population size is no guarantee of low risk of extinction (large e_s).

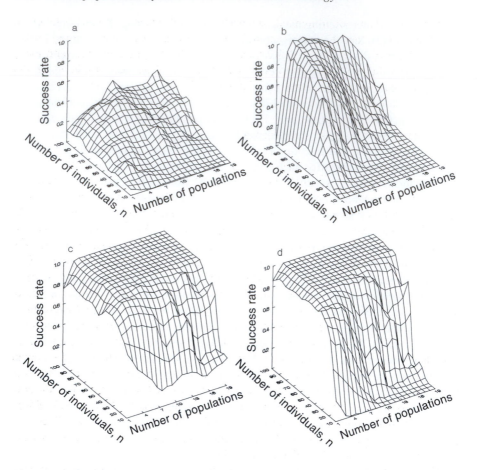

Fig. 10.7   The probability of successful metapopulation establishment (survival >100 years) in 100 replicate simulations using four combinations of parameters: (a) $\alpha = 3$, $e_s = 10$; (b) $\alpha = 0.1$, $e_s = 10$; (c) $\alpha = 3$, $e_s = 5$; and (d) $\alpha = 0.1$, $e_s = 50$. In all cases $r = 2$, $e_1 = 0.05$ and $K = 100$.

The effect of stochasticity is here similar to that in models examining the consequences of habitat subdivision on metapopulation persistence (Quinn and Hastings 1987). Subdivision might be advantageous if it does not create such small populations that demographic stochasticity becomes a prominent process, and provided that there is no strong regional stochasticity (Hanski 1991a).

   As a broad generalization, the multiple-site introduction is likely to be more successful in the case of relatively sedentary species with high rate of population increase but which are vulnerable to environmental stochasticity. Many, though not all, arthropods fit this description. In contrast, a single-site introduction is likely to be preferable for species with low population growth rate, species more affected by demographic than environmental stochasticity, and when a small number of individuals is available for introduction. This scenario applies to many vertebrates. It does not need to be emphasized that these considerations ignore genetics and many other factors that might be important in practice and must be considered in any particular case (see many chapters in Serena 1995).

10.4 The northern spotted owl

In this and the following sections I turn to a number of case-studies from conservation biology for which the metapopulation concept is relevant. The first species is the northern spotted owl (*Strix occidentalis caurina*). Very little was known about this subspecies only 20 years ago, other than that it occurs in the north-western United States, in the Douglas fir forests of Washington, Oregon and northern California. The spotted owl belongs to a genus with 18 species worldwide, including 3 species in North America and 4 species in Europe (Saurola 1995). Pioneering research by Eric Forsman (Forsman *et al.* 1984) showed that the northern spotted owl is unusually catholic in its habitat selection, with a pair of owls requiring a minimum of 1000 ha of old-growth forest within its home range. Today most people with any interest in conservation have heard about the northern spotted owl, and about the conflict of interest between the conservation and exploitation of its habitat, the temperate rainforests of North America. The President of the United States has been personally involved in the discussions seeking a resolution to the conflict, and the biology of the owl has been scrutinized in courtrooms (Harrison *et al.* 1993). Research on the spotted owl is supported at the level of several million dollars per year, equivalent to a few hundred dollars per individual owl. There are presently some 8000 individuals left (Gutiérrez *et al.* 1995), but the numbers have been decreasing rapidly (Burnham *et al.* 1994).

The conflict has forced ecologists to face an obvious but awkward question: How much old-growth forest should be preserved to give the owl a good chance of surviving in the future? The first attempts to find a quantitative answer employed linear Leslie matrix models (Marcot and Holhausen 1987; USDA 1988), refined by Boyce (1987) to include density dependence and an Allee effect. Even earlier, however, Shaffer (1985) had suggested that the Levins (1970) metapopulation model could be used to assess long-term persistence. He suggested that *s*potted *o*wl *m*anagement *a*reas (SOMAs) should have the same average size and median spacing as present-day occupied owl 'patches'. Lande's (1987, 1988a) seminal contribution was to take this approach further and to formulate a metapopulation model with suitable sites for individual territories interpreted as habitat patches, and including a description of how dispersing owls might be searching for vacant territories. Lande (1988a) arrived at a quantitative answer to the question about how much old-growth forest should be preserved—around 20% (Box 10.1), which happens to agree well with the percolation theory-based predictions and with empirical data for other forest-living birds and mammals (Fig. 10.2). Lamberson *et al.* (1992) studied how

Box 10.1 Lande's model of the northern spotted owl

Lande (1987, 1988a) constructed a metapopulation model from first demographic principles for territorial species such as the northern spotted owl. Assuming that the landscape is divided into patches corresponding to individual territories, let ε be the probability that a female offspring inherits the natal territory after the death of the mother. If the natal territory is not inherited, the juvenile is allowed to search m patches for vacancy. If no vacant territory is found the juvenile dies. Let h be the fraction of patches that are suitable, and denote by P the fraction of suitable patches that are currently occupied. At demographic equilibrium, the probability of successfully obtaining a territory times the lifetime production of female offspring conditional on the mother having found a territory, R_0, must equal unity:

$$[1 - (1 - \varepsilon)(\hat{P}h + 1 - h)^m]R_0 = 1,$$

where $\hat{P}$ is the equilibrium fraction of occupied patches. This equation can be solved for $\hat{P}$:

$$\hat{P} = 1 - (1 - k)/h,$$

where k is given by:

$$k = \left[\frac{1 - R_0^{-1}}{1 - \varepsilon}\right]^{1/m}$$

(if $h < 1 - k$ then $\hat{P} = 0$). k has been termed the 'demographic potential' of the population, and it is related to the life history of the species via R_0 (Lande 1987, 1988a). However, because $k = 1 - (1 - \hat{P})h$, the demographic potential can also be estimated from the knowledge of $\hat{P}$ and h only (Lande 1988a; Lawton *et al.* 1994). For the northern spotted owl, Lande (1988a) estimated that $h = 0.38$ and $\hat{P} = 0.44$, giving $k = 0.79$. From the second equation we can see that the critical amount of suitable habitat necessary for long-term persistence ($\hat{P} > 0$) is $h_c = 1 - k = 0.21$. This result also follows from the Levins rule (Section 4.4), which states that h is given by the fraction of empty patches out of all patches at equilibrium: $h - \hat{P}_{tot} = h - \hat{P}h = h(1 - \hat{P}) = 0.21$, where $\hat{P}_{tot}$ is the equilibrium fraction of occupied patches out of all patches, including those already destroyed. Note that Lande's P is the fraction of the suitable patches that are occupied, that is, Lande's $P = \hat{P}_{tot}/h$. Because one patch can be occupied by one individual at most in this model, and because the dynamics are fast and the rescue effect is not an issue, only some of the problems in the use of the Levins rule (Sections 4.4 and 10.2) affect Lande's result.

environmental stochasticity might change the conclusions based on deterministic models, and they also explored non-equilibrium situations. Lamberson *et al.* (1994) followed up their earlier study by considering how the spatial aggregation of suitable sites affects long-term persistence (see also Doak 1989; Thomas *et al.* 1990).

Lamberson *et al.* (1992, 1994) used lattice models. Assuming that suitable sites occur in clusters, they assumed that within-cluster migration follows a random walk, whereas between-cluster migration follows a straight line away from the cluster at a random azimuth. If no new cluster is intersected by a migrating owl in the entire landscape, the owl dies. If a new cluster is hit, the probability of actually reaching it declines exponentially with distance. The model was parameterized with data on

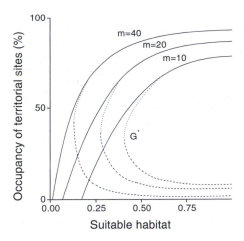

Fig. 10.8 Occupancy of territorial sites plotted against the proportion of landscape suitable for the northern spotted owl. Solid lines ignore pairing and sexual reproduction, dotted and dashed lines include mate search and show stable and unstable (dotted lines) equilibria, respectively, resulting from the Allee effect. Parameter *m* gives the juvenile search ability. (Based on Lamberson *et al.* 1992.)

demography and juvenile migration. The model parameters were set to correspond to the actual status of the forested landscape within the owl's range.

To start with modelling results ignoring the spatial arrangement of suitable sites, Fig. 10.8 shows the fraction of sites that are occupied as a function of the fraction of the landscape that remains suitable. The three lines correspond to three values of the migration parameter setting the number of sites that a migrating owl is able to search before perishing (thus large *m* corresponds to good migration ability). The results demonstrate the threshold amount of suitable habitat for persistence and how it increases with decreasing migration rate. Figure 10.8 also shows that taking explicitly into account the pairing of owls reduces the fraction of occupied sites, increases the threshold and creates an Allee effect with alternative stable equilibria. Stochasticity smoothens the threshold habitat requirement for persistence, increasing the amount of suitable habitat required for high probability of survival but also giving a low probability of 250-year survival in landscapes with less suitable habitat than predicted by the deterministic model (Lamberson *et al.* 1992). Numerical simulations have demonstrated that it takes tens or even hundreds of years before equilibrium is reached after habitat destruction, making it very risky to use occupancy data to predict long-term survival (Lamberson *et al.* 1992; see also Section 12.5).

In the version of the model with explicit space and clustered suitable sites (Lamberson *et al.* 1994), the average occupancy level of suitable sites increases with the number of sites in a cluster and with the percentage of the landscape in clusters (Fig. 10.9). The latter result can be expressed in terms of between-cluster

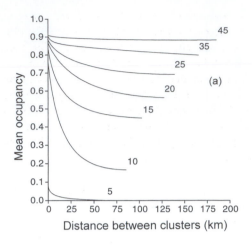

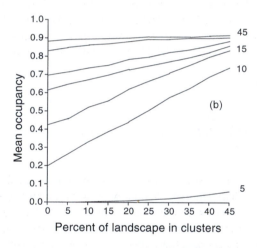

Fig. 10.9 (a) Mean occupancy versus edge-to-edge distance to the nearest cluster for seven different cluster sizes, of which 60% were assumed to be suitable habitat for the northern spotted owl. Occupancy rate was calculated for the last 10 years of simulation. (b) As (a) but now giving the mean occupancy against percentage of landscape in clusters for different cluster sizes. (Based on Lamberson *et al.* 1994.)

distances—mean occupancy increases with decreasing isolation (Fig. 10.9). The significance of cluster size is due to reduced mortality of migrating owls in large clusters, where they are likely to find an unoccupied site before leaving the cluster for risky long-distance migration. The amount of suitable habitat in the landscape and cluster spacing have especially great influence in the case of small clusters, with less than 20–25 sites per cluster. One conclusion emerging from this research is that the planned *h*abitat *c*onservation *a*reas (HCA) should be large enough to accommodate 15 to 25 pairs of owls, and that at least 15–30% of the forested land should be preserved in the HCAs (Lamberson *et al.* 1994), giving a between-cluster isolation of around 20 km (Thomas *et al.* 1990). In contrast with the northern spotted owl, the Californian spotted owl occurs in a discontinuous environment, with discrete local populations of fewer than 100 individuals. Lahaye *et al.* (1994) have applied a *n*-population simulation model to the Californian spotted owl. They found that the metapopulation would have a low risk of extinction if there was no deterministic decline in local populations. Empirical results had shown a decline in 1987–1993, but this might have been a temporary decline due to drought (Lahaye *et al.* 1994).

The great longevity of the spotted owl, with annual adult survival rate exceeding 0.9 (Gutiérrez *et al.* 1995), and the drastic changes that have occurred in the landscape, make it practically impossible to test model predictions with actual data. Using the models to devise management plans has encountered difficulties, and the current management plans contain no explicit reference to metapopulation models (Gutiérrez and Harrison 1996). The spatially explicit models of the spotted owl have nonetheless played a significant role in shifting the attention of conservation biologists to spatial dynamics more generally, and the results have helped to popularize the qualitative effects of habitat fragmentation on the survival of species.

10.5 Other case-studies

Bachman's sparrow

A long-term research project on Bachman's sparrow (*Aimophila aestivalis*), an endangered North American passerine, has employed metapopulation modelling as a research tool (Pulliam *et al.* 1992; Liu *et al.* 1995). Like the models developed for the northern spotted owl, the BACHMAP model is a lattice model including landscape variables that describe the extent and spatial arrangement of different types of habitat, habitat type-specific demographic variables, and behavioural variables describing migration. Migration is modelled as a series of discrete steps, ending either in an unoccupied suitable site or in death. A major focus of the modelling (Pulliam *et al.* 1992), and a difference to the models on the northern spotted owl, is an explicit description of the 20-year timber harvest rotation, which maintains forest succession that greatly affects the quality of the site for the sparrow. Pulliam *et al.* (1992) found that demographic parameters had the greatest influence on population size, whereas migration parameters had relatively little effect, most likely because the simulated landscape had a relatively large amount of more or less suitable habitat (see Section 10.2). Pulliam *et al.* (1992) identified the significance of stable source habitats, mature forests in the case of Bachman's sparrow, in lowering the risk of metapopulation extinction.

Red squirrel—habitat fragmentation in ordinary landscapes

Conservation efforts and public attention tend to be drawn to endangered large-bodied or otherwise appealing species in their last remaining populations. The threat that habitat loss and fragmentation pose is not restricted to these species, of course. Rampant habitat destruction goes on very close to most of us, undermining population diversity of a vast number of species (Ehrlich 1992). Many natural habitats are already fragmented to the extreme in large parts of Europe, North America, and other continents, though not being unique globally the species inhabiting these landscapes have not always received the attention that they deserve. Habitat destruction of these ordinary landscapes, and landscape-level restoration of the already lost habitats, is likely to be an area where the application of metapopulation concepts and models will be most productive in the future.

To take an example, the red squirrel (*Sciurus vulgaris*) is a common and widespread species in Europe, generally preferring coniferous to deciduous forests and woodlands. Working in central Sweden, Andrén and Delin (1994) found that 24% of the landscape was covered by the habitat preferred by the red squirrel; habitat that was used but not preferred accounted for 46%, whereas the rest consisted of avoided habitat. In this landscape, even the most isolated fragments of the preferred habitat were within the usual movement range of squirrels, and hence habitat fragmentation as such was no problem. In Belgium and the Netherlands, where habitat suitable for the red squirrel is much less common and is more fragmented than in Sweden, habitat fragmentation and the density of hedgerows (movement corridors) greatly affects the home range size and space use by individual squirrels (Wauters *et al.* 1994). The presence of squirrel dreys (nests) increases with woodland area and the cover of coniferous trees, and decreases with increasing distance to the nearest large permanently occupied woodland (Verboom and van Apeldoorn 1990). It is questionable whether individual woodlots have local populations (van Apeldoorn *et al.* 1994), but it seems that habitat destruction in Belgium and the Netherlands is close to the threshold level beyond which fragmentation effects become apparent (Fig. 10.2). In the Po Plain, in Italy, the threshold has already been exceeded for the red squirrel, and only those small woodlots that are located close to the more extensive riverine forests are occupied (Celada *et al.* 1994). The fraction of occupied patches was 50–70% in Belgium and the Netherlands (Apeldoorn *et al.* 1994; Wauters *et al.* 1994), but only 28% in the Po Plain (Celada *et al.* 1994). In the case of the red squirrel, the critical degree of isolation is 5–10 km (Celada *et al.* 1994). Other small vertebrates are less mobile and fragmentation effects may occur in less fragmented landscapes (Soulé *et al.* 1988, 1992—chaparral-requiring birds and rodents in California; Van Dorp and Opdam 1987; Verboom *et al.* 1991— passerine birds in Europe; Smith 1974, 1980—the American pika in California; Dickman and Doncaster 1989; van Apeldoorn *et al.* 1992—small mammals in Europe).

European butterflies

The past five years have seen an outburst of metapopulation studies on endangered European butterflies, including the Glanville fritillary (*Melitaea cinxia*; Part III), the bog fritillary (*Proclossiana eunomia*; Baguette and Nève 1994; Nève *et al.* 1996), the silver-spotted skipper (*Hesperia comma*; Thomas and Jones 1993), the silver-studded blue (*Plebejus argus*; Thomas and Harrison 1992; Thomas 1994b), the chequered blue (*Scolitantides orion*; Saarinen 1993; Hanski 1994a), the heath fritillary (*Mellicta athalia*; Warren 1987), and the false heath fritillary (*Melitaea diamina*; Wahlberg *et al.* 1996). For general reviews see Thomas (1994b,c), Hanski and Kuussaari (1995) and Thomas and Hanski (1997). There are at least three reasons for the explosion of studies on butterflies. First, there is a long tradition of ecological studies on butterflies, established by the classical works of Ford and Ford (1930), Ford (1945), Ehrlich *et al.* (1975) and Ehrlich (1983, 1984), and butterflies are in many ways convenient (and fun) to work with. Second, many butterflies have the kind of spatially structured populations that are assumed by classical metapopulation models (Ehrlich and Murphy 1987; Murphy *et al.* 1990; Thomas and Hanski 1997). And third, butterflies, high-profile species for conservation, have declined dramatically especially in northern and central Europe in recent decades, with a large fraction of species having become endangered or regionally extinct (van Swaay 1990; Warren 1993; Hanski and Kuussaari 1995; Pullin 1995). To take an admittedly extreme example, 17 of 71 species in the Netherlands have become extinct (van Swaay 1995).

The butterfly metapopulation studies, like the research on the northern spotted owl, have shifted attention from the conservation of small sites and local populations to regional issues (Murphy *et al.* 1990; Hanski and Thomas 1994; Thomas 1994b,c; Thomas and Hanski 1997). This is the second major reorientation in butterfly biology and conservation in 20 years, following the discoveries by Jeremy Thomas and others in the 1970s of the extremely specific habitat requirements of many species (Thomas 1984, 1991). The empirical evidence now available on the extinction-proneness of local butterfly populations even at protected small sites (Warren 1994) makes it clear that a regional perspective is needed. The butterfly metapopulation studies have employed spatially realistic metapopulation models (Chapter 5), such as the incidence function model and *n*-population simulation models, instead of the lattice models applied to the northern spotted owl and the Bachman's sparrow. This different modelling framework has been largely dictated by the different degrees of fragmentation of butterfly and bird landscapes.

What are the contributions that the metapopulation studies have made to butterfly conservation? It is too early to draw definite conclusions, given that most studies have been published since 1994 and that they have hardly had enough time to reach the management level, but here is a short list of achievements:

- greatly improved empirical knowledge of the spatial population structures and migration rate and distances of endangered butterflies (Section 8.1, Part III)

- successful application of spatially realistic metapopulation models to predict the spread of species into empty patch networks (Fig. 5.5) and the equilibrium distribution of species in patch networks (Fig. 12.4)
- the models have been used to make specific recommendations as to which patches in particular networks are especially critical for metapopulation survival (Wahlberg *et al.* 1996) and which sorts of networks are required for long-term survival (Thomas and Hanski 1997). In the case of *Melitaea diamina*, habitat management based on model predictions (Wahlberg *et al.* 1996) has been initiated (N. Wahlberg, personal comminication).

Furbish's lousewort

There are few detailed studies of plant metapopulations, although at least some plant communities include species with much local turnover (Grubb 1988; Ouborg 1993; Margules *et al.* 1994; Eriksson 1996; Husband and Barrett 1996; Fischer and Stöcklin 1997; Valverde and Silvertown 1997) and which hence might be persisting as classical metapopulations. Ouborg (1993) found that isolation affected both extinctions of and colonizations by several plant species occurring on dry grasslands along the IJssel and Rhine rivers in the Netherlands.

Menges's (1988, 1990) work on Furbish's lousewort (*Pedicularis furbishiae*) describes a fine example of a classical plant metapopulation. Furbish's lousewort is something of an exception, a species endemic to northern Maine, USA, where no endemics are expected to occur owing to the recent biological history spanning less than 12 000 years. The entire species has less than 10 000 individuals, distributed in some tens of local populations along a 250-km stretch of St. John River (Menges 1990). Waller *et al.* (1987) found no genetic variation in 22 allozyme loci in 28 individuals sampled from four populations. The plant has no seed bank nor seed dormancy, and it is propagated exclusively by sexual reproduction. Seeds lack adaptations for wind or animal dispersal, but they can float in water for several days (Menges 1990), hence long-distance migration can occur. Furbish's lousewort is a habitat specialist, a hemiparasite that flourishes on moist early-successional sites, created by physical disturbance such as ice scour and bank slumping along steep riverbanks. It is probably significant that St. John River is known for dramatic seasonal and long-term fluctuations in the water level, which cause substantial primary disturbance on the steep riverbanks where Furbish's lousewort is found.

Menges's (1990) demographic studies revealed the detailed environmental requirements for population growth. The finite rate of population increase (λ) in the entire study area was 0.77 in 1983–1984, 1.27 in 1984–1985 and 1.02 in 1985–1986 (Menges 1990). Multiplying these values gives the overall rate of 0.997 for the 3-year period, casting some doubt on the assumption that there is no density dependence in the dynamics (Menges 1990). Menges (1990) calculated λ values separately for 15 local populations for two time intervals. These results show year-to-year variation, as the populations performed generally better in 1984–1985 than in 1985–1996, but there was no strong overall synchrony in local dynamics (Menges

1990), suggesting that metapopulation dynamics might promote long-term persistence.

Menges's (1990) results suggest that the local extinction rate is around 5% per year. He recorded four new populations in 1981–1984; this was 3% of the extant populations. From these figures alone, however, one cannot draw any conclusions about long-term persistence at the metapopulation level, other than that there is substantial population turnover, and that Furbish's lousewort might indeed survive as a classical metapopulation. Population extinction was most likely at sites that otherwise were most favourable for population growth, with wet soils and low herb-dominated vegetation (Menges 1990). In other words, the Furbish's lousewort has become adapted to early-successional ephemeral sites. Sites with low probability of disturbance simply did not support viable local populations.

10.6 Spatially extended population viability analysis

Population viability analysis (PVA; Soulé 1980, 1987; Lande 1988b; Shaffer 1990; Boyce 1992) aims at assessing the risk of population extinction under specific environmental conditions, which in the conservation context often involve a particular management regime. PVA is typically based on a simulation model which includes the factors thought to pose the greatest threat to the long-term survival of the population, and of which suitable information is available. When applied to metapopulations, one needs to include the spatial population structure in the PVA. Metapopulation-level PVAs have typically employed models with great complexity, with tens or even hundreds of parameters, including patch-specific habitat quality and possibly changes in habitat quality, population age structure, details of migration, etc. (Akçakaya and Ferson 1992; Possingham *et al.* 1992; Burgman *et al.* 1993; Lindenmayer and Possingham 1994, 1995). Computer models available for metapopulation-level PVA include VORTEX (Lacy 1993), GAPPS (Harris *et al.* 1986), ALEX (Possingham *et al.* 1992) and RAMAS/Space (Akçakaya and Ferson 1992). For a review of these generic simulation models see Lindenmayer *et al.* (1995) and Table 5.1.

PVA generates a distribution of expected times to metapopulation extinction and thereby also the probability of metapopulation extinction during a given period of time. From a practical point of view, such predictions with their inevitable uncertainties (Wennergren *et al.* 1995) are not as useful as comparisons of alternative scenarios with appropriate sensitivity analyses (Possingham *et al.* 1992; Hanski and Simberloff 1997). In other words, although we might never be able to predict accurately the destiny of particular metapopulations in particular landscapes, we might well be able to rank alternative scenarios, possibly involving specific management actions, in terms of metapopulation viability. Metapopulation-level PVAs employing spatially realistic models transform the infamous debate about whether a manager should opt for a single large or several small (SLOSS) reserves (Soulé and Simberloff 1986) to a more meaningful comparison among different subdivisions of a given amount of habitat (Hanski and Simberloff 1997). The difference here is that SLOSS and other related 'rules' of reserve design (Simberloff 1988), although implicitly based on the dynamic theory of island biogeography (MacArthur and Wilson 1967), involve static comparisons of fixed alternatives. The fundamental concept is the species–area relationship, which in these applications means a fixed number of species within a fixed area. In contrast, spatially realistic metapopulation models explicitly address the dynamics of species survival in fragmented environments, and instead of contrasting fixed alternatives one is practically forced to make comparative predictions for specific fragmented landscapes. As has become apparent in the previous chapters, for this task my preference is for relatively simple models such as the incidence function model rather than for complex simulation models, especially when the habitat is highly fragmented. Relatively simple and general metapopulation models are also helpful in elucidating management principles ('rules of thumb') that would be hard to extract from complex models. Thus Drechsler and Wissel (1998) and Frank and Wissel (1998) discuss the relative advantages of managing (average) patch size or quality, patch number and patch connectivity in enhancing metapopulation persistence in the context of a general metapopulation model. They find that the kinds of management action that are likely to be most helpful depend on the environmental setting. Insight from relatively simple and general metapopulation models are likely to be more helpful for managers than uncertain predictions based on untested complex models.

Part III

THE GLANVILLE FRITILLARY—A CASE STUDY

Even the best reviews of ecological field studies suffer from the problem that the context of particular studies is generally lost—and the context usually matters in ecology. For this reason, I decided to complement the review of field studies in Part II with a more thorough account of the metapopulation ecology of a single species, the Glanville fritillary, *Melitaea cinxia* (L.) (Nymphalidae, Melitaeinae), a checkerspot butterfly ranging across the entire temperate region of the Old World (Tolman 1997). The Glanville fritillary project began in the spring 1991, when I searched for an insect species on which to launch a large-scale field project that would enable the testing of some metapopulation models I was studying at that time (Gyllenberg and Hanski 1992; Hanski and Gyllenberg 1993). In Finland, the Glanville fritillary has been extinct on the mainland since the late 1970s (Marttila *et al.* 1990; Kuussaari *et al.* 1995), but it survives on the Åland islands in northern Baltic. Like many other butterfly species (Thomas 1984, 1991; Warren 1993; Thomas 1994b; New *et al.* 1995; Pullin 1995), the Glanville fritillary has greatly declined in northern Europe in past decades, and its current distribution is highly fragmented on a large spatial scale (Hanski and Kuussaari 1995). By presenting a detailed account of this particular project I hope to demonstrate the value of the metapopulation approach as described in the two earlier parts of the book. Those readers who have already perused the previous chapters know that they include several examples on the Glanville fritillary, used in the absence of other suitable examples. I do not repeat here what has already been said, but I have included cross-references where appropriate.

Our checkerspot butterfly project has an illustrious predecessor in Paul Ehrlich's and his students' and collaborators' influential research on the population biology of checkerspot butterflies in California (Ehrlich 1961, 1965, 1984; Ehrlich and Birch 1967; Brussard and Ehrlich 1970; Singer 1971, 1972; Gilbert and Singer 1973; Ehrlich *et al.* 1975, 1980; Singer and Ehrlich 1979; Weiss *et al.* 1988, 1993; Ehrlich and Murphy 1987; Harrison *et al.* 1988; Harrison 1989; Murphy *et al.* 1990; Singer and Thomas 1996). In Europe, these butterflies are usually called fritillaries, but fritillaries also include other related taxa apart from Melitaeinae, which Americans call checkerspots. Melitaeinae are widely distributed in the boreal and temperate regions in America (39 species; Scott 1986) and in Europe (21 species; Tolman 1997), with *ca.* 250 species worldwide (Higgins 1981). The temperate species, which are best known and to which the Glanville fritillary belongs, have from one to several generations per year, except in the far north where larval development can take two or even more years (Ehrlich and Murphy 1981; Luckens 1985; Eliasson 1991; Wahlberg 1998). Most Melitaeinae use only one or a few larval host plants locally, but the same species may use different host plants in different localities (Singer 1983). Detailed studies by Michael Singer and his colleagues have revealed rapid evolution of larval host plant use in response to changes in the species composition and abundances of potential host plants (Singer 1983, 1994; Parmesan 1991; Singer *et al.* 1992, 1993, 1994; Thomas and Singer 1998). Such shifts have obvious relevance for metapopulation dynamics in heterogeneous patch networks, but little work has been done in this direction so far (Section 6.1; Singer and Thomas 1996).

One characteristic that apparently attracted Ehrlich to checkerspot butterflies is their spatial population structure (Ehrlich also wanted to study a 'normal' insect species at a time when insect ecology was excessively concerned with outbreak species; M. Singer, personal communication). Because the suitable habitat often occurs in discrete patches, local populations are relatively easy to delimit (Ehrlich 1961), a situation that is not uncommon in butterflies in general (Thomas and Hanski 1997). Pioneering studies by Ehrlich (Ehrlich *et al.* 1975; Singer and Ehrlich 1979) and others (Thomas 1984) established the concept of a 'closed' population, by which they meant very limited migration among, and demographic independence of, even closely situated local populations (Brussard and Ehrlich 1970). The three local populations of the Bay checkerspot butterfly (*Euphydryas editha*) on the Jasper Ridge in Stanford, California (Fig 1.1), attained the status of a classic example. Extending the work of Ehrlich and others, the next two chapters describe the most comprehensive study of classical metapopulation dynamics attempted so far.

11

Metapopulation patterns and processes

In ecology, the context matters but so does the scale. In a review of field studies on population responses to habitat fragmentation, Doak *et al.* (1992) found that the median number of habitat fragments per study was 34, the median distance among patches was 30 m, and the median duration of the study was less than one generation. Most of these studies were not concerned with metapopulations, but they reflect the general practical problems of conducting research across large areas. The results reviewed by Doak *et al.* (1992) also reflect, indirectly, the current emphasis on experimental studies, which are almost impossible to extend to most metapopulations in nature. However, accepting that a different approach is needed for ecological research on large scales, the difficulties need not be insurmountable. It also helps, of course, to select a study system with convenient attributes for field work. For instance, the plant–pathogen metapopulation studied by Antonovics and his colleagues occurs along roadsides in Virginia, USA (Antonovics 1994; Antonovics *et al.* 1997).

Our studies on the Glanville fritillary cover the entire geographical range of the species in Finland, which is currently restricted to the Åland islands in the northern Baltic, within an area of 50 by 70 km^2, of which some 1500 km^2 is land (Fig. 11.1). Dry meadows with the larval host plants, which are the suitable habitat patches in metapopulation parlance, are common on several large islands; there are altogether more than 1600 habitat patches in the entire system. We have been able to study the Glanville fritillary on such a large scale because of several helpful features in the butterfly's life cycle, described below, and because the some hundreds of farm houses and small villages scattered across the Åland are connected by a dense network of small roads, making movement easy. The local butterfly populations are mostly very small, which facilitates population censuses in the field and leads to high rate of population turnover. High turnover rate has, in turn, allowed us to accumulate much information about local extinctions and colonizations. Small population size is probably not an uncommon feature of many species, because most species are locally rare by any criteria we might wish to use (Gaston 1994). The large Glanville fritillary metapopulation serves, therefore, as an ideal model system, a kind of ecological field facility, with which one may address questions that are logistically difficult to study with most other species.

Over the years, the comment has frequently been made that our 'local populations' are not, really, populations at all but merely aggregates of individuals, and that we should expand the spatial scale of 'populations', for instance to a scale within which there is no significant variation in allele frequencies. I do not wish to

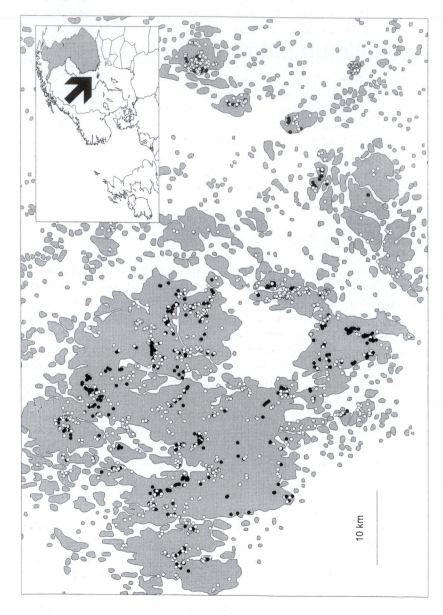

10 km

Fig. 11.1 A map of the Åland islands in SW Finland with the meadows suitable for the Glanville fritillary shown by small dots. Black dots represent meadows that were occupied in 1993.

promote an unhelpful semantic argument, but I want to make the point that ignoring the actual spatial population structure of any species, at whatever scale that structure might occur, would be an exceedingly silly thing to do. In the case of the Glanville fritillary with a large number of extinction-prone local populations (by my definition), pooling data for larger areas would yield better-behaved time-series of population density but at the great cost of losing much of the mechanistic understanding of the dynamics that the metapopulation approach may yield. The spatial population structure makes a difference because it restricts interactions among individuals. For instance, demographic stochasticity and inbreeding depression are significant mechanisms of population extinction in the Glanville fritillary. They would simply not play the same role if the population structure were not highly fragmented. The empirically-observed patterns in habitat occupancy discussed in this chapter could not be explained by pretending that we have a single panmictic population. It is not so important which nomenclature we use, but it is important that our description of populations reflects reality.

11.1 The Glanville fritillary on Åland islands

The Glanville fritillary has two larval host plants in Åland, the ribwort plantain, *Plantago lanceolata* L., and the spiced speedwell, *Veronica spicata* L. Out of a total of 3900 host plant records, only 0.3% are of other *Plantago* and *Veronica* species (Kuussaari *et al.* 1995). The common attributes of the two host plants include somewhat similar growth form, secondary chemistry (below) and habitat—dry meadows ranging from dry rocky outcrops with naturally sparse vegetation to more diverse grazed meadows on deeper soils. The rocky outcrops are permanent features of the landscape, but even here the long-term success of *P. lanceolata* and *V. spicata* is increased by disturbance. A detailed study of one part of Åland by Hering (1995) revealed a substantial decrease in the amount of suitable habitat for the butterfly with declining cattle and sheep grazing over a period of 20 years (Fig. 11.2). How representative this result is for the Åland islands as a whole is not known. Though the reduction in suitable habitat has been substantial, at least regionally, the rate of change has been relatively slow in comparison with the fast turnover rate of local populations—so slow that for many purposes we are justified in ignoring past changes in landscape structure.

Rearing experiments and field observations have not revealed systematic differences in larval performance on the two host plants (M. Kuussaari and M. Singer, personal communication), even if there is a distinct east-west gradient in female host plant preference, resembling a similar longitudinal gradient in the relative abundances of the two host plants (Kuussaari *et al.* 1998). In the well-studied *Euphydryas editha* populations in California, females typically agree on the rank order of oviposition preference (Thomas and Singer 1998). In contrast, many Glanville fritillary populations have both *Plantago*-preferring and *Veronica*-preferring females (Kuussaari *et al.* 1998), possibly due to the influence of the

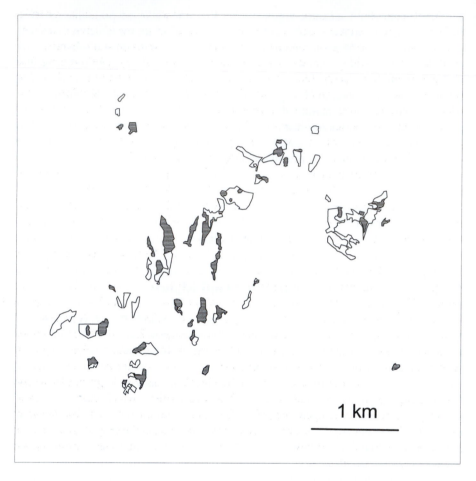

Fig. 11.2 The extent of suitable habitat for the Glanville fritillary within an area of 25 km²
in northern Åland *ca.* 20 years ago and today (shaded). During this period, the total area of
suitable habitat declined to one third of its original extent and the number of distinct habitat
  patches declined from 55 to 42, largely due to reduced grazing by cattle (Hering 1995).

extreme western and eastern ends of the study area, where females are more
systematically *Veronica*-preferring and *Plantago*-preferring, respectively, reflecting
the difference between the numerically dominant host plant species (Kuussaari *et al.*
1998). Apparently, host preference has not evolved in the Glanville fritillary in
Åland because of differences in larval performance on the two host plants but for
some other reasons, for instance local specialization might have been favoured to
improve host plant recognition from a distance by ovipositing females.

Apart from the larval host plants, flowers that provide nectar to adult butterflies affect movement patterns and thereby habitat selection of the Glanville fritillary (Kuussaari *et al.* 1996). In contrast to the catholic female oviposition habits, adult butterflies use a wide range of flowers (Wahlberg 1995; Pöyry 1996) during their flight season, from early June to early July. Males tend to emerge earlier than females (Hanski *et al.* 1994), as is usual in butterflies (protandry; Wiklund 1984). Females typically mate only once; Kuussaari *et al.* (1998) found that only 8% of 167 field-collected mated females had mated twice. Females lay several batches of 100–200 eggs at intervals of one or more days, when the weather is sunny, usually starting 2–3 days after mating (Wahlberg 1995; Kuussaari 1998). Females are choosy about the particular plant individual on which they oviposit, the selection involving a chain of behavioural decisions studied in detail with *Euphydryas editha* (Singer 1972, 1983, 1994). Eggs are laid on the underside of a leaf close to the ground. Eggs hatch in 10–14 days, and the newly-hatched larvae spin a web around the host plant. If and when the plant individual becomes defoliated, the larvae move on as a group to another plant, leaving a trail of silk behind them. During food shortages, the larval groups may split into two or even three subgroups (Kuussaari 1998). By late August or early September the larvae have completed the first three instars. They moult to the 4th instar, change colour from brown to black, and spin a dense 'winter nest' with a diameter of only 1–6 cm (usually 3–4 cm). All larvae in a group huddle into the nest, where they enter diapause. Larvae of the Glanville fritillary always diapause as a group, unlike the larvae of many related species, which hibernate individually (Kuussaari 1998). Diapause is terminated soon after the snow melts in April or even in late March. The black larvae bask in groups in the sun to raise their body temperature to 35°C (Kuussaari 1998), which enables them to develop in ambient air temperatures of 10°C or even less. Larvae grow fast on sunny days, and they are ready to pupate in early May. The pupa is the only stage in the life cycle of the Glanville fritillary which is hard to find; the pupa hangs from a plant stem in dense vegetation.

Spatial population structure

The gregarious larval habits of the Glanville fritillary have enabled us to census the sizes of local populations by counting the numbers of larval groups in late summer, when they are most conspicuous. There have been 300–500 local populations in Åland in 1993–1997, but most populations have only a few larval groups, each with usually 50 to 100 larvae in late summer (Hanski *et al.* 1995a; Kuussaari 1998). Much emphasis has been placed in the field work on ascertaining the presence or absence of larvae in individual meadows, because of the importance of this information for modelling (Section 12.2). In the larger populations, we estimate that some 30–50% of larval groups remain undiscovered (Hanski *et al.* 1995a; M. Nieminen and J. Pöyry, unpublished). Assuming that half of the larval groups have been detected, there are typically 2000–4000 larval groups in late summer in Åland, corresponding to some 10 000–30 000 adult butterflies.

The lowest distinct spatial unit in the Glanville fritillary system is thus the group of usually ful-sib larvae. The larval groups in the same meadow constitute a local population, with an average of 3.4 larval groups per occupied meadow (in autumn 1996). The third level has been dubbed SIN, a *semi-independent* patch *network* (Hanski *et al.* 1996c), of which there are 127, with an average of 12 habitat patches. Assemblages of local populations in individual SINs comprise largely independent metapopulations, because the different SINs are separated by at least 1.5 km from each other and usually by some physical barrier to migration (Hanski *et al.* 1996c). We have estimated that two neighbouring metapopulations in different SINs exchange less than 1%, and usually less than 0.1%, of individuals per generation (see Section 11.3 for migration distances). The metapopulations are, therefore, not expected to affect each other dynamically, but the low level of migration between SINs means that recolonization of an empty SIN is possible, which is helpful, because otherwise most or all SINs would have gone extinct a long time ago. Finally, the metapopulations in the 127 SINs constitute the fourth level in the spatial hierarchy, a metapopulation of metapopulations, which I have called a megapopulation.

Natural enemies

The two larval host plants, *Plantago lanceolata* (Plantaginaceae) and *Veronica spicata* (Scrophulariaceae), contain iridoid glycosides, which the larvae sequester into their bodies, apparently to defend themselves against potential natural enemies (Bowers 1980, 1983; M. Camara, personal communication). The defence is effective—although the black gregarious larvae are conspicuous in the spring, we have no records of birds or mammals attempting to attack them. On a few occasions, a sting bug or a spider has charged a larva, but even insect predators pose an insignificant threat to the larvae. Ants run away after stumbling on a larva (M. Camara, personal communication), although large *Formica* ants willingly attack adult butterflies, for instance ovipositing females (N. Wahlberg, personal communication). Nor have we observed any disease in the larvae, with one exception (in a population experimentally introduced to a small island; M. Kuussaari, personal communication). The small size of local populations makes it difficult for specific disease agents to persist.

The significant natural enemies are hymenopteran parasitoids (Fig. 11.3). These wasps include two specific larval parasitoids, *Cotesia* (*Apanteles*) *melitaearum* (Wilkinson) (Braconidae) and *Hyposoter horticola* (Gravenhorst) (Ichneumonidae), which in turn are attacked by four species of hyperparasitoids, of which *Gelis agilis* (Fabricius) (Ichneumonidae) attacking *C. melitaearum* and *Mesochorus* sp. cf. *stigmaticus* Brischke (Ichneumonidae) attacking *H. horticola* are common (Lei *et al.* 1997; Lei and Hanski 1997). We have also reared four species of generalist pupal parasitoids. Of the hyperparasitoids, *M. stigmaticus* is apparently a specialist in our study area, and hence we have an exceptional chain of specialist species consisting of the Glanville fritillary, *H. horticola* and *M. stigmaticus*. The three hyper-

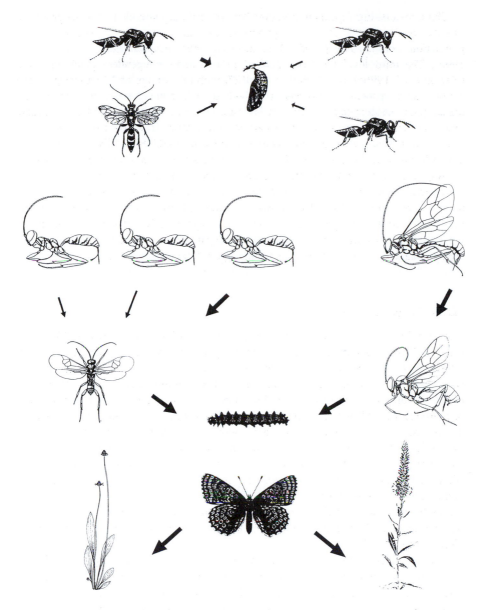

Fig. 11.3 The parasitoid assemblage associated with the Glanville fritillary in Åland. The two primary parasitoids are *Cotesia melitaearum* (Braconidae) and *Hyposoter horticola* (Ichneumonidae). The latter is attacked by *Mesochorus* sp. cf. *stigmaticus* (Ichneumonidae), *C. melitaearum* is attacked by *Gelis agilis*, *G. acarorum* and *G. ruficornis* (all Ichneumonidae). The four pupal parasitoids are all Ichneumonidae. Thick arrows indicate strong interactions (abundant species), thin arrows weak interactions (based on Lei *et al.* 1997; Lei and Hanski 1997). The two host plants are *Plantago lanceolata* and *Veronica spicata*.

parasitoids attacking *C. melitaearum* are, in contrast, all generalists (Lei *et al.* 1997; Lei and Hanski 1997).

Cotesia melitaearum is the parasitoid with the most dramatic impact on host dynamics (Sections 11.2 and 12.5). It develops three generations during one host generation (Lei *et al.* 1997), with most of the mortality in the host population being caused in the spring, when the wasps attack large 6th instar host larvae. A single parasitized host larva produces from a few up to some tens of *C. melitaearum* larvae, depending on the size of the host larva. In contrast, *H. horticola* is a solitary parasitoid with a single generation per year. The interactions between the host butterfly and the primary parasitoids are complicated by interspecific competition between the primary parasitoids (Lei and Hanski 1998) and by the presence of the hyperparasitoids, which often cause high mortality in, and even local extinctions of, the primary parasitoids (Lei and Hanski 1997). Comparable parasitoid assemblages are associated with other checkerspot butterflies, though comprehensive studies on other species are lacking. Typically, a local butterfly population is parasitized by one species of *Cotesia*, for example *C. acuminatus* on *Euphydryas maturna* in Sweden (Eliasson 1995), *C. bignellii* on *E. aurinia* in England (Porter 1981, 1983) and elsewhere in Europe (Komonen 1997), *C. euphydryas* on *E. phaeton* in Virginia, USA (Stamp 1981, 1982a,b), and *C. koebelei* on *E. editha* in California (Moore 1989). In all these cases, the primary parasitoid was attacked by several species of hyperparasitoids, often in the genus *Gelis* (Porter 1981; Stamp 1981). Occasionally, a local butterfly population might be parasitized by two *Cotesia* species, for instance *C. acuminatus* and *C. melitaearum* on *E. maturna* in SE Finland (Komonen 1997), but this is exceptional, and several lines of evidence suggest that two species of *Cotesia* might not be able to persist on a single host species (Komonen 1997). The coexistence of *C. melitaearum* and *H. horticola* on the Glanville fritillary in Åland is facilitated by the classical competition–colonization trade-off; *C. melitaearum* is a superior competitor but an inferior disperser to *H. horticola* (Lei and Hanski 1998).

11.2 Local dynamics and population extinction

Local populations of the Glanville fritillary often consist of a few butterflies only. Spatial and temporal variance in the agents of mortality are so huge in such small populations that a standard life-table analysis (Varley *et al.* 1973) would not be informative. For instance, larval mortality caused by the parasitoid *C. melitaearum* varies from 0 to 100% from population to population and from one year to another (Lei *et al.* 1997). Rather than focusing on the factors and processes that contribute to population regulation and persistence, one is naturally led to consider the factors and processes involved in local extinction. This is not to say that local populations would not be regulated at all—it becomes apparent below that they are. The point is that the local populations are mostly so small that regardless of potentially regulating density dependence the populations have a high risk of extinction for many reasons. But let us start with the dynamics of the smallest spatial units, the survival of larval groups.

Survival of larval groups

Newly-hatched larvae have no chance of surviving outside the communal web which the larvae spin around the leaves of the host plant. At this stage, entire larval groups often perish, while in the surviving groups larval mortality is relatively low. Kuussaari (1998) has estimated that mortality at the group level is 59% before winter diapause and that 34% of the larvae in the surviving groups die. Survival of larval groups, and survival of larvae in groups, is positively group size-dependent in summer, but especially so during the winter diapause, when only rarely groups with less than 15–20 larvae have survived until spring (Kuussaari 1998). Group-living probably facilitates feeding in the earliest larval instars, and the web that the larvae spin seems to provide substantial protection against the parasitoid *Cotesia melitaearum* (G. Lei and M. Camara, personal communication).

The average rate of overwinter survival of larval groups has been around 80%. For instance, in autumn 1994, there were 400 local populations in the entire study area, with a total of 1665 larval groups recorded, of which 83% survived until the following spring. Using these data we can assess the impact of demographic, environmental and regional stochasticities in overwinter survival. Of the 164 populations which had just one larval group in the autumn, 79% survived until the spring, which is close to the overall mean rate. Using binomial probabilities, I found that overwinter mortality was significantly higher than expected in 19 populations, for instance in one population 8 of 10 groups disappeared. The population sizes are so small that in only 2 cases a significantly lower rate of mortality than expected could be scored; in one population 67 of 69 groups survived. The populations that deviated from the average rate exemplify the action of 'environmental stochasticity'. The causes of this extra variance in mortality are largely unknown, but one likely factor is among-population variation in group size, which would make within-population probabilities of group survival correlated, as overwinter survival is strongly group-size dependent (Kuussaari 1998). On the other hand, there was no sign of regional stochasticity in these data, in other words no spatial correlation in the occurrence of high-survival and low-survival populations. Lack of regional stochasticity in overwinter survival contrasts with the observed spatially correlated changes in population size from one year to another (Fig. 11.7 below). Evidently, the factors responsible for regional trends in population change do not operate during winter diapause.

Mortality of entire larval groups reduces effective population sizes, because variance in the fitness of reproducing females is increased (Hedrick 1983), but of greater concern here are the straightforward demographic repercussions—among-group variance in survival contributes an extra component of variance to the population growth rate, which increases the risk of population extinction (Lande 1993; Foley 1994, 1997). This result goes some way towards explaining the very high extinction rates observed in local populations (below). More generally, it might not be a coincidence that the larvae of many checkerspot butterflies live in groups at least during the early instars and that they are among the most threatened butterfly species in Europe (Warren *et al.* 1984; Eliasson 1991, 1995; Warren 1994).

Contributing to the among-group variance in survival is parasitism by *C. melitaearum*. Parasitism in the first two generations increases with increasing larval group size, because *C. melitaearum* females are attracted to the larger larval groups and tend to stay with one such group, once detected, for several days (Lei 1997). Similar results have been found by Stamp (1982) for the related American species *Cotesia euphydryidis* parasitizing *Euphydryas phaeton*. The level of parasitism is, however, so low in the first two parasitoid generations (<20%; Lei 1997) that it poses no great selection pressure against large group size. What is less clear is to what extent among-group variance in parasitism in the third generation in the spring, when parasitism is often very high (Lei 1997), is still affected by the spatial distribution of parasitoids among the larval groups in the earlier generations. In contrast to the often long stay-times of *C. melitaearum* with individual larval groups, the other primary parasitoid of the Glanville fritillary, *Hyposoter horticola*, moves frequently among larval groups, apparently parasitizing only a small number of larvae in a group (Lei and Hanski 1998). The comparable species in Stamp's (1982) study, *Benjamina euphydryalis*, also travelled frequently among larval groups, spending only less than 1 min per host group. At the population level, the incidence of parasitism by *H. horticola* tends to increase with isolation of the larval group, whereas groups parasitized by *C. melitaearum* tend to occur close to other larval groups, in the center of the population (Lei and Hanski 1998). The same pattern is observed also at the metapopulation level, as the effect of host population isolation on parasitoid presence is more pronounced for *C. melitaearum* than for *H. horticola* (Lei and Hanski 1998). The greater mobility of *H. horticola*, which leads to parasitism of relatively isolated larval groups and populations, apparently helps it to survive competition with the superior competitor, *C. melitaearum* (Lei and Hanski 1998).

Density dependence in local dynamics

The early results of Ehrlich *et al.* (1975) on population fluctuations in *Euphydryas editha* were taken as evidence for weak or no density dependence in classical checkerspot populations (Ehrlich 1984). Indeed, Fig. 1.1 does not create the impression of strongly regulated populations, although without appropriate statistical analyses the case remains open. In Åland, the populations of the Glanville fritillary are mostly so small that they do not persist for many years, and hence density dependence cannot be analysed for individual populations. Instead, I examine whether the relative change in population size from year t to year $t + 1$ depends on the corresponding change from year $t - 1$ to year t, comparing populations which had more than five larval groups in year t (this restriction produced a data set with very few zeros). In this analysis, I included as an explanatory variable the regional trend in population size, to account for any general environmental effects on the change in population size (N_{trend}, Box 11.1). In both 1994–1995 and 1995–1996, much of the variance in the relative change in population size was explained by the regional trend, but in both years also the relative change in population size in the previous year had a significant negative effect (Fig. 11.4). Therefore, although there is much fluctuation in population sizes,

Box 11.1 Measurement of patch isolation

A measure of population/patch isolation should reflect the expected numbers of immigrants arriving at the patch in unit time. Immigrants may arrive from several populations, and generally we expect that the contribution of population j to the pooled numbers of immigrants to patch i increases with the size of population j but decreases with its distance from patch i. The following measure of isolation, denoted by S_i, takes into account these effects:

$$S_i = \sum_{j \neq i} \exp(-\alpha d_{ij}) N_j.$$

Note that large values of S correspond to small isolation, hence S really measures connectivity, but the term isolation is used here to accord with common practice in the literature. The constant α describes how fast the numbers of migrants from patch j decline with increasing distance. The value of α can be estimated with mark–recapture data (Box 2.1) or an informed guess can be made; the ranking order of the S_i values in a data set is not sensitive to the exact value of α used. d_{ij} can be the straight-line (Euclidian) distance between patches j and i or some 'corrected' distance, taking into account the quality of the intervening habitat for migration. Often the population sizes N_j are not known, in which case one may assume that $N_j \approx p_j A_j$, where p_j equals 1 for occupied and 0 for empty patches and A_j is the area of patch j. If known, a more accurate scaling of N_j with patch area may be used. For instance, in the Glanville fritillary N_j scales roughly as $\approx A_j^{0.5}$. S_i is also a convenient relative measure of the numbers of individuals in the populations surrounding patch i. The ratio $S_i(t)/S_i(t-1)$ is then a measure of the regional change in population sizes around patch i. The logarithm of this ratio, denoted by N_{trend}, is used in Table 11.1.

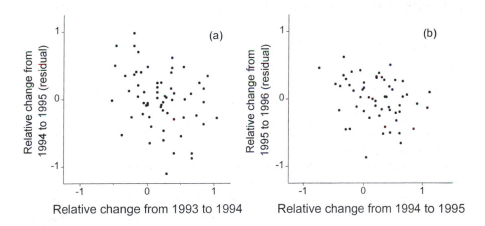

Fig. 11.4 (a) Relationship between the relative change in population sizes from 1994 to 1995 (vertical axis) against the corresponding change from 1993 to 1994 (including populations with at least five larval groups in 1994; the regression slope is significant, $P = 0.01$). The vertical axis gives the residual from a regression model in which the change in population size was explained by the regional trend in population sizes (N_{trend}; Box 11.1). (b) Similar analysis for the relative change in population sizes from 1995 to 1996 ($P = 0.03$).

and even large populations can go extinct in a matter of few years, it is nonetheless clear that some local density-dependent processes operate in these populations.

What are these processes? The two main candidates are natural enemies and resource limitation. The two primary parasitoids, *C. melitaearum* and *H. horticola*, have been observed to inflict heavy mortality on the host populations, and in extreme cases *C. melitaearum* has killed every host larva in the population (Lei *et al.* 1997; Lei and Hanski 1997). The substantial differences between the biologies of the two parasitoids (Lei 1997) are reflected in a much more stable level of parasitism by *H. horticola* than by *C. melitaearum* in both space and time (Lei *et al.* 1997), but both species contribute to density dependence in parasitism, as they both have higher incidence of presence in larger host populations (Lei and Hanski 1997). Interestingly, the same pattern is generated by somewhat different processes in the two species, as host population size had the strongest effect on the extinction rate in *H. horticola* but on the colonization rate in *C. melitaearum* (Lei and Hanski 1998).

Turning to resource limitation, in a few instances population densities have been so high that larvae have defoliated all available host plants in a habitat patch and starved to death in large numbers. Although this is exceptional, resource limitation as such is not exceptional, because what really matters is the availability of host plants in close proximity to larval groups, not in the meadow as a whole. A female might have unluckily oviposited on an isolated host plant and thereby greatly increased the risk of eventual starvation of her progeny. In the spring, when large post-diapause larvae forage on the still small leaves of their host plants, local food scarcity is the rule rather than the exception. Density dependence at the level of larval groups is however complicated by the necessity of the larvae to spend at least the early instars in a group until the spring, and by overwinter survival increasing with group size (Kuussaari 1998).

Population extinction

In the years 1993–1995, more than 400 population extinctions were recorded. The overall extinction rate is high, 50 and 42% in 1993–1994 and 1994–1995, a reflection of the generally very small population sizes. The most consistent predictor of local extinction has been the size of the population in the previous year (Table 11.1). Several other factors have either increased or reduced the extinction risk, at least in some parts of the study area. High density of butterflies in the neighbourhood of the focal population tends to reduce extinction risk (Table 11.1), representing the 'rescue effect' (for a more detailed analysis see Table 8.2). Extinction risk is also related to regional changes in density (Fig. 11.7), apparently because extinctions reflect regional changes in population sizes, driven by some spatially correlated environmental factors. The immediate effect of grazing by cattle is negative (Table 11.1), but the long-term effect is positive, because grazing keeps the vegetation relatively open and thereby helps the larval host plants in competition with other plant species (Hanski *et al.* 1996c). The smallest isolated populations are

Table 11.1 Logistic regression results showing factors affecting population extinction and colonization of empty habitat patches in the Glanville fritillary in 1994–1995

Explanatory variable	Extinctions			Colonizations		
	Odds ratio (95% bounds)	G^2	P	Odds ratio (95% bounds)	G^2	P
Constant		40.0	<0.0001		156.1	<0.0001
Log N_{t-1}	(–)0.295 (0.46 .. 0.19)	37.5	<0.0001			
N_{neigh}	(–)0.744 (0.97 .. 0.57)	4.8	0.0277	2.645 (3.35 .. 2.09)	80.2	<0.0001
N_{trend}	(–)0.485 (0.78 .. 0.30)	9.1	0.0025	1.986 (3.37 .. 1.17)	6.6	0.0099
Log area				1.381 (1.62 .. 1.18)	16.6	<0.0001
Plantago	(–)0.506 (0.69 .. 0.37)	20.5	<0.0001	2.207 (2.75 .. 1.49)	21.3	<0. 0001
Grazing	2.473 (5.48 .. 1.12)	5.0	0.0243	(–)0.496 (0.94 .. 0.26)	5.0	0.0243
Dryness				(–)0.988 (1.00 .. 0.98)	28.4	<0.0001
Correctly predicted cases	61%			79%		
Extinctions/colonizations	163			122		
Populations/empty patches	392			831		

Regional butterfly density (N_{neigh}) was measured by the logarithm of the index S (Box 11.1), and regional trend in population sizes (N_{trend}) by log (S_t/S_{t-1}). There were no significant interaction terms.

affected by the Allee effect, inverse density dependence due to high emigration rate and reduced mating frequency at the lowest densities (Fig. 2.4 and below).

In summary, the Glanville fritillary study has produced evidence of demographic and environmental stochasticity in various disguises increasing the risk of extinction, immigration reducing it (the rescue effect), and inverse density dependence at low density again increasing it (the Allee effect). This is not the full story, however, as three additional factors have substantially influenced the risk of local extinction: inbreeding depression, female host plant preference and parasitism. Inbreeding depression and female host preference have been discussed in Sections 6.2 and 6.1, respectively, and I do not repeat that discussion here. Below, I elaborate briefly on parasitism and host population extinction.

Guangchun Lei conducted a comprehensive study of parasitism in a 50-patch network in Åland in 1993–1996 (Lei 1997; Lei and Hanski 1997). Lei's results showed that parasitism by *C. melitaearum* significantly increased the likelihood of host extinction (Lei and Hanski 1997). In an experiment in which six parasitoid females and six males were introduced into four of six experimental butterfly populations, each originally established with six groups of 100 caterpillars, we found that all four populations into which the parasitoid was introduced went extinct in one year (Lei and Hanski 1997). Lei's results throw light also on the factors that lead to the extinction of parasitoid populations. Clearly, not all host extinctions are caused by the parasitoid, but high rate of population extinction in the host inevitably forces a high rate of extinction in the specialist parasitoids. In the period 1993–1995, Lei observed 23 local extinctions of *C. melitaearum*, of which 14 were caused by host extinction (to which the parasitoid might have contributed). Three extinctions were assigned to hyperparasitism, on the basis of direct observations, while the remaining six extinctions were due to other causes, typically involving a small parasitoid population (Lei and Hanski 1997). The host populations in which *C. melitaearum* is absent tend to be small, implying that small parasitoid populations have a high risk of extinction even when the host population survives.

The list of causes and mechanisms of local extinction detected in the Glanville fritillary practically exhausts the list of plausible processes of population extinction (Caughley 1994). This is at first surprising in view of the generally very small sizes of local populations, on which basis one could expect that demographic stochasticity is the overridingly dominant mechanism of extinction (MacArthur and Wilson 1967; Goodman 1987a; Lande and Orzack 1988; Lande 1993). The large number of extinctions recorded for the Glanville fritillary has facilitated the detection of many mechanisms of extinction, but ample statistical power alone does not explain these results as the effects have often been large. More to the point, the fact that extinctions are recurrent in the metapopulation context allows a wider range of processes to influence extinctions than might be expected in isolated populations.

11.3 Migration and population establishment

The idea that many species of butterflies live in 'closed' populations with little migration between the populations became firmly established in the literature towards the end of the 1980s (Section 8.1). The Glanville fritillary has been classified among the species with closed populations (Thomas 1984), but when we conducted our first mark–recapture study in 1991, we found that there was substantial movement among populations (Hanski *et al.* 1994). We estimated that 15% of males and 30% of females moved to another habitat patch during their lifetime. Though the exact figures must vary from one year to another and from one patch network to another (Kuussaari *et al.* 1996), this species is not all that sedentary. In this section, I summarize what is now known about the factors that affect emigration and immigration in the Glanville fritillary, emphasizing those factors that are most likely to influence metapopulation dynamics. I summarize results on migration and colonization distances and explore what determines the rate of establishment of new local populations in presently empty meadows.

Factors affecting emigration and immigration

An experiment conducted in 1992 on the small island of Husö (1.6 km^2) with 64 small meadows yielded much information on migration in the Glanville fritillary

Table 11.2 Factors affecting emigration and immigration rates in males and females

Explanatory variable	Emigration rate		Immigration rate	
	Males	Females	Males	Females
Patch area	– – –	–	+	+++
Density of flowers	– – –		+	
Butterfly density	–	– –		
Open patch boundary	+++	+		

The number of symbols indicates the statistical significance of the effect (based on the results of Kuussaari *et al.* 1996).

(Kuussaari *et al.* 1996). Husö island, located some 20 km south-east of the main Åland island, did not have the butterfly in 1991. In the spring of 1992, we reared large numbers of larvae collected from the main island, and were thereby able to release 882 newly-enclosed marked butterflies on 16 meadows in Husö, selected to cover a range of habitat patch sizes and release densities. We followed the movements of these butterflies in the patch network for the next two weeks (Kuussaari *et al.* 1996).

Emigration rate was higher than in our previous mark–recapture study, with at least 40% of butterflies moving to another patch during their lifetime. High emigration rate is largely or entirely explained by the small size of the habitat patches in Husö, mostly less than 0.1 ha. Patch area-dependent emigration rate has been reported for *Hesperia comma* (Hill *et al.* 1996) and for *Melitaea diamina* (Hanski *et al.* 1999). The latter study employed the model described in Box 2.1 and showed that emigration rate scaled to power 0.2 of patch area. In the Husö experiment, high density of butterflies and great abundance of nectar flowers, apart from large patch area, were the factors that tended to reduce emigration rate, whereas open landscape around the patch increased emigration (Table 11.2). The numbers of immigrants to a patch increased with the abundance of flowers and with patch area (Table 11.2). The negative effect of butterfly density on emigration is especially interesting, because it leads to inverse density dependence at low density (the Allee effect), and thereby makes the smallest populations even more vulnerable to extinction. Apparently density-dependent emigration rate has been reported for other butterflies, including *Euphydryas editha* (Gilbert and Singer 1973) and *E. chalcedona* (Brown and Ehrlich 1980), but it was not possible to ascertain in these earlier studies whether increased emigration rate was due to low density of butterflies or to low habitat quality, which are often correlated (Brown and Ehrlich 1980; M. Singer, personal communication). The significant influence of the openness of the surrounding landscape on emigration rate is another result with important implications for metapopulation dynamics and conservation. Populations in small habitat patches might go deterministically extinct due to high rate of

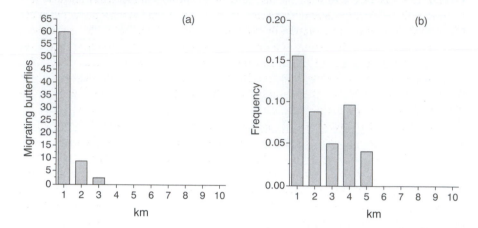

Fig. 11.5 (a) Frequency distribution of movement distances in a mark–recapture study conducted within an area of 4 by 5 km^2 (Hanski *et al.* 1994). 1737 butterflies were marked, of which 741 were recaptured and 72 had moved to another meadow. (b) Frequency distribution of the observed colonization distances in the pooled data for 1993–1996 (distance from the meadow that became colonized to the nearest local population).

emigration especially if the patches are surrounded by open landscape (Thomas and Hanski 1997).

Migration and colonization distances

In summer 1991 we marked 1737 butterflies in a 50-patch network within an area 4 by 5 km^2 (Hanski *et al.* 1994). We obtained a total of 741 recaptures, of which 72 were butterflies that had moved from one meadow to another. Migration distances were generally short, less than 500 m; the longest distance was 3.1 km (Fig. 11.5). These data have not been properly modelled to correct for variation in population and sample sizes, but it is clear that butterfly movements were affected by distance and that most migrating butterflies moved only a few hundred meters. In *Melitaea diamina*, a close relative of the Glanville fritillary, only 1% of migrating individuals were estimated to move more than 1 km in one day (Hanski *et al.* 1999; using the model in Box 2.1). In the Glanville fritillary, males moved more than females within habitat patches, but females moved more and somewhat longer distances than males between patches (Hanski *et al.* 1994; Kuussaari *et al.* 1996).

The observed migration distances suggest that an empty patch has a small chance of becoming colonized if isolated by more than 2–3 km from existing populations. Using the 240 colonization events recorded in 1993–1996, we can examine the effect of isolation on colonization. In these data the longest distances from empty patches to the nearest population are 13 km. The longest observed colonization distance was 4.7 km, but most of the observed colonization distances were less than 3 km, in agreement with direct observations on migration distances (Fig. 11.5). The effect of isolation on colonization is even clearer when isolation is measured not by the distance to the nearest population but by a measure that takes into account the distances to and the sizes of all nearby populations (Box 11.1). The colonization distances observed for the Glanville fritillary agree well with the results of Nève *et al.* (1996) for the morphologically similar *Proclossiana eunomia* in Morvan, central France, where the species was introduced in 1970. The butterfly has been spreading at an average rate of 0.4 km year^{-1} over the past 25 years, with several colonization jumps of 3–4 km across unsuitable habitat. On the other hand, suitable habitat isolated by >10 km has remained unoccupied (Nève *et al.* 1996), suggesting that, just as for the Glanville fritillary (Fig. 11.5), a jump longer than 5 km is very unlikely. As predicted by basic diffusion theory (Section 1.2), the speed of invasion of *P. eunomia* seems to have been relatively constant (Nève *et al.* 1996: fig. 5).

Population establishment

The annual colonization rate of empty patches was only around 15%, largely because many empty patches are currently so isolated that their chances of becoming immediately recolonized are small. In three parts of Åland, the annual colonization rate has ranged from 9 to 34%. As expected, the most significant factor affecting colonizations has been the density of butterflies in the neighbourhood of the focal patch (Table 11.1). The abundance of the more abundant larval host plant, *Plantago lanceolata*, has had a strong and consistent effect on colonization in the pooled data for the study area, most probably because the density of host plants in a patch affects the probability that an immigrant female perceives it to be a habitat patch in the first place. A more detailed analysis of the effects of the two host plants on colonization revealed that habitat patches dominated by the regionally more abundant host plant have substantially higher colonization rate than patches dominated by the regionally scarce host plant (Section 6.1). This result can be explained by regional variation in female host plant preference, for which there is experimental evidence (Kuussaari *et al.* 1998), and by preference-influenced movement behaviour, for which there is evidence for the Glanville fritillary and related species (M. Singer, personal communication).

Guangchun Lei has been able to accumulate data on colonization of host populations by the parasitoid *C. melitaearum* in a 50-patch network. Nine new host populations were established in 1994–1995, of which four became colonized by *C. melitaearum* within one year. Of the older but previously unparasitized host populations, three of ten possible populations became colonized in the same period (Lei and Hanski 1997). The history of the host population did not seem to affect colonization by the parasitoid, and the overall annual colonization rate was 37% in this network.

11.4 Four necessary conditions for metapopulation-level persistence

The previous sections have documented a high rate of population turnover in the Glanville fritillary metapopulations in Åland. Population turnover is the hallmark of classical metapopulation dynamics (Levins 1969), but turnover occurs also in species which are thought to persist owing to the persistence of one or a few large 'mainland' populations (Schoener and Spiller 1987; Harrison 1991; Harrison and Taylor 1997). To establish that long-term persistence of the Glanville fritillary is based on classical metapopulation dynamics in the sense of Levins (1969, 1970), I show in this section that the following four conditions are met: (1) habitat patches support local breeding populations, (2) no single population is large enough to ensure long-term persistence, (3) the patches are not too isolated to preclude recolonization; and (4) local dynamics are sufficiently asynchronous to make simultaneous extinction of all local populations unlikely. Fulfilment of conditions 1 to 3 eliminates the possibilities of 'patchy' populations (Harrison 1991, 1994), mainland–island metapopulations (Hanski 1991a; Schoener 1991; Harrison 1994) and non-equilibrium metapopulations (Harrison 1994), respectively.

Condition 1—discrete breeding populations

The dry meadows used by the Glanville fritillary occur as discrete small habitat patches in Åland. The mean, median and maximum patch areas are 0.13, 0.03 and 3.0 ha (Hanski *et al.* 1995a). Given that roughly 60–80% of butterflies spend their entire lifetime in the natal meadow (Hanski *et al.* 1994; Kuussaari *et al.* 1996), it is fair to conclude that the meadows support local breeding populations. Migration, when it occurs, takes place among nearby populations and habitat patches (Section 11.3). In these respects the Glanville fritillary metapopulation is similar to many other butterfly metapopulations (Ehrlich 1984; Baguette and Nève 1994; Warren 1994; Hanski and Kuussaari 1995; Wahlberg *et al.* 1996; Thomas and Hanski 1997).

Condition 2—all populations have a high risk of extinction

The very largest local populations in 1993–1997 had 100–200 larval groups, corresponding to roughly 1000–2000 adult butterflies. For instance, in the autumn 1995 there were only 16 populations with more than 20 larval groups recorded (the true population sizes are roughly twice the recorded size). In many species, a population of 1000 individuals has a long expected time to extinction (Shaffer 1981; Diamond 1984; Goodman 1987a; Akcakaya and Ginzburg 1991; Foley 1994), but the Glanville fritillary does not belong to these species. A population of *ca.* 2000 butterflies in 1991 (Hanski *et al.* 1994) went extinct by 1994, to become re-established the following year. Another population with nearly 100 larval groups in 1993–1994 crashed close to extinction in the spring 1996, and three other populations with 20–40 larval groups in autumn 1993 had been reduced to only 1–2 groups by autumn 1995. In the first two cases at least, the parasitoid *C. melitaearum* played a decisive role in bringing the population down (Lei 1997). *Cotesia melitaearum* soon colonizes large host populations and quickly builds up to a level where it can cause massive host mortality, which occurs when the phenological synchrony in the development of the host and the parasitoid is good in the spring (Lei 1997). Parasitism is thus a process which tends to reduce variance in host population sizes. Results for *Euphydryas editha* in California also show striking population fluctuations (Fig. 1.1; Harrison *et al.* 1991; Foley 1994), although parasitoids seem to play a relatively insignificant role in this species (S. Weiss, personal communication). In the Glanville fritillary in Åland, there simply are no populations having an insignificant risk of extinction.

Condition 3—recolonization possible

The rate of recolonization of empty meadows decreases rapidly with increasing distance from the nearest extant populations, but empty patches up to 3–4 km from existing populations have been colonized (Fig. 11.5). The mean nearest-neighbour distance among the patches is only 240 m (median 128 m, maximum 3870 m), which implies that most patches in Åland are unlikely to remain empty for a long time because of isolation. There are some that will, however. For instance, a network of 17 patches in the island of Sottunga 10 km south-east from the main island was empty in 1991. Mikko Kuussaari and Marko Nieminen transferred 62 larval groups to the island in autumn 1991, and thereby established a metapopulation that has functioned as any respectable metapopulation ever since, with 23 turnover events recorded so far (Fig. 10.5).

Condition 4—asynchrony in local dynamics

Results on extinction and colonization rates in Table 11.1 clearly demonstrate that what happens in a particular population is related to what happens in other populations nearby—extinctions have been common in regions where butterflies were generally declining, colonizations have been common in regions where densities were on increase. It is hence clear that some degree of spatial synchrony is present in the dynamics of the metapopulation, contrary to the assumption of most metapopulation models (Chapter 4). The important questions include how strong is spatial synchrony, over what spatial scale does it occur, and what are the causes.

To answer the first two questions, I calculated correlation coefficients for abundance changes in pairs of populations and regressed the average value of the correlation coefficient against the pairwise distance of the two populations. The results show a moderate level of spatial synchrony, with the largest value at the shortest distances (Fig. 11.6). The overall positive correlation reflects the large-scale synchrony evident in Fig. 11.7 and is most likely due to some spatially correlated weather conditions. The scale of the strongest spatial synchrony up to 500 m corresponds to the scale of butterfly movements, which are common up to this

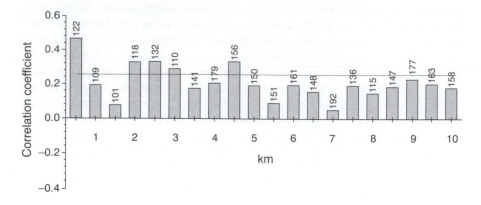

Fig. 11.6 Strength of spatial synchrony in the dynamics of the Glanville fritillary. The vertical axis gives the average value of the correlation coefficient of temporal fluctuations calculated for pairs of populations as a function of their pairwise distance (horizontal axis; data for 1993–1997, including populations with at least 10 larval groups in the pooled data). Sample size in each distance class is given by the number above the bar. The horizontal line gives the critical value of the correlation coefficient at 1% level for $n = 100$.

distance (Fig. 11.5). Unexpectedly, the average value of the correlation coefficient was low from 1.0 to 1.5 km, suggesting that spatial dynamics might have a more complex structure, with nearby populations tending to oscillate in synchrony but populations at somewhat greater distances being often out of synchrony. Predator–prey dynamics can create such spatial patterns (Section 7.3). In the case of the Glanville fritillary, the interaction between the butterfly and its specialist parasitoid *Cotesia mealitaearum* is strong enough to potentially generate complex spatial dynamics (Section 12.5).

Figure 11.7 shows the actual data, relative changes in population sizes from one year to another, for the period 1993–1997. For clarity, and to emphasize the large-scale patterns, the results have been pooled for metapopulations living in different semi-independent patch networks (SINs). These results demonstrate spatially correlated changes at a large scale, of the order of 10 km, but there is no temporal continuity in these patterns, hence these changes did not generate high overall spatial correlation in the previous analysis. Lack of temporal continuity, the scale of the correlated patterns, and field observations strongly suggest that the large-scale synchrony in Fig. 11.7 is due to the effects of weather on larval survival. Thus late summer 1993 was very dry and host plants withered especially in NW Åland, causing high larval mortality and apparently the large-scale decline shown in Fig. 11.7. Showers are often restricted to one part of Åland only, and as a heavy shower at a critical time can make a substantial difference to plant condition, spatially correlated rainfall is likely to explain at least partly the patterns in population change in Fig. 11.7. There is also some regional variation in the type of meadows, which will interact with the prevailing weather conditions. Thus dry meadows are favourable in most summers but can be catastrophically bad in drought years. Interactions between weather conditions and habitat quality have been well documented for the American checkerspot species (Ehrlich *et al.* 1975; Ehrlich and Murphy 1987; Weiss *et al.* 1993) and other butterflies (Sutcliffe *et al.* 1997). Large-scale spatial synchrony has been reported for many butterfly (Pollard and Yates 1993; Sutcliffe *et al.* 1996) and moth species (Hanski and Woiwod 1993b), and these patterns are generally attributed to spatially correlated weather conditions (Pollard and Yates 1993; Thomas and Hanski 1997).

In summary, in any given year there is substantial spatial correlation in the dynamics of the Glanville fritillary. Small-scale positive correlation up to 500 m is probably augmented by migration, whereas spatially correlated but temporally haphazard changes at an order of magnitude larger scale are most likely due to some effects of prevailing weather conditions on adults and pre-diapause larvae. It is noteworthy that so far populations across the entire study area have not all changed simultaneously in the same direction, which highlights the great overall stability of the Åland 'megapopulation' despite drastic regional changes. Considering the modelling of megapopulation dynamics, it is clear that the assumption of completely independent local dynamics is violated, though not so drastically (Fig. 11.6) that the standard models would become obsolete. Nonetheless, it is desirable to consider also models which incorporate regional stochasticity (spatially correlated environmental stochasticity). This will be discussed in Section 12.2.

11.5 Patch occupancy patterns

A fundamental prediction of classical metapopulation models with population turnover is that not all suitable habitat is occupied all the time (Chapter 4). For the Glanville fritillary, we have several lines of evidence demonstrating the presence of empty but suitable habitat. First, hundreds of meadows are empty every year, though by our best judgement these meadows are suitable for local populations. Second, we have observed roughly 400 natural colonizations of meadows that were unoccupied in the previous year (Section 11.3). And third, we have successfully established new populations experimentally, the most striking example being the island of Sottunga with a network of 17 meadows (Fig. 10.5). There is thus no doubt that a great deal of suitable habitat remains unoccupied. This section explores the actual patterns of habitat use.

Patch area and isolation effects

The risk of local extinction decreases with increasing patch area and the probability of colonization increases with decreasing patch isolation (Table 11.1). Therefore, one expects that small and isolated patches have a smaller probability of being occupied at any one time than large and well-connected patches. To visualize the effects of patch area and isolation on occupancy, and to assess the consistency of

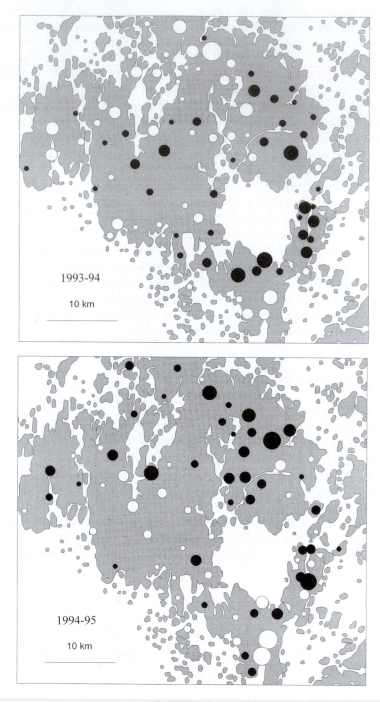

Fig. 11.7 Relative changes in metapopulation sizes in 1993–1994, 1994–1995, 1995–1996 and 1996–1997. Open symbols indicate metapopulation declines, black symbols metapopulation increases. The size of the symbol is proportional to relative change in metapopulation size.

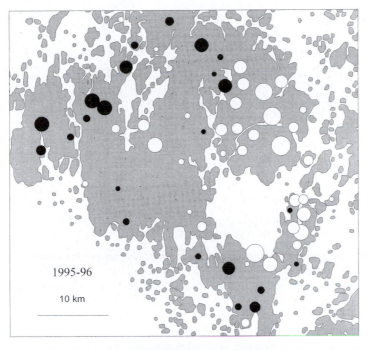

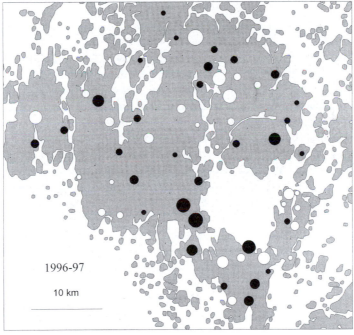

Fig. 11.7   continued

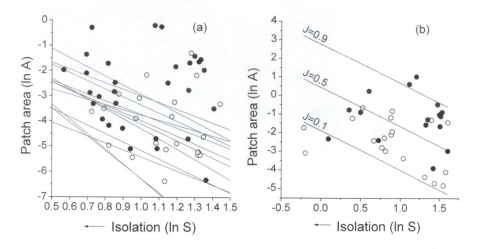

Fig. 11.8 (a) A plot of the logarithm of patch area (in ha) against the logarithm of isolation (measured by S, Box 11.1, isolation increases to the left) for one patch network of the Glanville fritillary: open symbols represent empty patches and black dots occupied patches during one survey. The incidence lines give the 0.5 probability of occupancy, estimated separately for 12 different patch networks (Hanski *et al.* 1996c). (b) Area-isolation plot for the parasitoid *Cotesia melitaearum*, with open symbols representing empty meadows (only the host butterfly present). This panel shows the 0.1, 0.5 and 0.9 incidence lines for this metapopulation.

these patterns in sub-sets of the data, I have plotted in Fig. 11.8a the logarithm of patch area against the logarithm of isolation, the latter measured by the index S described in Box 11.1. The open and black dots in this figure represent one snapshot of patch occupancy in one patch network. The lines give the fit of the logistic regression to 12 similar data sets from 12 other networks. The lines show the 0.5 incidence level—above the lines the predicted probability of patch occupancy is greater than 0.5, below the lines it is less than 0.5. Different networks show consistent effects of patch area and isolation on occupancy (the lines have a negative slope), though clearly there is also substantial variation in the parameter values, reflected in the position of the lines. The dynamic consequences of these patterns are explored in Section 12.2.

Patch occupancy is naturally also affected by factors other than area and isolation. Table 11.3 indicates that the abundance of the larval host plant *P. lanceolata* has a positive effect and grazing by cattle has a negative effect on occupancy. As expected, the same factors also affect rates of population extinction and recolonization (Table 11.1).

The effects of habitat patch area and isolation have been detected in the two specialist parasitoids of the Glanville fritillary, *Cotesia melitaearum* and *Hyposoter horticola* (Lei and Hanski 1997, 1998). In the case of the parasitoids, empty but

Table 11.3 Logistic regression results showing factors affecting patch occupancy by the Glanville fritillary in 1995

Explanatory variable	Odds ratio (95% bounds)	G^2	P
Constant		348.8	<0.0001
N_{neigh}	2.488 (2.89 .. 2.14)	177.6	<0.0001
Log area	1.347 (1.48 .. 1.23)	41.7	<0.0001
Plantago	2.082 (2.52 .. 1.72)	61.7	<0.0001
Grazing	(–)0.361 (0.56 .. 0.24)	23.6	<0.0001
Correctly predicted cases		69%	

suitable patches are meadows where there is a population of the host butterfly but no parasitoid population. As with the butterfly, small and isolated patches are the ones that tend to be empty at any given time (Fig. 11.8b). The effect of isolation is less marked in *Hyposoter horticola*, consistent with the greater mobility of this species than *C. melitaearum* (Lei and Hanski 1998).

Patterns at the network level

Patch-level predictions about habitat occupancy extend to the level of patch networks—the fraction of occupied patches in a network is predicted to decline with decreasing average patch size and with increasing average patch isolation (Chapter 4). Results on the Glanville fritillary confirm both predictions (Table 9.2). In networks with a small number of habitat patches, extinction–colonization stochasticity plays a great role and a metapopulation can go extinct owing to such stochasticity (Section 4.5). This prediction too is supported by empirical results for both the butterfly (Fig. 9.3) and the parasitoids (Table 7.2).

11.6 Summary

The Glanville fritillary butterfly (*Melitaea cinxia*) inhabits a network of *ca.* 1600 small meadows with one or both of the larval host plants, *Plantago lanceolata* and *Veronica spicata*, in Åland, SW Finland. Females typically mate once and lay several large batches of eggs; larvae live gregariously and diapause in a silken nest on the ground. Local populations are mostly very small, with only a few larval groups, and even the largest populations with 100–200 larval groups have a substantial risk of extinction. Extinction risk increases with decreasing population size, decreasing number of butterflies in the neighbouring populations (the rescue effect), increasing density of a specialist parasitoid (*Cotesia melitaearum*) and grazing by cattle. Sixty to eighty per cent of butterflies stay at their natal meadow during their lifetime. Emigration rate is increased by small patch size and by low density of conspecifics, leading to an Allee effect; immigration rate is reduced by low density of nectar flowers and by small patch size. The abundance of larval host plants in an empty meadow affects colonization probability, as does the local host plant composition in relation to which host plant is most commonly used in the neighbourhood, reflecting regional evolution in female host plant preference. The observed colonization distances are in good agreement with direct estimates of migration distances; patches located more than 3–4 km away from the nearest population have only a small chance of being colonized.

The large Glanville fritillary metapopulation in Åland satisfies the four necessary conditions for metapopulation-level persistence of a species consisting entirely of extinction-prone local populations. Patterns of habitat patch occupancy are in agreement with predictions of classical metapopulation models—patch occupancy increases with increasing patch size and with decreasing isolation; the fraction of occupied patches in well-connected patch networks increases with average patch size and increasing connectivity; and the butterfly is frequently absent from small networks with <15 patches due to extinction–colonization stochasticity leading to metapopulation extinction. Similar conclusions apply to the specialist larval parasitoid *Cotesia melitaearum*, which occurs as a classical metapopulation in a classical metapopulation of the host butterfly.

12

The incidence function model—applications

This chapter has a dual aim, to describe the dynamics in the Glanville fritillary metapopulation and to illustrate, with this and a few other examples, the application of the incidence function model (IFM) to real metapopulations. The Glanville fritillary metapopulation in Åland, SW Finland, is exceptional in terms of the quantity and quality of data available, a great advantage when parameterizing and testing models. Fortunately, the IFM can also be used to study metapopulations for which much less information is available, as Section 12.4 will demonstrate. My hope is that this chapter will stimulate others to explore the dynamics of other metapopulations using the IFM and related modelling approaches. Below, I first describe how the model is parameterized and how the estimated parameter values can be used to simulate metapopulation dynamics in real and hypothetical patch networks. One can also simulate the model with hypothetical parameter values to study more general questions (Moilanen and Hanski 1995; Doncaster *et al.* 1996; Hanski *et al.* 1996b). In the remaining sections, I analyse metapopulation dynamics in the Glanville fritillary and some other species, moving from basic to more complex dynamics with alternative quasi-equilibria and to transient dynamics away from the equilibrium.

12.1 Parameter estimation and model simulation

The incidence function model is described in Section 5.3 in the context of other spatially realistic metapopulation models. The IFM is attractive because of its relative simplicity and because it can be parameterized with data that are frequently available to ecologists. The following list summarizes the kinds of situation to which the IFM can be applied and the type of data that are required.

- The suitable habitat occurs in discrete patches which together constitute only a small fraction of the total landscape, say less than 20%. This condition is necessary because the model ignores patch shapes and assumes that pair-wise isolations can be approximated by the distances between the midpoints of the patches. Because the IFM is a patch model with no account of local dynamics, it is most applicable to highly fragmented landscapes with no very large patches.
- There is substantial variation in patch areas and/or isolations, which provides information for parameter estimation. In practice, this is not a restrictive condition, because real patch networks seldom consist of equal-sized and equally connected patches.

- The patches are occupied by local breeding populations which may persist at least for a few generations in the absence of migration. This condition excludes panmictic populations with patchy within-generation distribution of individuals, in other words patchy distributions at scales smaller than the metapopulation scale.
- At least one complete survey of patch occupancy has been conducted, in which all existing patches within some relatively large area were included and their areas and spatial coordinates were measured. If some patches and populations were not included in the survey, the true connectivity of the surveyed patches becomes underestimated. Although one 'snapshot' of patch occupancy is sufficient for parameter estimation, it is preferable to have data for several years for more accurate parameter estimates and/or to test predicted population turnover.
- The survey must include a sufficient number of patches (say 30 or more), a sufficient number of occupied patches (say 10 or more), and a sufficient number of empty patches (say 10 or more). If there are very few occupied or empty patches, there is insufficient information for parameter estimation.
- The metapopulation which is used for parameter estimation should occur at a stochastic extinction–colonization quasi-equilibrium, which in practice means that there is no strong increasing nor decreasing trend in the fraction of occupied patches. This condition is hard to establish rigorously, but often there is sufficient knowledge about the landscape and the species to make a well informed judgement.

Let us recall that the IFM is a stochastic patch model, with each patch having two possible states, presence or absence of the species. Assuming that the transition probabilities between the two states, E_i (extinction) and C_i (colonization), are functions of patch area and isolation as defined by eqns 5.9 and 5.12, and assuming the rescue effect as described in Section 5.3, we arrive at the following expression for the long-term probability of occupancy of patch i, called the incidence J_i,

$$J_i = \left[1 + \frac{ey}{S_i^2 A_i^x}\right]^{-1}, \tag{12.1}$$

where A_i is the area of patch i, S_i is a measure of patch isolation (Box 11.1), and x, e and y are three model parameters. Equation 12.1 can be rewritten:

$$J_i = \frac{1}{1 + \exp[\ln(ey) - 2\ln S_i - x\ln A_i]}, \tag{12.2}$$

which can be linearized to:

$$\ln\left(\frac{J_i}{1 - J_i}\right) = -\ln(ey) + 2\ln S_i + x\ln A_i. \tag{12.3}$$

Here is a very explicit connection between the IFM and empirical field studies. Equation 12.3 defines a logistic regression model, which has become the preferred method of data analysis in empirical metapopulation studies (Sjögren 1991; Verboom *et al.* 1991; Thomas *et al.* 1992; Eber and Brandl 1994; Hanski *et al.* 1995b). The theory behind the IFM suggests that, in logistic regression, patch area should be log-transformed and that isolation (connectivity) should be measured by S. Instead of logistic regression, one may use general non-linear regression or some other method to estimate the parameters of eqn 12.1, which is also necessary for the version of the model without the rescue effect (Section 5.3), because it cannot be linearized with the logit-transformation.

Parameter estimation is accomplished in the following steps:

- Determine the value of α, which gives the spatial scale of migration in the expression for S (Box 11.1). If available, use mark–recapture data to estimate α (Hanski 1994a; Box 2.1); this is preferable to reduce the number of parameters estimated from patch occupancy data. If no independent estimate of α is available, one can fit eqn 12.1 with several values of α and select the best-fitting value (below), or one can use some global optimization method to estimate the values of all parameters at the same time (Moilanen 1995, 1998). If α is estimated from patch occupancy data, check that the value obtained is biologically feasible. If not, use the best-fitting value that is consistent with the biology of the species. Studies on the Glanville fritillary (Hanski *et al.* 1996c; Moilanen 1998) and the grasshopper *Oedipoda caerulescens* (Appelt and Poethke 1997) found that the α values estimated from patch occupancy data were similar to independent estimates based on mark–recapture studies.
- Calculate S_i for each patch, which requires knowledge of patch areas A_i and their spatial coordinates, the latter to calculate the pair-wise distances d_{ij}.
- The IFM is then fitted to the data by regressing the observed patch occupancies p_i against the predicted incidences J_i. If only one snapshot of patch occupancies is available, $p_i = 1$ for occupied patches and 0 for empty patches. If several snapshots are available, p_i is the average occupancy of patch i in the data. Note that maximum likelihood parameter estimates obtained by fitting eqn 12.3 to the data are not true maximum likelihoods, because no allowance is made for spatial and temporal correlations in patch occupancy (ter Braak *et al.* 1998). Although this is unlikely to lead to severely biased parameter estimates, I discuss below (p. 237) an alternative approach to parameter estimation which does not suffer from this problem.
- The above procedure will give an estimate of the product ey in eqn 12.1. Not knowing the values of e and y independently means, in biological terms, that we cannot distinguish patterns of patch occupancy produced by high rates of population turnover from patterns produced by low rates. Teasing apart the values of e and y is nonetheless essential if we want to simulate the extinction–colonization dynamics in some patch network as will be described below. The simplest method is to estimate (or guess) the threshold patch area for which

$E = eA_0^{-x} = 1$, from which one can calculate the value of e having obtained an estimate of x in the previous step. Alternatively, if several consecutive surveys of patch occupancy have been conducted, one can use the observed population turnover rate to estimate the values of e and y. Denoting by T the total number of turnover events (extinctions and colonizations) per unit time, we can use the equality:

$$T = \sum (1 - C_i)E_i p_i + C_i(1 - p_i), \tag{12.4}$$

with appropriate expressions for C_i and E_i (Section 5.3) to obtain:

$$T = \sum \frac{1}{S_i^2 + y^2}\left[S_i^2(1 - p_i) + \frac{eyp_i}{A_i^x} \right], \tag{12.5}$$

from which the value of y can be found numerically, given that the values of x and ey have been obtained from eqn 12.1.

A few additional comments are appropriate. The predicted rates of extinction and colonization are affected by the value of A_0 used, but the spatial pattern of patch occupancy at quasi-equilibrium is little affected. If nothing better is available, one can use the area of the smallest occupied patch as a rough estimate of A_0, although it should be noted that a patch smaller than A_0 can be occupied for one time unit. If turnover data are available, one should use the second method to separate the values of e and y. The amount of turnover data required for this purpose is modest, as the purpose is not to model the factors affecting the extinction and colonization probabilities, as in state transition models (Section 5.4), but to use eqn 12.5 merely to fix the absolute rate of population turnover. Nonetheless, the more data there are available the more accurate are the parameter estimates one obtains (Hanski *et al.* 1996c), and the best estimates are obtained using the Monte Carlo approach described below. Finally, two additional parameters can be added to the model. First, one can add a parameter to describe the scaling of population size by patch area in the calculation for S (Box 11.1). And second, one can make the coefficient of $\ln S$ in eqn 12.3 a free parameter and estimate its value from the data. If this parameter is allowed to vary between 1 and 2, one can estimate, in principle, the shape of the relationship between the colonization probability C_i and the number of immigrants M_i arriving at patch i in unit time (Section 5.3). My preference has been to keep the number of parameters to a minimum, and hence I have assumed a value of 2 for this parameter; this is often justified on biological grounds (Section 5.3; Hanski 1994a).

Parameter estimation with extra environmental factors

The incidence function model can be generalized to include the effects of other environmental factors, apart from patch area and isolation, on the rates of extinction and colonization. The quality of habitat patches may affect the expected population sizes and thereby the extinction rate, while other factors might affect emigration and immigration rates. For instance, we know that the abundance of nectar flowers reduces emigration from, and increases immigration to, a habitat patch in the Glanville fritillary (Section 11.3).

Moilanen and Hanski (1998) modified the IFM to include extra environmental factors. The idea is to add to the model multipliers which modify patch areas and isolations. The extended model is parameterized with some global optimization method such as simulated annealing (Moilanen 1995). Specifically, Moilanen and Hanski (1998) modified eqns 5.11 and 5.14 as:

$$S'_i = V_i \sum_{j \neq i} p_j V_j \exp(-\alpha d_{ij}) U_j A_j \qquad (12.6)$$

$$J_i = \left[1 + \frac{ey}{S'^2_i (U_i A_i)^x} \right]^{-1}, \qquad (12.7)$$

where:

$$U_i = \prod_{r=1}^{R} G_r(g_{ri}),$$

$$V_i = \prod_{s=1}^{S} H_s(h_{si}),$$

and $G_r(r = 1..R)$ and $H_s(s = 1..S)$ are functions that map R environmental factors affecting patch area and S factors affecting isolation to the corresponding modifiers of patch area and isolation. g_{ri} and h_{si} are the values of these factors for patch i. For instance, one can use third-degree polynomials to model G_r and H_s, thus, e.g., $G_r(g_{1i}) = ag_{1i}^3 + bg_{1i}^2 + cg_{1i} + d$.

Monte Carlo approach to parameter estimation

Parameter estimation based on eqn 12.1 ignores spatial and temporal correlations in patch occupancy. Moilanen (1998) has developed a parameter estimation method based on Monte Carlo inference for implicit statistical models, which enables one to parameterize the IFM without making these simplifying assumptions. The basic idea is to consider patterns of patch occupancy in the entire patch network, instead of patch occupancy of individual patches, and to calculate transition probabilities between the empirically observed occupancy patterns at the network level. Let $O(0)$, $O(1), \ldots, O(n)$ be the patch occupancy patterns in the network at times $0, 1, \ldots, n$. The aim of the maximum likelihood parameter estimation is to find the set of parameter values that maximizes the probability of observing the sequence $O(0), O(1), \ldots, O(n)$. As a Markov chain expression, the overall probability can be decomposed to the probability of arriving at the first pattern, $O(0)$, multiplied by the probabilities of the subsequent transitions:

$$\max P[O(0)] \times P[O(1)|O(0)] \times , \ldots, \times P[O(n)|O(n-1)]. \qquad (12.8)$$

The simplest task is to calculate the probability of transition between two consecutive occupancy patterns if any such patterns occur in eqn 12.8. The exact probability of such a transition is simply the product of the probabilities of the respective transitions for individual patches. If two occupancy patterns $O(t)$ and $O(t+n)$ are separated by two or more time units ($n \geq 2$), one may use a Monte Carlo approach. With a set of parameter values, simulate the IFM to obtain $O(t+n-1)$ starting from $O(t)$, after which the probability of transition from $O(t+n-1)$ to $O(t+n)$ can be calculated exactly. This step is repeated for a large number of replicate simulations, and the results are averaged to obtain an approximation for the transition probability from $O(t)$ to $O(t+n)$. The first probability in eqn 12.8 can be obtained by simulating metapopulation dynamics with a set of parameter values for a long period of time and calculating the one-step probability of transition to $O(t)$ from many points along the long simulated time series of patch occupancy. Finally, the set of parameter values that maximize eqn 12.8 is obtained with some numerical optimization method, for instance simulated annealing, and one may also calculate confidence intervals for the parameter estimates (Moilanen 1998).

One important consideration has to be added. The method as described above tends to select parameter values that lead to high population turnover rate, because such values tend to increase the transition probabilities. To eliminate unrealistically high turnover rates, Moilanen (1998) used the observed turnover rate as a ceiling which the predicted turnover rates were not allowed to exceed. In practice, therefore, this method can only be used for data sets with at least two snapshots of patch occupancy patterns. Moilanen (1998) has tested the Monte Carlo method with simulated data. The Monte Carlo method gives more accurate parameter estimates than the simple method based on eqn 12.1, especially when one estimates α and A_0 as well as x, e and y from the occupancy data. The parameter estimates obtained with eqn 12.1 turned out not to be badly biased, however, and hence the simple method can be used to obtain reasonably good estimates especially in situations when only one snapshot of data is available.

Using the IFM to simulate metapopulation dynamics

The parameters of the IFM can be used to calculate the extinction (E_i) and colonization (C_i) probabilities for given patch areas and isolations using eqns 5.9 and 5.12 in Section 5.3. This enables one to simulate metapopulation dynamics in real and hypothetical patch networks using empirically estimated or hypothetical parameter values. The IFM can therefore be used, for instance, to predict the consequences of increased fragmentation of a particular landscape.

Simulation of the IFM includes the following steps:

- Assign an area and spatial coordinates for each habitat patch, using empirically measured values for real patch networks, and calculate all pair-wise distances between the patches.

- Define the initial pattern of patch occupancies, which might range from just one occupied patch to all patches occupied. For the purpose of calculating population turnover during one time interval, calculate the S_i values (Box 11.1) for the current pattern of patch occupancy.

- Assign stochastic extinctions and colonizations to the patches. If patch i is occupied, draw a uniformly distributed random number and allow patch i to go extinct (set $p_i = 0$) if that number is less than $E_i(1 - C_i)$ (assuming the rescue effect; Section 5.3). If patch i is empty, draw a random number and allow the patch to become colonized (set $p_i = 1$) if that number is less than C_i. Only one turnover event is allowed per patch in one time unit. Move to the next patch and repeat the above calculations. After possible turnover events for each patch have been assigned, return to the previous step and recalculate the S_i values using the current set of patch occupancy. Repeat these two steps for as long as needed.

- One can add regional stochasticity (spatially correlated environmental stochasticity) to the model by multiplying all patch areas in each time interval with a lognormally distributed random variable with unit mean and variance σ^2 (Hanski *et al.* 1996c). Regional stochasticity is likely to be influential for instance in insect metapopulations (Sutcliffe *et al.* 1997; Thomas and Hanski 1997; Section 11.4; Fig. 11.7). If a large amount of data are available, the value of σ^2 can be estimated from empirical data with the Monte Carlo method (Moilanen 1998).

As the networks that can be simulated with the IFM have a finite number of patches all with a finite probability of extinction, sooner or later all the patches become unoccupied at the same time and the metapopulation has gone extinct. However, for large patch networks and for species which have a high colonization rate in relation to the extinction rate, time to metapopulation extinction approaches infinity and meanwhile the metapopulation settles down to a stochastic quasi-equilibrium with no long-term increasing nor decreasing trend in the number of occupied patches, though local populations continue to turn over. The equilibrium may be calculated either by simulating the actual patch dynamics or by iterating the equations for C_i as explained in Section 5.2.

An important point to realize is that although one must make the assumption of metapopulation equilibrium for the purpose of parameter estimation, one can subsequently use the estimated (or hypothetical) parameter values to simulate the dynamics in arbitrary patch networks starting from arbitrary patterns of patch occupancy. Thus the model can be used to make predictions about both the transient (non-equilibrium) dynamics as well as the eventual stochastic steady state, whether it is a positive quasi-equilibrium or metapopulation extinction. For instance, one can assess whether a metapopulation is likely to survive in a patch network after a reduction in patch number or in patch areas (Hanski 1994a,c; Wahlberg *et al.* 1996). One can assess the significance of particular habitat patches to the dynamics of the metapopulation as a whole (Hanski 1994c; Moilanen *et al.* 1998). One can also predict the likely consequences of a perturbation that kills all but a few local populations. The following sections present a range of applications of the IFM to real metapopulations.

12.2   Predicted dynamics and steady states

In this section I use the IFM to predict the stochastic quasi-equilibrium and population turnover in the Glanville fritillary metapopulation in Åland. Parameter values were estimated using data collected in 1991 from a 50-patch network, within an area of 25 km², in northern Åland (Hanski *et al.* 1994). In that year, 42 of the 50 patches were occupied, and exactly which patches were occupied with regard to their areas and isolations from the existing populations provides the information needed to parameterize the model. Equation 12.1 was fitted to these data, and the observed nine population turnover events between 1991 and 1992 were used to obtain independent estimates of *e* and *y* (Hanski *et al.* 1996c). Parameter α was estimated using independent data on migration distances (Hanski *et al.* 1996c).

With parameter values thus obtained, we proceeded to iterate the dynamics of the butterfly in the entire Åland islands, which was divided into 4 by 4 km² squares ($n = 100$) for the purpose. Many of the squares had so few patches that the respective metapopulation had a high risk of extinction, which was prevented by including a small probability of outside colonization for each patch (Hanski *et al.* 1996c). This is not unrealistic, because in reality butterflies may occasionally move long distances. The empirical data to test model predictions were collected in 1993. I compare here the predicted and observed fractions of occupied patches, *P*, in the 4 by 4 km² squares, although an even more informative measure to compare would be the entire pattern of patch occupancies. The observed *P* value was within two standard deviations of the predicted value in most (74%) squares. In five squares, the observed value was significantly (>2 SD) smaller, and in 21 squares it was significantly greater, than the predicted value (Hanski *et al.* 1996c). As a comparison, we repeated the above exercise with logistic regression and basing the prediction only on the effect of patch area on occupancy. In these results, 73% of the observed values were smaller than predicted, 2% were greater than predicted, and 25% matched the prediction. Thus predictions generated by the non-dynamic logistic regression model not including the effect of population isolation (*S*) tended to be higher than the observed values. As it happened, discrepancies between the observed and the IFM-predicted *P* values occurred mostly in one part of Åland, whereas the model predicted the *P* values remarkably successfully for squares with more than 15 habitat patches across most of Åland (Fig. 12.1).

These results are encouraging, but it is also clear that the distribution of the butterfly on the meadows cannot be predicted accurately with information only on patch areas, their spatial locations, and one snapshot of patch occupancies. There are several possible reasons why the model failed in parts of the study area (Hanski *et al.* 1996c). First, the habitat patches vary to some extent in quality, and this variation affects extinction and colonization rates (Table 11.1). I return below to model predictions incorporating extra environmental variables. Second, the assumption that the metapopulations occur in a stochastic quasi-equilibrium in the patch networks might not be valid. Apart from changes in environmental conditions, the butterfly metapopulations might not be locally at equilibrium because of a delay caused by

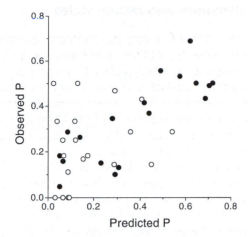

Fig. 12.1 Comparison of the predicted and observed fractions of occupied patches (*P*) in 4 by 4 km² areas in western Åland. Black dots represent squares with more than 15 habitat patches, open symbols are for squares with fewer patches. Statistical tests and further details in Hanski *et al.* (1996c).

the interaction with the specialist parasitoid *Cotesia melitaearum* (Section 11.2), which might generate large-scale oscillations in both species. I return to such non-equilibrium dynamics in Section 12.5. Third, even though the metapopulations were at equilibrium, there might be multiple equilibria which would limit our ability to predict the distribution of the species at a given point in time. This possibility will be investigated in Section 12.3.

The incidence function model can be used to predict the rate of population turnover as well as the distribution of species in fragmented landscapes. Using the parameter values estimated from the 50-patch network in 1991, I predicted the extinction and colonization events in Åland in 1993–1995. The observed turnover rate (26%) was close to the predicted rate (23%), although the predicted extinction rate was somewhat higher, and the predicted colonization rate was lower, than the observed rate (Table 12.1). The discrepancy between predicted and observed rates may reflect regional stochasticity; this will be discussed on p. 244.

Consistency of parameter estimates

Let us return to the empirically-observed patterns of patch occupancy depicted in Fig. 11.8. The lines in that figure illustrate the effects of patch area and isolation on patch occupancy as estimated from 12 different patch networks in Åland. The effects of patch area and isolation seemed to be reasonably similar in these networks, with a few exceptions, but the question remained what are the metapopulation dynamic consequences of the observed differences among the networks? This question can be

Table 12.1 Comparison of the numbers of predicted and observed turnover events between 1993 and 1995 in 13 large patch networks in Åland

	Extinctions	Colonizations	Turnover (rate)
Predicted	127	141	268 (23 ± 12%)
Observed	183	94	277 (26 ± 9%)

The turnover rate was calculated as the number of turnover events (extinctions and colonizations) per year divided by the number of patches in the network. (The total number of habitat patches was 555.) The figures in brackets give the average ± standard deviation in 13 patch networks (Hanski *et al.* 1996c).

answered by parameterizing the IFM separately with data from different networks, then simulating the model with the different sets of parameter values but in the same network, for instance that depicted in Fig. 11.8. Figure 12.2 shows the results. In no case is the metapopulation predicted to go extinct. In three cases the predicted fraction of occupied patches (*P*) is consistently greater than the observed *P*, in two cases it is lower, whereas in the remaining seven cases the predicted *P* crosses the

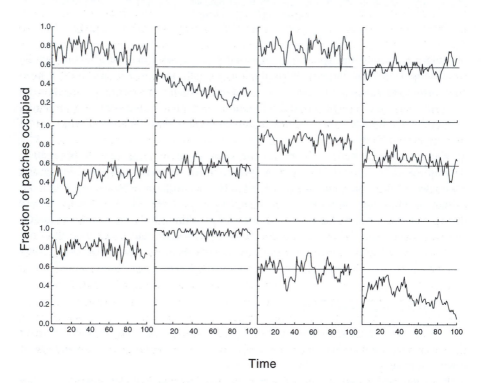

Fig. 12.2 Predicted metapopulation dynamics in the network depicted in Fig. 11.8a using the 12 sets of parameter values for which the lines in Fig. 11.8a were drawn (the actual values are given in Table 3 in Hanski *et al.* 1996c). The horizontal line gives the observed fraction of occupied patches in this particular network.

observed value, $P = 0.58$, at least once in 100 years. There is much variation around the average P value in the model predictions due to inevitable stochasticity associated with extinctions and colonizations (Fig. 12.2). Exactly how much variation there is depends both on the parameter values and on the structure of the fragmented landscape.

I draw two conclusions from these results. First, parameter values estimated from different patch networks tended to predict similar dynamics, though not without exceptions. This is important in demonstrating satisfactory repeatability of the results. Second, the predicted long-term dynamics show much variation in the fraction of occupied patches owing to extinction–colonization stochasticity. This variation puts a definite limit to the possible accuracy of our predictions: we can only predict the behaviour of the metapopulation in statistical terms even if the metapopulation is at a 'steady' state.

Influence of other factors apart from patch area and isolation

Additional environmental factors, apart from patch area and isolation, can be added to the IFM as described in Section 12.1. In brief, this is done by assuming a set of modifiers for patch area and isolation, which are functions of the extra environmental factors. With sufficient data, which we have available for the Glanville fritillary, one can parameterize quite complex models. Whether a particular factor should be included or not in the model is assessed by examining the improvement in model fit. Using this approach, Moilanen and Hanski (1998) found that grazing by cattle reduces effective patch area. The biological mechanism is clear—grazing increases larval mortality and hence the risk of population extinction (Table 11.1). It is worth adding, though, that the long-term effect of grazing is positive, because grazing keeps vegetation sufficiently open to allow *Plantago* to persist despite superior plant competitors (Hanski *et al.* 1996c). The amount of low vegetation in a patch tended to reduce effective patch area, which was not expected, because low vegetation generally means a better chance for *Plantago* to persist. However, meadows with much low vegetation are also the driest, and during dry summers the larval host plants might completely wither on the generally dry meadows, often on rocky outcrops, leading to widespread larval mortality.

A substantial improvement in model fit was obtained by adding the cover of the vegetation type with abundant nectar flowers for butterflies, and assuming that abundant nectar flowers increase immigration but reduce emigration (Moilanen and Hanski 1998). This is another encouraging result, because this is exactly how the Glanville fritillary responds to flowers on the meadows (Table 11.2; Kuussaari *et al.* 1996). The behavioural response of butterflies to flowers is strong enough to affect their movement behaviour and thereby also population turnover and the pattern of patch occupancy.

In summary, several aspects of habitat quality improved model fit and produced results with meaningful biological interpretation. On the other hand, the improvement in model fit was not very great, and Moilanen and Hanski (1998) concluded that, in this metapopulation, the effects of patch area and isolation clearly dominate. The result will depend greatly on the types of patches one has included in the study network in the first place, and how much variation there is generally in patch quality. Grazing by cattle greatly increased the extinction risk of local populations, but because the fraction of meadows with grazing mammals was small, the effect remained relatively small for overall metapopulation dynamics.

Dynamics with regional stochasticity

Fluctuations in the fraction of occupied patches in Fig. 12.2 are due to spatially uncorrelated stochasticity in the extinction and colonization events. In reality, environmental stochasticity affecting extinction rate is often more or less correlated in space. The prime example is year-to-year variation in weather conditions, which have a great impact on most insect populations (Andrewartha and Birch 1954). Spatially correlated environmental stochasticity, or regional stochasticity for short (Hanski 1991a), is expected to reduce metapopulation size and lifetime (Levins 1969; Gilpin 1990; Harrison and Quinn 1990; Hanski 1991a).

The incidence function model is a patch model and hence does not include an explicit description of local dynamics. However, regional stochasticity can be modelled by assuming temporal variation in patch areas, which affect extinction and colonization rates in the model (Hanski *et al.* 1996c). Increasing regional stochasticity reduces the average fraction of occupied patches (P) and increases the variance of P (Hanski *et al.* 1996c). In model simulations, the distribution of P values became often bimodal, suggesting the presence of alternative equilibria. In the following I discuss the occurrence of such more complex spatial dynamics in the Glanville fritillary in Åland.

12.3 Alternative equilibria

Dynamic systems such as natural populations may have two or even more locally stable attractors for a given set of parameter values. The attractors may be point equilibria, limit cycles, or more complex attractors, but here it is sufficient to consider just point equilibria, which are predicted by deterministic metapopulation models (Section 4.2). In the real world, of course, the point equilibria are not points (constant P values), as the metapopulation is constantly perturbed by external environmental forces and dragged back towards the equilibrium by the metapopulation dynamic forces. As a result, in the long course of time the metapopulation settles down to a stationary distribution of P values. However, if a system has alternative equilibria, around which it spends most of the time, the stationary distribution is bimodal, and we cannot predict the state of the system without knowing its history. A familiar example from population dynamics is the Allee effect, which implies that below a threshold population size the local population growth rate becomes negative, above the threshold it is positive. The threshold population size represents an unstable equilibrium, the boundary between the domains of attraction of two stable equilibria which, in the case of local dynamics with the Allee effect, are population extinction and some positive population size. How commonly natural populations exhibit alternative equilibria is a matter of long-standing debate (Connell and Sousa 1983).

Metapopulation models demonstrate that the dynamics can exhibit alternative equilibria when immigration from large populations sufficiently reduces the likelihood of small populations going extinct: the rescue effect (Sections 4.2 and 4.3). We have seen that small populations of the Glanville fritillary have a high risk of extinction (Table 11.1), that migration rate among nearby populations is relatively high (Section 11.3), and that the smallest populations are indeed affected by the rescue effect (Table 8.2). We can therefore expect alternative equilibria in the metapopulation dynamics of the Glanville fritillary.

Alternative equilibria in the Glanville fritillary metapopulations

The most convincing evidence for alternative equilibria would come from experimental perturbations, in which the metapopulation is pushed from the domain of attraction of one equilibrium to the domain of attraction of the alternative equilibrium. Unfortunately, replicated field manipulations involving hundreds of local populations are practically impossible to perform in real metapopulations. Our approach has been to take advantage of the 127 semi-independent patch networks (SIN) into which the entire habitat patch network in Fig. 11.1 has been divided (Section 11.1), and to use the different SINs as replicates (Hanski *et al.* 1995b). The SINs are largely independent, because they are separated by at least 1.5 km from each other and usually by some physical barrier to migration (Hanski *et al.* 1996c). The metapopulations are not expected to affect each other dynamically, but because of low level of migration between SINs recolonization of an empty SIN is possible. The idea here is that if one observes a large sample of independent replicates of a system with alternative equilibria, one would expect that, at any given time, some replicates would occur at or close to one of the equilibria, while others would occur at or close to the alternative equilibrium, and only few replicates would be found at intermediate states. Of the 127 SINs into which the *ca.* 1600 patches were divided (Hanski *et al.* 1996c), 65 SINs have at least five habitat patches and were included in the following analysis.

Figure 12.3a gives a theoretical example of alternative equilibria from the model of Hanski and Gyllenberg (1993; Section 4.3). In this example, all parameters except one were kept constant, and metapopulation size, as measured by the fraction of occupied patches (P) was plotted against the colonization rate parameter. One could as well choose some other parameter for the analysis (Hanski and Gyllenberg 1993). In this example, metapopulation extinction is the only stable equilibrium when colonization rate is very low, and a positive equilibrium with large P is the only stable equilibrium when colonization rate is very high. But for intermediate colonization rates there are two alternative equilibria (Fig. 12.3a). In the empirical

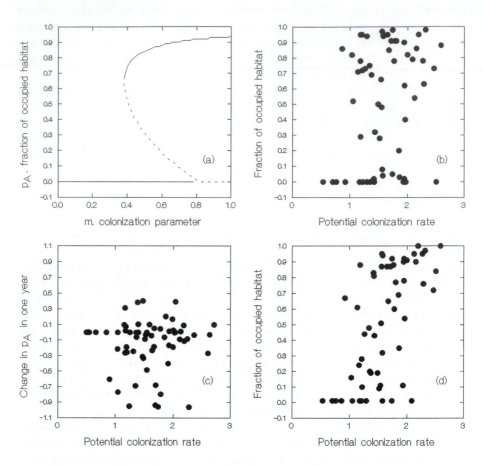

Fig. 12.3 Bifurcation diagrams for the fraction of occupied habitat in patch networks (P_A) as a function of the colonization rate (m). Panel (a) gives a theoretical example of alternative equilibria from Hanski and Gyllenberg (1993). Continuous lines represent stable equilibria, broken lines are unstable equilibria. Panel (b) gives an empirical example from 65 patch networks of the Glanville fritillary. The potential rate of colonization, M, on the horizontal axis, corresponding to the colonization parameter m in panel (a), gives the expected rate of exchange of butterflies among the patches on the assumption that all patches are occupied. Panel (c) shows the change in P_A from 1993 to 1994 as a function of M. Some very large negative changes were observed in networks which had very low density of larvae in 1993. In general, large (>0.15) absolute changes were mostly observed in networks with intermediate M between 1 and 2 ($\chi^2 = 9.40$, $P = 0.002$). Panel (d) gives the pattern of P_A against M as predicted by the IFM parameterized for the Glanville fritillary and simulated in the same networks as in panel (b). Note the correspondence between the predicted (panel d) and observed patterns (panel b).

example, the colonization rate parameter as such is constant, reflecting a fixed biological attribute of the species, but what varies is the structure of the landscape. The potential rate of exchange of butterflies among the habitat patches is different in different SINs—high where patches are large and located close to each other, low where patches are small and located far apart from each other. An analogous parameter to the colonization rate parameter (m) in the theoretical example is, therefore, a measure of the potential colonization rate, say M, which has a specific value for each network. Potential colonization rate was calculated as the familiar measure of patch connectivity, S (Box 11.1), but now assuming that all patches are occupied, because M characterizes the network and is independent of how common the species happens to be at some particular point in time. As the dependent variable, we used the occupied fraction of the pooled patch area, P_A, which gives more weight to the larger patches and thereby eliminates much of the stochasticity due to erratic occurrence of the species in the many very small patches.

The relationship between M and P_A (Fig. 12.3b) resembles greatly the predicted bifurcation pattern (Fig. 12.3a), with most networks with intermediate M having either a large or a small fraction of the suitable habitat occupied. The seemingly good quantitative correspondence is largely coincidental, however, as the model of Hanski and Gyllenberg (1993) cannot be used to make quantitative predictions; what matters is the qualitative correspondence. The marginal distribution of the P_A values in the pooled material is strikingly bimodal (Fig. 9.5b), which gives further support to the hypothesis of alternative stable states. The existing metapopulations with less than *ca.* 70% of the habitat occupied are predicted to be in transit towards one of the two stable equilibria (Fig. 12.3b). This prediction is supported by the observed changes in P_A between 1993 and 1994—large (>0.15) absolute changes were mostly observed in networks with intermediate M between 1 and 2 (Fig. 12.3c).

One additional test of the hypothesis of alternative equilibria in this system can be accomplished by using the IFM to predict long-term dynamics in the different SINs. Using the set of parameter values estimated from the 50-patch network in 1991 (Section 12.2), I simulated the dynamics in each SIN for 2000 generations and sampled the final pattern of patch occupancy as one data point for the respective SIN. These data points (snapshots) were subsequently treated in exactly the same manner as the observed data points. The relationship between M and P_A in the IFM-generated data (Fig. 12.3d) is very similar to the observed pattern (Fig. 12.3b). This correspondence provides further quantitative support for alternative equilibria in the Glanville fritillary metapopulations.

Clusters of patches winking in and out

Deterministic models such as the one studied by Hanski and Gyllenberg (1993) may capture key features of metapopulation dynamics, but whenever the number of patches is small, either locally or in the entire network, stochasticity becomes important. Spatially restricted movements among nearby populations and patches are likely to create correlated changes in patch occupancy—most patches in a cluster of well-connected patches are likely to be in the same state, either occupied or empty. Such correlations and the consequent stochasticity in metapopulation dynamics do not disappear even in large networks, because interactions among populations and patches continue to be localized. Here is an apparent difference between extinction–colonization stochasticity in metapopulations and demographic stochasticity in local populations, though various forms of heterogeneity might amplify demographic stochasticity in real local populations (Ebenhard 1991). Clusters of patches may wink in and out, from being well occupied to being empty or practically empty. This sort of behaviour is analogous to spatial aggregation observed in interacting particle system models (Section 5.1). Though this behaviour is essentially due to stochastic population turnover and spatially restricted movements, the resulting patterns are reminiscent of alternative equilibria in deterministic models, which do not involve patch aggregation and restricted movements, but are based on the mean-field approximation (all patches equally connected). Given the independent evidence for the rescue effect in the Glanville fritillary (Table 8.2), it seems probable that both deterministic and stochastic mechanisms contribute to the bifurcation pattern in Fig. 12.3b.

12.4 Predictions for other species

Ecologists working with unique biological species often envy scientists who can reach universally valid conclusions with a single critical experiment. Though such sentiments might reflect an excessively rosy picture about other fields of science, they do underscore an important problem ecologists have to face. Our results are qualified by the study species, by the study site (population), and by the time when the study was conducted (environmental conditions). The following examples illustrate the application of the IFM to other species than the Glanville fritillary, with the purpose of demonstrating that this modelling approach can be usefully applied to a wide range of taxa living in fragmented landscapes. Other recent metapopulation studies employing the IFM have been conducted on a gall-forming fly (Eber and Brandl 1996), a grasshopper (Appelt and Poethke 1997), the European nuthatch (ter Braak *et al.* 1998) and plant assemblages (Quintana-Ascencio and Menges 1996).

The false heath fritillary and other classical metapopulations

In northern Europe, many species of butterflies have become endangered or have already gone regionally extinct (New 1991; Thomas 1994b,c; Hanski and Kuussaari 1995; New *et al.* 1995; Pullin 1995). The checkerspot butterflies (Melitaeinae) have done especially poorly. For instance, all three species occurring in the UK and five of the six species in Finland have either a very restricted remnant distribution or are endangered otherwise. The rarest of the Finnish species is the false heath fritillary (*Melitaea diamina*), a close relative of the Glanville fritillary. This species has only two remaining metapopulations in Finland, of which one is located, without any obvious reason, in one small area in southern central Finland. Very little was known about *M. diamina* before 1995, when Niklas Wahlberg set out to study its biology. Wahlberg found that the caterpillars feed exclusively on *Valeriana sambucifolia* and that the butterflies hence occur on the moist meadows where this plant grows (Wahlberg 1997). With the help of two other students, Wahlberg mapped all the 94 suitable meadows for the butterfly within an area of 600 km².

Having the previously estimated parameter values for the Glanville fritillary (Hanski *et al.* 1996c) as well as the patch areas and spatial locations in the *M. diamina* network, Wahlberg *et al.* (1996) predicted the patch-specific incidences for the latter species using the parameter values for the Glanville fritillary. This prediction was subsequently tested with empirical data on the actual pattern of patch occupancy in the *M. diamina* metapopulation. The match between the predicted incidences and the observed snapshot of patch occupancy was good (Fig. 12.4). The model correctly predicted that the three main clusters of patches have many or most patches occupied, whereas the level of occupancy was low in the more isolated patches. The observed few isolated local populations in the NW corner of the study area may represent relict populations in a region where many habitat patches have become recently overgrown (Wahlberg *et al.* 1996). As might be expected from the results in Fig. 12.4, the parameter values subsequently estimated for *M. diamina* turned out to be very similar to the parameter values for the Glanville fritillary (Wahlberg *et al.* 1996).

The incidence function model has been parameterized for two other species of butterfly, for the silver-spotted skipper (*Hesperia comma*), studied in the UK by

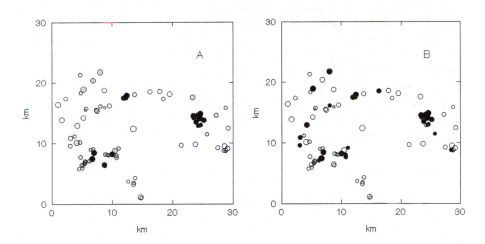

Fig. 12.4   (a) A map of the patch network occupied by *Melitaea diamina* showing the relative patch sizes, their spatial locations and the predicted incidence of occupancy based on parameter values estimated for the congeneric Glanville fritillary. Higher incidences are shown by darker shading. (b) A snapshot of patch occupancy in 1995. (Based on Wahlberg *et al.* 1996.)

Chris Thomas (Thomas *et al.* 1992; Thomas and Jones 1993), and the chequered blue (*Scolitantides orion*), studied in Finland by Pekka Saarinen (1993; Hanski 1994a). The parameter values for the chequered blue are similar to those for the two fritillaries, but the skipper tended to occupy somewhat larger patches in its native network (Wahlberg *et al.* 1996). Why the skipper is different is not clear, but Hanski and Thomas (1994) suggested that it experiences higher level of stochasticity, and has a lower intrinsic rate of population increase, than the Glanville fritillary, which factors are expected to increase the risk of local extinction (Lande 1993; Foley 1994, 1997). Not all species are evidently similar, but the results of Wahlberg *et al.* (1996) raise the interesting possibility that one set of parameter values could provide a good approximation for a set of species sharing the same basic biology and living in similarly fragmented landscapes, such as many butterflies living on meadows. Of course, you never know when another exception turns up, but even so much can be accomplished with parameter values that are not too much off. To predict the absolute rate of population turnover requires accurate parameter values but to rank alternative landscape structures in terms of metapopulation persistence is much less sensitive to small changes in parameter values. And the latter is the kind of application for which spatially realistic metapopulation models such as the IFM are most likely to be needed in practice.

The next example takes us from butterflies to mammals. A unique 20-year record of classical metapopulation dynamics in the American pika (*Ochotona princeps*) in a highly fragmented landscape in California (Smith 1974, 1980; Smith and Gilpin 1997) provides an interesting example of spatially correlated changes in patch occupancy. Moilanen *et al.* (1998) parameterized the IFM for the pika using four snapshots of patch occupancy recorded in 1972, 1977, 1989 and 1991 (Fig. 12.5). The model was tested by comparing the observed 49 turnover events with the predicted turnover rate. In each period between two censuses, the observed number of turnover events was within one standard deviation of the model-predicted turnover number (Moilanen *et al.* 1998). Thus the model parameterized with spatial data only successfully predicted the temporal dynamics.

In 1972, a large fraction of habitat patches throughout the network was occupied, but unexpectedly the pika had practically disappeared from the southern part of the network by 1991 (Fig. 12.5). Using the estimated parameter values, and adding a realistic level of regional stochasticity to the model, Moilanen *et al.* (1998) simulated the pika dynamics in the real patch network. It turned out that the northern network remained relatively stable and constantly occupied in the simulations, but the southern network was predicted to periodically collapse to regional extinction (Fig. 12.6). The middle network, consisting of relatively small patches, showed a low level of occupancy both in model predictions (Fig. 12.6) and in reality (Fig. 12.5), but it functioned as a stepping stone in the model, through which the southern network was soon recolonized after regional extinction. These modelling results strongly suggest that the observed collapse of the southern pika meta-population between 1972 and 1991 can be explained by classical extinction–colonization dynamics with some regional stochasticity (Moilanen *et al.* 1998). In

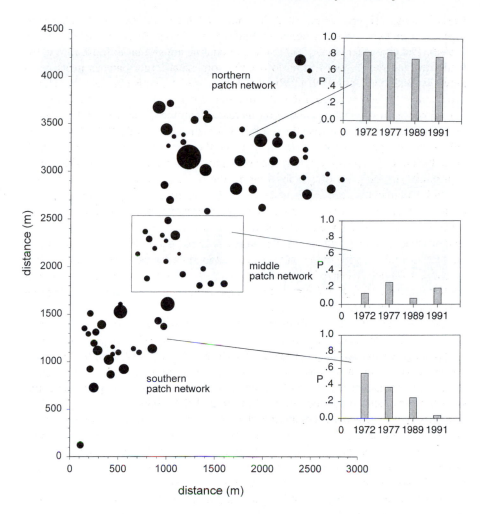

Fig. 12.5 The patch network of the American pika in Bodie, California (Smith 1974, 1980, personal communication). The size of the dot is proportional to the estimated carrying capacity of the habitat patch. The figure shows the division of the patches into northern, middle and southern patch networks, including 36, 16 and 24 patches, respectively. The small panels show the observed proportions of occupied patches in the three parts of the network in four surveys. Occupancy has remained stable in the northern network, whereas a distinct decline has occurred in the southern network. (Based on Moilanen *et al.* 1998.)

particular, there is no need to assume any environmental change affecting the southern network to explain the observed change in patch occupancy. Another good example of classical metapopulation dynamics in mammals is the Vancouver Island marmot (*Marmota vancouverensis*) living on small and patchily distributed sub-alpine meadows on Vancouver Island (Bryant and Janz 1996).

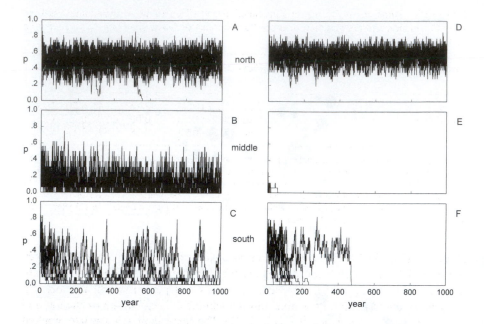

Fig. 12.6 Patch occupancy in 10 replicate simulations of the pika metapopulation using parameter values estimated from the four snapshots in Fig. 12.5. Each simulation was started by assuming the 1972 patch-occupancy pattern. In panels (a) to (c), the entire metapopulation was simulated, in panels (d) to (f) the three parts of the network (Fig. 12.5) were simulated in isolation. Note that only the northern network supports a metapopulation with a positive quasi-equilibrium. The pika frequently goes extinct in the southern network (c), which becomes recolonized via the middle network, although the middle network has low occupancy (b) and would go quickly extinct in isolation (e). (Based on Moilanen *et al.* 1998.)

Mainland–island metapopulations of small mammals and birds

The examples discussed so far represent classical metapopulations in which all local populations have a high risk of extinction. It is even more straightforward to apply the IFM to mainland–island metapopulations, because in this case both extinction and colonization probabilities remain constant and there is no doubt about the validity of the basic expression for incidence (eqn 5.8).

A good example of mainland–island metapopulations is the occurrence of small mammals on islands in lakes (Hanski 1986; Hanski and Kuitunen 1986; Peltonen and Hanski 1991; Lomolino 1993) and in the sea (Pokki 1981; Crowell 1986). In an early study, the IFM was parameterized for three species of shrew using a snapshot of their occurrence on 68 islands (Hanski 1992). A simplified version of the model was used, with the colonization probability for different islands assumed to be the same, as island isolations did not vary greatly. Using the estimated parameter values, I predicted the annual colonization and extinction probabilities for another set of

Table 12.2 Parameter estimates for a mainland–island IFM and the predicted and observed annual extinction and colonization rates for three species of *Sorex* shrew on small islands in lakes

Species	Body size (g)	Parameter estimates			Predicted		Observed	
		x	(SE)	e	C	E	C	E
S. araneus	9	2.30	(0.68)	0.20	0.26	0.04	0.20	0.04
S. caecutiens	5	0.91	(0.24)	0.53	0.03	0.28	0.05	0.33
S. minutus	3	0.46	(0.16)	0.73	0.18	0.53	0.13	0.46

e and x are the parameters of the annual extinction probability E as a function of island area A, $E = e/A^x$, and C is the annual colonization probability, assumed to be the same constant for all islands. To tease apart the values of e and C, the minimum island area for occupancy, A_0, was assumed to be 0.5 ha. The predicted extinction probability was calculated for an island of 1.6 ha, which is the average size of the 17 islands from which the observed colonization and extinction rates were measured in a 5-year study (Peltonen and Hanski 1991; Hanski 1992).

17 islands (Hanski 1992), on which a census had been taken for 5 years (Peltonen and Hanski 1991). The observed rates matched the predicted rates well in all three species (Table 12.2), which represent practically independent replicates for this analysis, as interspecific interactions did not significantly affect the extinction and colonization rates (Peltonen and Hanski 1991). The same approach was later applied to two North-American species of small mammal (Hanski 1993), based on long-term island studies by Crowell (1986) and Lomolino (1993).

The shrew example can be developed further to elucidate additional biological inference based on the value of parameter x in the IFM. Using data on population densities and the estimated x values (Table 12.2), I calculated the relationship between the expected time to population extinction and the expected population size in the three shrew species (Fig. 12.7). A rapidly increasing time to extinction with expected population size is associated with a large x value, which indicates weak environmental stochasticity (Box 5.1). Note that large species of shrew had the largest x values (Fig. 12.7). In another similar study, Cook and Hanski (1995) used the mainland–island IFM to estimate the value of x for species of birds in four sets of data which were considered to fit well the model assumptions—mountain tops in the Great Basin (Behle 1978), and oceanic islands in the western Torres Strait (Draffan *et al.* 1983), around New Zealand (Diamond 1984) and in the Sea of Cortez (Cody 1983). Excluding large carnivores, represented by raptors and the pheasant coucal in the Torres Strait, which behaved differently as a group, we found that, just like in the shrews, the value of x was positively correlated with body size (Cook and Hanski 1995). Recalling that large x values indicate weak environmental stochasticity, these results imply that large-bodied species are less affected by environmental stochasticity than small-bodied species. Such a difference is actually expected, because the small body reserves of small vertebrates make them vulnerable to temporal changes in food availability (Krebs and Davies 1984; Hanski 1992; Pimm 1992). In invertebrates, we would not expect a simple relationship between starvation time and body size, hence it is not surprising that Nieminen (1996c) found

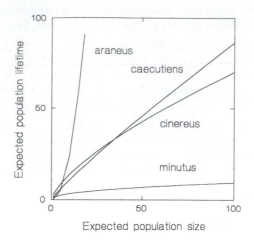

Fig. 12.7 Relationship between the expected population lifetime and the expected population size (island area times average density) in four species of *Sorex* shrews on islands. The results were obtained by fitting the IFM to data on island occupancy (Hanski 1993). The North American *S. cinereus* is similar in size to *S. caecutiens*.

no relationship between body size and the x value in herbivorous moths. These examples illustrate how estimation of parameter x can be helpful in assessing broad patterns in population extinction. One prediction that follows but has not been tested is a negative correlation between the value of x and the amplitude of population fluctuations.

12.5   Non-equilibrium dynamics

It is instructive to visit places in nature which one has not seen for two or more decades (I realize that some readers are in a better position to do that than others). Not always, but often enough, ones principal sensation is change. What used to be a meadow 20 years ago is a thicket today. The rare landscapes on earth that are not much affected by humans, and which are not just sand or ice, might have reached a steady state at a large spatial scale, but in our familiar landscapes such an equilibrium is unlikely to occur, and changes in landscape structure are likely to be directional even at large scales.

Metapopulation dynamics occur on the time-scale of local extinction and population establishment, which is generally a slow time-scale in comparison with the time-scale of local dynamics. Changes in landscape structure are often likely to occur on about the same time-scale than extinctions and recolonizations. In such situations, metapopulations are not at stochastic equilibrium with respect to the prevailing structure of the landscape, but they are in the process of tracking, with a delay, the changing environment. Given the ongoing destruction and fragmentation of natural habitats, the most likely scenario is the one in which metapopulations are 'too large'—larger than their respective equilibrium size—and are hence approaching the new equilibrium from 'above'. In the worst case, which might, unfortunately, be a common one, the new equilibrium is metapopulation extinction. Habitat destruction has levied a 'debt of extinctions' and created an illusory excess of rare species (Hanski 1994b, 1996; Tilman *et al.* 1994; Tilman and Lehman 1997). The best-known examples of non-equilibrium metapopulations include large-scale dynamics in which species number is declining in remnant habitat fragments on mountain tops (Brown 1971) and on land-bridge islands (Diamond 1984). But there is no reason to believe that similar scenarios would not occur on much smaller scales as well.

An almost authentic example

Working within an area of 25 km^2 in northern Åland, Hering (1995) mapped the present extent of the suitable habitat for the Glanville fritillary. He then proceeded to reconstruct the extent of the suitable habitat some 20 years ago, using the results of his field work, old aerial photographs and interviews with local people. During the 20-year period, the total area of suitable habitat had become reduced to one third of its original extent and the number of distinct patches had declined from 55 to 42 (Fig. 11.2), largely due to reduced grazing pressure. To assess the metapopulation dynamic consequences of this change in landscape structure, we did the following (Hanski *et al.* 1996b). For each patch that had lost some area we assumed that the loss had occurred linearly over a period of T years, where T is a random variable uniformly distributed between 1 and 20 years. We then used the previously parameterized IFM (Hanski *et al.* 1996c) to predict the fraction of occupied patches, P, during and following the 20-year period of habitat destruction. Using the model, we also calculated the equilibrium value of P, $\hat{P}$, at each point in time. The difference between P and $\hat{P}$ reflects the magnitude of the delay in metapopulation dynamics in a declining patch network.

The results in Fig. 12.8a suggest that the butterfly has tracked the amount of suitable habitat with only a small delay. The delay is small because the metapopulation occupied most of the habitat at the beginning of the study period and because the turnover rate is high (Table 12.1) in comparison with the rate of environmental change. But what would happen if the same trend in the amount of habitat continued for another 20 years? To take an example, let us assume that each of the present patches would lose a further 50% of its area in another 20 years. Figure 12.8b shows that this further loss of habitat would soon lead to a patch network that is not adequate for long-term metapopulation persistence, that is, the equilibrium $\hat{P}$ drops to zero in less than 20 years (Fig. 12.8b). However, the actual metapopulation extinction is predicted to take tens or even hundreds of years, with the largest remaining local populations dwindling to extinction slowly. The inevitable decline to extinction might be temporarily halted for long periods, with

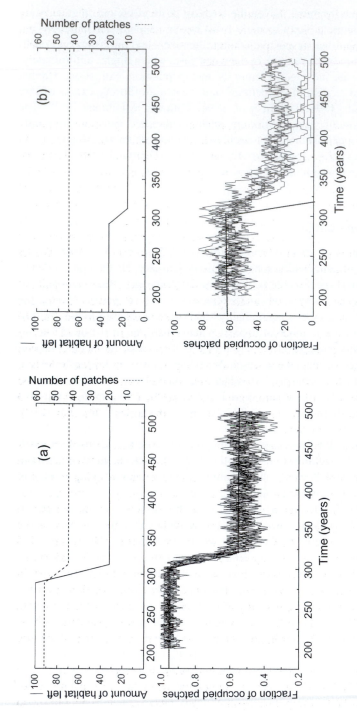

Fig. 12.8 Metapopulation dynamics of the Glanville fritillary in the landscape shown in Fig. 11.2. The size of the metapopulation is measured by the fraction of occupied patches (P). The results were obtained with the IFM using parameter values estimated in Section 12.1. Panel (a, lower) gives the predicted equilibrium metapopulation size (thick line) with ten replicate predicted trajectories before, during and following an observed reduction in habitat area over a 20-year period (shown in the upper panel). Panel (b) shows similar results for a hypothetical scenario of further loss of 50% of the area of each of the remaining patches (upper panel). In this case the equilibrium moves to metapopulation extinction even though a substantial amount of habitat remains. The simulated trajectories show a slow decline with much variation.

the number of occupied patches fluctuating without any obvious trend. The delay would be smaller if the dynamics were greatly affected by regional stochasticity, but the point is clear—many small metapopulations (rare species) may be around only because they have not yet had time to go extinct.

Host–parasitoid dynamics

Spatially explicit host–parasitoid metapopulation models generate spatially correlated abundance patterns in the host and in the parasitoid (Section 7.3). As there is definite spatial correlation in the abundance changes in the Glanville fritillary (Fig. 11.6), and as the butterfly has two specialist parasitoids (Section 11.1), of which at least one, the braconid *Cotesia melitaearum*, increases the rate of local host extinctions (Section 11.2), it is natural to inquire whether host–parasitoid metapopulation dynamics might drive spatially correlated changes in the butterfly metapopulations. There are some anecdotal data suggesting that *Cotesia* parasitoids attacking other fritillaries (Porter 1981) and other butterflies (Fulto 1940; Biever 1992) may cause local extinctions of their host populations.

Guangchun Lei's (1997) and Saskya van Nouhuys's (personal communication) results provide a unique field record of the dynamics of the Glanville fritillary and *C. melitaearum* in a patch network in northern Åland over a period of five years. The network itself has remained relatively constant during this period, but the set of patches suitable for the parasitoid has varied greatly with host population turnover. During the period 1993–1995, there was a declining trend in the number of host populations, from 35 to 13, at least to some extent caused by parasitism (Lei and Hanski 1997). The number of parasitoid populations declined only little from 1993 to 1994, but then dramatically to only 2 populations in the spring 1995 (Fig. 12.9). The scene is now set for the host to recover.

The near extinction of the parasitoid metapopulation from this network suggests that the network is too small to reveal the entire picture, for which we have to turn to the scale of the entire Åland islands. If a survey is conducted in the spring soon after the second generation of *C. melitaearum* larvae have emerged from the host caterpillars, it is possible to get a good idea of the presence of the parasitoid populations by counting their cocoons in the host larval webs. A large-scale operation of this sort is tricky, however, because timing is critical. In spring 1995 Juha Pöyry managed to get everything right, and with the help of 10 students he surveyed all 392 butterfly populations recorded the previous autumn. The map in Fig. 12.10 shows the result. In these data, both patch area and isolation from the existing parasitoid populations had strong effects on the presence of the parasitoid (Lei and Hanski 1997), just as for the host butterfly. The parasitoid populations have a markedly clumped spatial distribution. Does this pattern represent a stochastic quasi-equilibrium?

We used the IFM to answer the question about metapopulation-level equilibrium (Lei and Hanski 1997). Model parameters were estimated in the same manner as for the butterfly (Section 12.1), using the snapshot of presence/absence data collected in

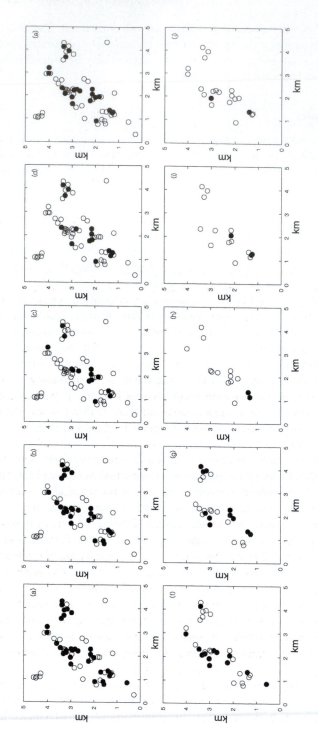

Fig. 12.9 Metapopulation patterns in the Glanville fritillary (upper panels) and the specialist parasitoid *Cotesia melitaearum* (lower panels) from 1993 (a and f) until 1997 (e and j). Open circles indicate empty patches suitable for colonization (meadows for the butterfly, meadows occupied by the butterfly for the parasitoid), black dots show occupied patches.

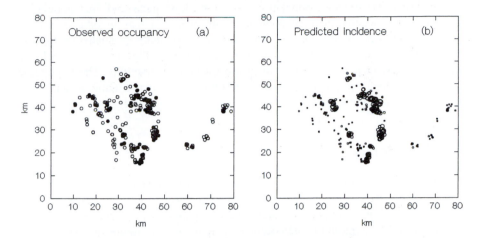

Fig. 12.10 The observed and predicted occupancy of host populations by *Cotesia melitaearum* in Åland in the spring 1995. (a) The dots show the habitat patches occupied by the host butterfly; the ones for which the parasitoid was recorded are shown by a black dot. In (b), the size of the dot indicates the predicted incidence based on fitting the IFM to the data in (a). (Based on Lei and Hanski 1997.)

the spring 1995. With the parameterized IFM, we then predicted the incidence of the parasitoid in the suitable patches, that is meadows with a butterfly population, in the entire study area. The predicted incidence tended to be high in the areas where the parasitoid had been recorded, essentially in the regions with high density of host populations (Fig. 12.10). However, some regions had clearly fewer while others had more parasitoid populations than predicted. Such discrepancies suggest that the observed pattern of patch occupancy in 1995 does not represent an equilibrium, but to an unknown extent reflects the past history of changes in the populations. Given the observed high turnover rate in the host, this is not a surprising conclusion. The non-equilibrium situation is reflected by the simulations of the parasitoid in the 1995 network of host populations using the parameter values estimated from the same data. If the system were in equilibrium, the predicted fraction of occupied patches should closely match that observed. In the present case, however, the predicted fraction of occupied patches (41%) was twice the observed fraction (20%; Lei and Hanski 1997), a discrepancy which can arise if the metapopulation from which the parameter values were estimated is severely out of equilibrium. In other words, the parasitoid was absent from many patches in which it should have occurred based on the inferred effects of area and isolation on occupancy. It was most likely absent due to a delay in colonizing newly established host populations.

The predicted parasitoid incidences in Fig. 12.10 are very small in large areas of Åland, where the density of host populations is apparently too low to enable persistence of the parasitoid. It is likely that in those parts of Åland where the potential density of host populations is high, host–parasitoid metapopulation dynamics significantly increase population turnover, whereas in other areas the parasitoid plays no significant role. The large-scale spatially correlated changes in host population size do not show temporal correlation (Fig. 11.7), as would be expected if the changes were due to host–parasitoid interactions. The large-scale changes are likely to reflect spatially correlated weather conditions and their interaction with habitat quality.

12.6 Summary

The incidence function model can be parameterized with data that are available for many species living in highly fragmented landscapes—a description of the landscape structure in terms of the areas and the spatial locations of discrete habitat patches, and a description of patch occupancy by the species in one or several years. One advantage of the incidence function model is that it can be parameterized with spatial data only, yet the model can be used to make predictions about metapopulation dynamics. This chapter has described the kinds of situation to which the incidence function approach can be applied and has summarized the steps in parameter estimation and model simulation. The use of the model is illustrated with examples on classical and mainland–island metapopulations of butterflies, small mammals and birds. Apart from being useful in predicting the occurrence of species in specific fragmented landscapes, the incidence function model can be used to study complex spatial dynamics with alternative equilibria and transient dynamics following changes in the structure of the landscape or perturbations to local populations.

13

Epilogue

13.1 The achievements

Viewed from outside, metapopulation ecology represents a particular brand of spatial ecology. Ecologists are used to thinking about assemblages of interacting entities, such as individuals in populations, but usually without an explicit reference to space. The novelty of spatial ecology is in the claim that the spatial locations of individuals, populations and communities can have equally significant consequences on dynamics as birth and death rates, competition and predation. Spatial ecology is one of the most visible developments in ecology and population biology in recent years; some regard it as a new paradigm arising towards the end of the twentieth century. New paradigm or not, it is true that never before has space been considered to be so pivotal to so many biological phenomena as today.

Paradoxically, the emphasis on space challenges the age-old habitat–organism relation. Species have adopted to their environments and, according to common wisdom, the spatial distribution of individuals reflects the match between the environment and the species' ecological requirements. Spatial ecology, in contrast, predicts that species might exhibit complex spatial patterns even in uniform environments; that species might be absent where environmental conditions are favourable; and that species might be present where the conditions are not favourable (sink populations). As with other challenges to common wisdom, for instance the challenge to the diversity–stability paradigm, one should recognize the variegated nature and scale dependence of the phenomena of interest. Spatial ecology does not claim that elephants are more numerous in Africa than in Europe because of some subtle population dynamic process, but spatial ecology makes, for instance, the claim that with gradually increasing habitat loss and fragmentation species might drop to extinction rather abruptly. In other words, the mapping of species abundance on environmental conditions is modified by population dynamics.

Spatial ecology includes metapopulation ecology but also other approaches, most notably landscape ecology and diffusion-based and lattice-based models of spatially structured populations in uniform environments. In comparison with these two other approaches, one heavily empirical, the other one heavily theoretical, metapopulation ecology strives towards the middle ground between modelling and field studies. This position has been bought with the cost of certain simplifying assumptions. For landscape ecologists, the major simplifying assumption in metapopulation ecology is the vision of space as a network of discrete and ecologically distinct habitat patches. This is clearly not a sensible assumption for all landscapes nor for all species, but where the assumption can be defended it has enabled the construction of effective theory. For theoretical ecologists, the prominent simplification in classical metapopulation ecology is the use of patch models; other models of spatial dynamics depict a much more fine-grained picture of populations. The patch-network assumption about the structure of the environment and the patch-occupancy description of populations obviously match each other, and they are at the heart of the interplay between theory and field studies in classical metapopulation ecology. The lesson to be drawn is not that it always pays to compromise; the lesson is that it is necessary to make simplifying assumptions, and that the assumptions you make largely determine whether or not your work is going to be successful.

Successful theory is born from successful applications, which in ecology often means a successful research tradition focused on particular taxa. The dynamic theory of island biogeography might not exist without knowledge of bird communities on islands. Although here I am blatantly partial, I suggest that the study of classical metapopulation biology has been similarly stimulated and advanced by empirical studies on butterflies as a model group of species.

Classical metapopulation ecology

The gradual unfolding of the theory of classical metapopulation ecology and its implications for empirical research is the most visible achievement of current metapopulation ecology. I distinguish two stages in its formation. The first stage involved the conception of the general notion of extinction–colonization dynamics, first in the writings of Nicholson and Andrewartha and Birch, subsequently in the model of Levins. This stage directly paved the way to theoretical studies based on the assumption of regional stability of assemblages of unstable local populations. The second stage in the advance of the classical metapopulation theory occurred when habitat patch area and isolation effects were included in the models—area-dependent extinction and isolation-dependent colonization rates. This stage was critical for empirical studies, because it led to straightforward instructions to field ecologists about the type of information that should be collected. Furthermore, the area and isolation effects are so general that their inclusion into models does not compromise their generality. This is important for forging a strong link between models and field studies. We now have a theoretical concept and a practical modelling framework to address the spatial dynamics of species living in highly fragmented landscapes. The patch area-isolation framework has been helpful for the study of single-species dynamics but also for studies of individual movement behaviour (Box 2.1) and community structure (Box 7.1). Research approaches based on the patch area and isolation effects comprise an auxiliary paradigm in population biology of individuals, populations and communities living in highly fragmented landscapes, adding a significant new element to the basic model of extinction–colonization dynamics.

The concept of populations persisting in a stochastic balance between local extinctions and recolonizations of empty habitat patches has become well established. There is no need to reiterate here the results and conclusions discussed in the previous chapters about the dynamics of species in fragmented landscapes. I emphasize just two issues. First, it is appropriate to conclude that classical metapopulation persistence involves the 'risk-spreading' process (den Boer 1968)— different local populations oscillate to some extent independently, which will reduce variance in the growth rate of the metapopulation as a whole, thereby reducing the risk of extinction (Goodman 1987a; see eqn 2.3—independent local dynamics reduce v when the equation is applied to the metapopulation as a whole). However, it clearly does not follow from this observation that species would persist as metapopulations without density dependence. Second, the fundamental threshold condition for metapopulation persistence deserves extra emphasis, as it underscores most other theoretical conclusions. Here we have a fundamental difference to traditional non-spatial population dynamics. Space has added another requirement, apart from positive local growth rate, for long-term persistence.

Metapopulations and conservation

Metapopulation ecology has largely replaced the dynamic theory of island biogeography as the population-dynamic doctrine for conservation. The general message for conservation is that spatial dynamics often matter, not only in explaining the occurrence of species in distant archipelagos but also in explaining the abundance of species in our ordinary landscapes. Even if most species are not structured as classical metapopulations, the rise of the metapopulation paradigm has served and continues to serve the useful function of forcing conservation biologists to gather data that are likely to be important for effective conservation strategies, such as data on migration rates, spatially varying rates of reproduction and mortality, and the like.

With the growing number of empirical studies, it has become possible to construct and test spatially realistic models, capable of making quantitative predictions about the dynamics of particular species in particular landscapes. My own preference is for relatively simple models, which do not attempt to model everything but instead focus on the key processes, such as area-dependent extinction and isolation-dependent colonization in highly fragmented landscapes. Similarly, Drechsler and Wissel (1998) analyse the relative advantages of managing average patch size or quality, patch number and patch connectivity, in enhancing metapopulation persistence in the context of a general metapopulation model. It is unlikely that we will ever have enough data and knowledge of the dynamics of most species to warrant the application of models with tens of parameters. Often the best we can hope to achieve with spatially realistic models is comparative predictions, for instance, contrasting the likelihood of persistence under alternative scenarios of landscape change or under alternative management actions. Luckily, these are exactly the kinds of predictions that are likely to be most useful for conservation and management.

There are two reasons to expect predictions of spatially realistic metapopulation models to be more helpful for conservation and management than the well-known rules of reserve design based on the island theory (Terborgh 1974; Diamond 1975; Wilson and Willis 1975). First, the island biogeographic rules of reserve design were in practice static, although justified by the dynamic theory of island biogeography. In contrast, metapopulation predictions explicitly address the dynamics of survival in space. Second, the reserve-design rules contrasted fixed alternatives, whereas with the spatially realistic metapopulation models one is practically forced to make predictions for specific fragmented landscapes. It is to be hoped that these metapopulation dynamic considerations will be incorporated into the currently popular reserve-selection algorithms (e.g. Pressey *et al.* 1996), which aim at finding representative areas for conservation with a constraint on the total protected area. These algorithms, like the old island theory-based rules, pay no attention to spatial dynamics. An apparent drawback of metapopulation models is that they deal with single species, not with entire communities, but in practice it might be sufficient to consider only a few types of species with different responses to habitat fragmentation. The really important issue is whether spatial dynamics are considered at all in landscape management and conservation.

13.2 Old and new challenges

I have alluded to the preoccupation of landscape ecologists with ever more sophisticated descriptions of landscape patterns, with the apparent cost that theories of landscape ecology have remained largely verbal. Many theoretical models of spatial dynamics have the opposite problem. A key mission of theoretical spatial ecology has been to demonstrate the emergence of complex dynamics and complex spatial patterns due to population dynamics in homogeneous space (Tilman and Kareiva 1997). The formidable empirical challenge of relating these modelling results to empirical studies has only just begun. For all of us, regardless of which brand of spatial ecology we are attached to, the grand challenge is to forge a conceptual and theoretical synthesis of spatial ecology, embracing in one manner or another individual responses and population dynamics, and explaining patterns in species abundance caused by complex landscapes and patterns driven by complex dynamics. This synthesis is needed for the progress of our science and for its application to conservation and management, but which form the synthesis will take remains one of the unknowns of the next millennium.

Another general challenge for both theoretical and empirical metapopulation studies is to come to terms with non-equilibrium dynamics at the regional scale. Classical metapopulation models deal with unstable local populations but assume regional stability. But we know that most environments are currently changing so fast that the assumption about regional stability is not generally met. There is a need to develop approaches that are focused on regional non-equilibrium situations.

Many of the fundamental ecological issues about the persistence of populations in highly fragmented landscapes have been settled, for the time being, but the same cannot be said about metapopulation genetics and adaptive evolution in fragmented landscapes. The classical genetic models of spatially structured populations, such as the island model, are two steps behind ecology. Current and forthcoming work can be expected to change our views. Sewall Wright's shifting balance theory has remained a powerful metaphor of adaptive evolution in metapopulations, but there might not be much more to it. What is the next vision of evolution in fragmented environments? It is customary to plea for the unification of population ecology, genetics and evolutionary biology. Progress does not take place because such pleas are made, progress happens when the time is right or a new angle is found. The prospects are bright for increasingly comprehensive biological studies of populations in fragmented environments.

References

Abbot, I. and Black, R. 1988. Changes in species composition of floras on islands near Perth, Western Australia. *Journal of Biogeography* 7: 399–410.

Åberg, J., Jansson, G., Swenson, J. E. and Angelstam, P. 1995. The effect of matrix on the occurrence of hazel grouse (*Bonasa bonasia*) in isolated habitat fragments. *Oecologia* 103: 265–269.

Adler, F. R. and Nürnberger, B. 1994. Persistence in patchy irregular landscapes. *Theor. Pop. Biol.* 45: 41–75.

Ågren, G. I. and Fagerström, T. 1984. Limiting dissimilarity in plants: randomness prevents exclusion of species with similar competitive abilities. *Oikos* 43: 369–375.

Akçakaya, H. R. and Atwood, J. L. 1997. A habitat-based metapopulation model of the California gnatcatcher. *Cons. Biol.* 11: 422–434.

Akçakaya, H. R. and Ferson, S. 1992. RAMAS/Space user manual: spatially structured population models for conservation biology. Applied Biomathematics, New York.

Akçakaya, H. R. and Ginzburg, L. R. 1991. Ecological risk analysis for single and multiple populations. In *Species conservation: a population-biological approach* (ed. A. Seitz and V. Loeschcke), pp. 73–87. Birkhäuser, Basel.

Akçakaya, H. R., McCarthy, M. A. and Pearce, J. L. 1995. Linking landscape data with population viability analysis: Management options for the helmeted honeyeater *Lichenostomus melanops cassidix. Biol. Cons.* 73: 169–176.

Allee, W. C., Emerson, A. E., Park, O., Park, T. and Schmidt, K. P. 1949. *Principles of animal ecology*. Saunders, Philadelphia.

Allen, J. C., Schaffer, W. M. and Rosko, D. 1993. Chaos reduces species extinction by amplifying local population noise. *Nature* 364: 229–232.

Andersen, N. M. 1993. The evolution of wing polymorphism in water striders (Gerridae): a phylogenetic approach. *Oikos* 67: 433–443.

Anderson, R. M. and May, R. M. 1991. *Infectious diseases of humans: dynamics and control*. Oxford University Press.

Andow, D. A., Kareiva, P. M., Levin, S. A. and Okubo, A. 1990. Spread of invading organisms. *Landscape Ecology* 4: 177–188.

Andrén, H. 1994. Effects of habitat fragmentation on birds and mammals in landscapes with different proportions of suitable habitat: a review. *Oikos* 71: 355–366.

Andrén, H. 1996. Population responses to habitat fragmentation: Statistical power and the random sample hypothesis. *Oikos* 76: 235–242.

Andrén, H. 1997. Habitat fragmentation and changes in biodiversity. *Ecological Bulletins* 46: 171–181.

Andrén, H. and Delin, A. 1994. Habitat selection in the Eurasian red squirrel, *Sciurus vulgaris*, in relation to forest fragmentation. *Oikos* 70: 43–48.

Andrewartha, H. G. and Birch, L. C. 1954. *The distribution and abundance of animals*. The University of Chicago Press.

Angelstam, P. 1997. Landscape analysis as a tool for the scientific management of biodiversity. *Ecological Bulletins* 46: 140–170.

Antonovics, J. 1994. The interplay of numerical and gene-frequency dynamics in host–pathogen systems. In *Ecological genetics* (ed. L. A. Real), pp. 129–145. Princeton University Press.

Antonovics, J., Thrall, P., Jarosz, A. and Stratton, D. 1994. Ecological genetics of metapopulations: The *Silene–Ustilago* plant–pathogen system. In *Ecological genetics* (ed. L. A. Real), pp. 146–170. Princeton University Press.

Antonovics, J., Thrall, P. H. and Jarosz, A. M. 1997. Genetics and the spatial ecology of species interaction. In *Spatial ecology* (ed. D. Tilman and P. Kareiva), pp. 158–183. Princeton University Press.

van Apeldoorn, R. C., Celada, C. and Niuwenhuizen, W. 1994. Distribution and the dynamics of the red squirrel (*Sciurus vulgaris* L.) in a landscape with fragmented habitat. *Landscape Ecology* 9: 227–235.

van Apeldoorn, R. C., Oosterbrink, W. T., van Winden, A. and van der Zee, F. F. 1992. Effects of habitat fragmentation on the bank vole, *Clethrionomys glareolus*, in an agricultural landscape. *Oikos* 65: 265–274.

Appelt, M. and Poethke, H. J. 1997. Metapopulation dynamics in a regional population of the blue-winged grasshopper (*Oedipoda caerulescens*; Linnaeus, 1758). *Journal of Insect Conservation* 1: 205–214.

Armstrong, D. P. and McLean, I. G. 1995. New Zealand translocations: theory and practice. *Pacific Conservation Biology* 2: 29–54.

Armstrong, R. A. 1976. Fugitive species: experiments with fungi and some theoretical considerations. *Ecology* 57: 953–963.

Arneberg, P. 1996. Commonness and rarity among mammalian nematodes: A comparative study of parasite abundance. Ph.D. Dissertation, University of Tromsö, Norway.

Arrhenius, O. 1921. Species and area. *Journal of Ecology* 9: 95–99.

Atkinson, W. D. and Shorrocks, B. 1981. Competition on a divided and ephemeral resource: a simulation model. *J. Anim. Ecol.* 50: 461–471.

Atkinson, W. D. and Shorrocks, B. 1984. Aggregation of larval Diptera over discrete and ephemeral breeding sites: the implications for coexistence. *Am. Nat.* 124: 336–351.

Avery, P. J. and Hill, W. G. 1977. Variability in genetic parameters among small populations. *Genet. Res.* 29: 193–213.

Back, C. E. 1988. Effects of host plant size on herbivore density: patterns. *Ecology* 69: 1090–1102.

Baguette, M. and Nève, G. 1994. Adult movements between populations in the specialist butterfly *Proclossiana eunomia* (Lepidoptera, Nymphalidae). *Ecol. Entomol.* 19: 1–5.

Baker, R. D. 1978. *The evolutionary ecology of animal migration*. Hodder and Stoughton, London.

Barbault, R. and Sastrapradja, S. D. 1995. Generation, maintenance and loss of biodiversity. In *Global biodiversity assessment* (ed. W. H. Heywood), pp. 193–274. Cambridge University Press.

Barkai, A. and McQuaid, C. 1988. Predator–prey role reversal in a marine benthic ecosystem. *Science* 242: 26–64.

Barton, N. and Clark, A. 1990. Population structure and processes in evolution. In *Population biology* (ed. K. Wohrmann and S. K. Jain), pp. 115–173. Springer, New York.

Barton, N. H. and Rouhani, S. 1987. The frequency of shifts between alternative equilibria. *J. Theor. Biol.* 125: 397–418.

Barton, N. H. and Rouhani, S. 1991. The probability of fixation of a new karyotype in a continuous population. *Evolution* 45: 499–517.

Barton, N. H. and Rouhani, S. 1993. Adaptation and 'shifting balance'. *Genet. Res.* 61: 57–74.

Barton, N. H. and Whitlock, M. C. 1997. The evolution of metapopulations. In *Metapopulation biology* (ed. I. A. Hanski and M. E. Gilpin), pp. 183–214. Academic Press, San Diego.

Bascompte, J. and Solé, R. V. 1994. Spatially induced bifurcations in single-species population dynamics. *J. Anim. Ecol.* 63: 256–264.

Bascompte, J. and Solé, R. V. 1995. Rethinking complexity: modelling spatiotemporal dynamics in ecology. *Trends Ecol. Evol.* 10: 361–366.

Bascompte, J. and Solé, R. V. 1996. Habitat fragmentation and extinction thresholds in spatially explicit models. *J. Anim. Ecol.* 65: 465–473.

Bascompte, J. and Solé, R. V. 1998. *Modeling spatiotemporal dynamics in ecology.* Springer, New York.

Baur, B. and Bengtsson, J. 1987. Colonizing ability in land snails on Baltic uplift archipelagos. *Journal of Biogeography* 14: 329–341.

Begon, M., Harper, J. L. and Townsend, C. R. 1996. *Ecology.* Blackwell, Oxford.

Behle, W. H. 1978. Avian biogeography of the Great Basin and intermountain region. *Great Basin Naturalist Memoirs* 2: 55–80.

Beier, P. 1993. Determining minimum habitat areas and habitat corridors for cougars. *Conserv. Biol.* 7: 94–108.

Bengtsson, J. 1988. Life histories, interspecific competition and regional distribution of three rockpool *Daphnia* species. Ph.D. Dissertation, University of Uppsala, Sweden.

Bengtsson, J. 1989. Interspecific competition increases local extinction rate in a metapopulation system. *Nature* 340: 713–715.

Bengtsson, J. 1991. Interspecific competition in metapopulations. In *Metapopulation dynamics: empirical and theoretical investigations* (ed. M. E. Gilpin and I. Hanski), pp. 219–237. Academic Press, London.

Bengtsson, J. 1993. Interspecific competition and determinants of extinction in experimental populations of three rockpool *Daphnia* species. *Oikos* 67: 451–464.

Bennett, A. F. 1990. Habitat corridors and the conservation of small mammals in a fragmented forest environment. *Landscape Ecology* 4: 109–122.

Berg, L. M., Lascoux, M. and Pamilo, P. 1998. The infinite island model with sex-differentiated gene flow. *Heredity* in press.

Berger, J. 1989. *Environmental restoration.* Island Press, Washington, D.C.

Berryman, A. A. 1992. Intuition and the logistic equation. *Trends Ecol. Evol.* 7: 316.

Biever, K. D. 1992. Distribution and occurrence of *Cotesia rubecula* (Hymenoptera: Braconidae), a parasite of *Artogeia rapae* in Washington and Oregon. *J. Econ. Entom.* 85: 739–742.

Blondel, J. 1993. Habitat heterogeneity and life-history variation of Mediterranean blue tits (*Parus caeruleus*). *Auk* 110: 511–520.

Boecklen, W. J. and Mopper, S. 1998. Local adaptation in specialist herbivores: theory and evidence. In *Genetic structure and local adaptation in natural insect populations* (ed. S. Mopper and S. Y. Strauss), pp. 64–88. Chapman and Hall, New York.

den Boer, P. J. 1968. Spreading of risk and stabilization of animal numbers. *Acta Biotheor.* 18: 165–194.

den Boer, P. J. 1987. Detecting density dependence. *Trends Ecol. Evol.* 2: 77–78.

den Boer, P. J. 1991. Seeing the tree for the wood: random walks or bounded fluctuations of population size? *Oecologia* 86: 484–491.

Boughton, D. A. 1998. Ecological and behavioural mechanisms of colonisation in a metapopulation of the butterfly *Euphydryas editha*. Ph.D. Dissertation, University of Texas, Austin.

Bowers, M. A. and Matter, S. F. 1997. Landscape ecology of mammals: relationships between density and patch area. *J. Mamm.* 78: 999–1013.

Bowers, M. D. 1980. Unpalatability as a defence strategy of *Euphydryas phaeton* (Lepidoptera: Nymphalidae). *Evolution* 34: 586–600.

Bowers, M. D. 1983. Iridoid glycosides and larval hostplant specificity in checkerspot butterflies (*Euphydryas*, Nymphalidae). *J. Chem. Ecol.* 9: 475–493.

Boyce, M. S. 1987. A review of the U.S. Forest Service's viability analysis for the spotted owl. Final report to the national council of the paper industry for air and stream improvement. Unpublished.

Boyce, M. S. 1992. Population viability analysis. *Ann. Rev. Ecol. Syst.* 23: 481–506.

Boycott, A. E. 1930. A re-survey of the fresh-water mollusca of the parish of Aldenham after ten years with special reference to the effect of drought. *Trans. Hertf. Nat. Hist. Soc.* 19: 1–25.

ter Braak, J. F., Hanski, I. A. and Verboom, J. 1998. The incidence function approach to modeling of metapopulation dynamics. In *Modeling spatiotemporal dynamics in ecology* (ed. J. Bascompte and R. V. Solé), pp. 167–188. Springer, New York.

Brakefield, P. and Saccheri, I. J. 1994. Guidelines in conservation genetics and the use of the population cage experiments with butterflies to investigate the effects of genetic drift and inbreeding. In *Conservation genetics* (ed. V. Loeschcke, J. Tomiuk and S. K. Jain), pp. 165–179. Birkhäuser, Basel.

Briggs, C. J. and Godfray, H. C. J. 1996. The dynamics of insect–pathogen interactions in seasonal environments. *Theor. Pop. Biol.* 50: 149–177.

Bright, P. W., Mitchell, P. and Morris, P. A. 1994. Dormouse distribution: survey techniques, insular ecology and selection of sites for conservation. *J. Appl. Ecol.* 31: 329–339.

Brinck, P. 1948. Coleopteran from Tritan da Cunha. Results of the Norwegian Scientific Expedition to Tritan da Cunha 17: 1–121.

Brodmann, P. A., Wilcox, C. V. and Harrison, S. 1997. Mobile parasitoids may restrict the spatial spread of an insect outbreak. *J. Anim. Ecol.* 66: 65–72.

Brooks, T. M., Pimm, S. L. and Collar, N. J. 1997. Deforestation predicts the number of threatened birds in insular southeast Asia. *Cons. Biol.* 11: 382–394.

Brown, E. S. 1951. The relation between migration rate and type of habitat in aquatic insects, with special reference to certain species of Corixidae. *Proc. Zool. Soc. Lond.* 121: 539–545.

Brown, I. L. and Ehrlich, P. R. 1980. Population biology of the checkerspot butterfly, *Euphydryas chalcedona*: structure of the Jasper Ridge colony. *Oecologia* 47: 239–251.

Brown, J. H. 1971. Mammals on mountaintops: nonequilibrium insular biogeography. *Am. Nat.* 105: 467–478.

Brown, J. H. 1984. On the relationship between abundance and distribution of species. *Am. Nat.* 124: 255–279.

Brown, J. H. 1995. *Macroecology*. Chicago University Press.

Brown, J. H. and Kodric-Brown, A. 1977. Turnover rates in insular biogeography: effect of immigration on extinction. *Ecology* 58: 445–449.

Brown, J. S. and Pavlovic, N. B. 1992. Evolution in heterogeneous environments: effect of migration on habitat specialization. *Evol. Ecology* 6: 360–382.

Brussard, P. F. and Ehrlich, P. R. 1970. Contrasting population biology of two species of butterfly. *Nature* 227: 91–92.

Bryant, A. A. and Janz, D. W. 1996. Distribution and abundance of Vancouver island marmots (*Marmota vancouveriensis*). *Can. J. Zool.* 74: 667–677.

Bryant, E. H., McCommas, S. A. and Combs, L. M. 1986. The effect of an experimental bottleneck upon quantitative genetics variation in the housefly. *Genetics* 114: 1191–1211.

Bryant, E. H., Meffert, L. M. and McCommas, S. A. 1990. Fitness rebound in serially bottlenecked populations of the house fly. *Am. Nat.* 136: 542–549.

Buckley, P. 1989. *Biological habitat reconstruction*. Belhaven, London.

Buechner, M. 1987. A geometric model of vertebrate dispersal: tests and implications. *Ecology* 68: 310–318.

Burdon, J. J., Ericson, L. and Muller, W. J. 1995. Temporal and spatial changes in metapopulation of the rust pathogen *Triphragmium ulmariae* and its host, *Filipendula ulmaria*. *J. Ecol.* 83: 979–989.

Burgman, M. A., Ferson, S. and Akçakaya, H. R. 1993. *Risk assessment in conservation biology*. Chapman and Hall, New York.

Burkey, T. V. 1989. Extinction in nature reserves: the effect of fragmentation and the importance of migration between reserve patches. *Oikos* 55: 75–81.

Burkey, T. V. 1995. Extinction rates in archipelagos: implications for populations in fragmented habitats. *Cons. Biol.* 9: 527–541.

Burkey, T. V. 1997. Metapopulation extinction in fragmented landscapes: using bacteria and protozoa communities as model ecosystems. *Am. Nat.* 150: 568–591.

Burnham, K. P., Anderson, D. R. and White, G. C. 1994. Estimation of vital rates in northern spotted owl. In *Final supplemental environmental impact statement*, pp. 1–26. U.S. Department of Agriculture and U.S. Department of the Interior, Portland, Oregon.

Burt, A. 1995. The evolution of fitness. *Evolution* 49: 1–8.

Bush, A. O. and Holmes, J. C. 1986. Intestinal helminths of lesser scaup ducks: an interactive community. *Can. J. Zool.* 64: 142–152.

Buys, M. H., Maritz, J. S., Boucher, C. and van der Walt, J. J. 1994. A model for species–area relationships in plant communities. *J. Veget. Sc.* 5: 63–66.

Carlquist, S. 1966. The biota of long-distance dispersal II. Loss of dispersability in the Pacific Compositae. *Quarterly Reviews in Biology* 41: 247–270.

Caro, T. M. and Laurenson, M. K. 1994. Ecological and genetic factors in conservation: a cautionary tale. *Science* 263: 485–486.

Case, T. 1991. Invasion resistance, species build-up and community collapse in metapopulation models with interspecies competition. *Biol. J. Linn. Soc.* 42: 239–266.

Caswell, H. and Cohen, J. E. 1991. Disturbance, interspecific interaction and diversity in metapopulations. *Biol. J. Linn. Soc.* 42: 193–218.

Caswell, H. and Cohen, J. E. 1993. Local and regional regulation of species–area relations: a patch-occupancy model. In *Species diversity in ecological communities* (ed. R. E. Ricklefs and D. Schluter), pp. 99–107. The University of Chicago Press.

Caughley, G. 1976. Plant–herbivore systems. In *Theoretical ecology* (ed. R. M. May), pp. 94–113. Blackwell, Oxford.

Caughley, G. 1994. Directions in conservation biology. *J. Anim. Ecol.* 63: 215–244.

Celada, C., Bogliani, G., Gariboldi, A. and Maracci, A. 1994. Occupancy of isolated woodlots by the red squirrel *Sciurus vulgaris* L. in Italy. *Biol. Conserv.* 69: 177–183.

Charlesworth, D. and Charlesworth, B. 1987. Inbreeding depression and its evolutionary consequences. *Ann. Rev. Ecol. Syst.* 18: 237–268.

Chesson, P. 1991. A need for niches? *Trends Ecol. Evol.* 6: 26–28.

Chesson, P. 1998. Making sense of spatial models in ecology. In *Modeling spatiotemporal dynamics in ecology* (ed. J. Bascompte and R. V. Solé), pp. 151–166. Springer, New York.

Chesson, P. L. and Case, T. J. 1986. Overview: nonequilibrium community theories: chance, variability, history, and coexistence. In *Community ecology* (ed. J. Diamond and T. J. Case), pp. 229–239. Harper and Row, New York.

Chesson, P. L. and Warner, R. R. 1981. Environmental variability promotes coexistence in lottery competitive systems. *Am. Nat.* 117: 923–943.

Christiansen, F. B. and Fenchel, T. M. 1977. *Theories of populations in biological communities*. Springer, Berlin.

Clark, C. W. and Rosenzweig, M. L. 1994. Extinction and colonisation processes: parameter estimates from sporadic surveys. *Am. Nat.* 143: 583–596.

Clark, J. S., Fastie, C., Hurtt, G., Jackson, S. T., Johnson, C., King, *et al.* 1998. Reid's Paradox of rapid plant migration. *BioScience* 48:13–24.

Cody, M. L. 1983. The land birds. In *Island biogeography in the Sea of Cortez* (ed. T. J. Case and M. L. Cody), University of California Press, Berkeley.

Cody, M. L. and Overton, J. 1996. Short-term evolution of reduced dispersal in island plant populations. *Journal of Ecology* 84: 53–61.

Colas, B., Olivieri, I. and Riba, M. 1997. *Centaurea corymbosa*, a cliff-dwelling species tottering on the brink of extinction: a demographic and genetic study. *Proceedings of the National Academy of Sciences, USA* 94: 3471–3476.

Coleman, B. D. 1981. On random placement and species–area relations. *Math. Biosc.* 54: 191–215.

Collins, S. L. and Glenn, S. M. 1990. A hierarchical analysis of species' abundance patterns in grassland vegetation. *Am. Nat.* 135: 633–648.

Collins, S. L. and Glenn, S. M. 1991. Importance of spatial and temporal dynamics in species regional abundance and distribution. *Ecology* 72: 654–664.

Comins, H. N., Hamilton, W. D. and May, R. M. 1980. Evolutionary stable dispersal strategies. *J. Theor. Biol.* 82: 205–230.

Comins, H. N., Hassell, M. P. and May, R. M. 1992. The spatial dynamics of host-parasitoid systems. *J. Anim. Ecol.* 61: 735–748.

Condit, R. 1995. Research in large, long-term tropical forest plots. *Trends Ecol. Evol.* 10: 18–22.

Condit, R. 1996. Defining and mapping vegetation types in mega-diverse tropical forests. *Trends Ecol. Evol.* 11: 4–5.

Connell, J. H. 1971. On the role of natural enemies in preventing competitive exclusion in some marine animals and in rain forest trees. In *Dynamics of populations* (ed. P. J. den Boer and G. R. Gradwell), pp. 298–312. Pudoc, Wageningen.

Connell, J. H. 1978. Diversity in tropical rain forest and coral reefs. *Science* 199: 1302–1310.

Connell, J. H. 1983. On the prevalence and relative importance of interspecific competition: evidence from field experiments. *Am. Nat.* 122: 661–696.

Connell, J. H. and Slatyer, R. O. 1977. Mechanisms of succession in natural communities and their role in community stability and organization. *Am. Nat.* 111: 1119–1144.

Connell, J. H. and Sousa, W. P. 1983. On the evidence needed to judge ecological stability or persistence. *Am. Nat.* 121: 789–824.

Connor, E. F. and McCoy, E. D. 1979. The statistics and biology of the species–area relationship. *Am. Nat.* 113: 791–833.

Cook, R. R. and Hanski, I. 1995. On expected lifetimes of small and large species of birds on islands. *Am. Nat.* 145: 307–315.

Cornell, H. V. 1985. Local and regional richness of cynipine gall wasps on California oaks. *Ecology* 66: 1247–1260.

Coyne, J. A., Barton, N. H. and Turelli, M. 1997. A critique of Wright's shifting balance theory of evolution. *Evolution* 51: 643–671.

Crawley, M. J. 1993. *Natural enemies*. Blackwell, London.

Crawley, M. J. and May, R. M. 1987. Population dynamics and plant community structure: competition between annuals and perennials. *J. Theor. Biol.* 125: 475–489.

Crone, E. E. 1997. Delayed density dependence and the stability of interacting populations and subpopulations. *Theor. Pop. Biol.* 51: 67–76.

Crow, J. F. 1993. Mutation, mean fitness, and genetic load. *Oxford Surveys in Evolutionary Biology* 9: 3–42.

Crowell, K. L. 1973. Experimental zoogeography: introduction of mice to small islands. *Am. Nat.* 107: 535–558.

Crowell, K. L. 1986. A comparison of relict versus equilibrium models for insular mammals of the Gulf of Maine. In *Island biogeography of mammals* (ed. L. R. Heaney and B. D. Patterson), pp. 37–64. Academic Press, London.

Crowley, P. H. 1981. Dispersal and the stability of predator–prey interactions. *Am. Nat.* 118: 673–701.

Curnutt, J. L., Pimm, S. L. and Maurer, B. A. 1996. Population viability in sparrows in space and time. *Oikos* 76: 131–144.

Cutler, A. 1991. Nested faunas and extinction in fragmented habitats. *Cons. Biol.* 5: 496–505.

Danielson, B. J. 1991. Communities in a landscape: The influence of habitat heterogeneity on the interactions between species. *Am. Nat.* 138: 1105–1120.

Danielson, B. J. 1992. Habitat selection, interspecific interactions and landscape composition. *Evol. Ecology* 6: 399–411.

Darlington Jr, P. J. 1943. Carabidae on mountains and islands: data on the evolution of isolated faunas, and on atrophy of wings. *Ecol. Monogr.* 13: 37–61.

Darwin, C. 1859. *On the origin of species by means of natural selection*. John Murray, London.

Davis, G. J. and Howe, R. W. 1992. Juvenile dispersal, limited breeding sites, and the dynamics of metapopulations. *Theor. Pop. Biol.* 41: 184–207.

Debinski, D. M. 1994. Genetic diversity assessment in a metapopulation of the butterfly *Euphydryas gillettii*. *Biol. Conserv.* 70: 25–31.

Dempster, J. P. 1983. The natural control of populations of butterflies and moths. *Biol. Rev.* 58: 461–481.

Dempster, J. P. 1991. Fragmentation, isolation and mobility of insect populations. In *Conservation of insects and their habitats* (ed. N. M. Collins and J. A. Thomas), pp. 143–154. Academic Press, London.

Dempster, J. P., King, M. L. and Lakhani, K. H. 1976. The status of the swallowtail butterfly in Britain. *Ecol. Entomol.* 1: 51–56.

Dempster, J. P., Atkinson, D. A. and French, M. C. 1995. The spatial population dynamics of insects exploiting a patchy food resource: II. Movements between patches. *Oecologia* 104: 354–362.

Dennis, R. L. H. and Shreeve, T. G. 1996. Diversity of butterflies on British Isles: ecological influence underlying the roles of area, isolation and the size of the faunal source. *Biol. J. Linn. Soc.* 60: 257–275.

Dennis, B. and Taper, M. L. 1994. Density dependence in time series observations of natural populations: estimating and testing. *Ecol. Monogr.* 64: 205–224.

Denno, R. F. and Peterson, M. A. 1995. Density-dependent dispersal and its consequences for population dynamics. In *Population dynamics: New approaches and synthesis* (ed. N. Cappuccino and P. W. Price), pp. 113–130. Academic Press, London.

Denno, R. F., Roderick, G. K., Peterson, M. A., Huberty, A. F., Dödel, H. G., Eubanks, M. D., *et al.* 1996. Habitat persistence shapes the interspecific dispersal strategies of planthoppers. *Ecol. Monographs* 66: 389–408.

Deutschmann, D. H., Bradshaw, G. A., Chidress, W. W., Daly, K. L., Grunbaum, D., Pascual, M., *et al.* 1993. Mechanisms of patch formation. In *Patch dynamics* (ed. S. A. Levin, T. M. Powell and J. H. Steel), pp. 184–209. Springer, Berlin.

Diamond, J. M. 1969. Avifaunal equilibria and species turnover rates in the Channel islands of California. *Proceedings of the National Academy of Sciences, USA* 64: 57–63.

Diamond, J. M. 1971. Comparison of faunal equilibrium turnover rates on a tropical and temperate island. *Proceedings of the National Academy of Sciences, USA* 68: 2742–2745.

Diamond, J. M. 1975. The island dilemma: lessons of modern biogeographic studies for the design of natural reserves. *Biol. Cons.* 7: 129–146.

Diamond, J. M. 1979. Population dynamics and interspecific competition in bird communities. *Fortschr. Zool.* 25: 389–402.

Diamond, J. M. 1984. 'Normal' extinction of isolated populations. In *Extinctions* (ed. M. H. Nitecki), pp. 191–246. University of Chicago Press.

Diamond, J. M. and May, R. M. 1977. Species turnover rates on islands: dependence on census interval. *Science* 197: 266–270.

Diamond, J. M. and May, R. M. 1981. Island biogeography and the design of natural reserves. In *Theoretical ecology* (ed. R. M. May), pp. 228–252. Blackwell, Oxford.

Dias, P. C. 1996. Sources and sinks in population biology. *Trends Ecol. Evol.* 11: 326–330.

Dias, P. C. and Blondel, J. 1996. Local specialization and maladaptation in Mediterranean blue tits *Parus caeruleus*. *Oecologia* 107: 79–86.

Dias, P., Verheyen, G. R. and Raymond, M. 1996. Source–sink populations in Mediterranean blue tits: evidence using single-locus minisatellite probes. *Journal of Evolutionary Biology* 9: 965–978.

Dickman, C. R. and Doncaster, C. P. 1989. The ecology of small mammals in urban habitats. II. Demography and dispersal. *J. Anim. Ecol.* 58: 119–128.

Dingle, H. 1996. *Migration: the biology of life on the move*. Oxford University Press.

Dingle, H., Blackley, N. R. and Miller, E. R. 1980. Variation in body size and flight performance in milkweed bugs (*Oncopeltus*). *Evolution* 34: 371–385.

Dixon, A. F. G. 1990. Population dynamics and abundance of deciduous tree-dwelling aphids. In *Population dynamics of forest insects* (ed. M. Hunter, N. Kidd, S. R. Leather and A. Watt), pp. 11–23. Intercept, Andover.

Dixon, A. F. G. and Kindlmann, P. 1990. Role of plant abundance in determining the abundance of herbivorous insects. *Oecologia* 83: 281–283.

Doak, D. F. 1989. Spotted owls and old growth logging in the Pacific Northwest. *Cons. Biol.* 3: 389–396.

Doak, D. F. 1995. Source–sink models and the problem of habitat degradation: general models and applications to the Yellowstone Grizzly. *Cons. Biol.* 9: 1370–1379.

Doak, D. F. and Mills, L. S. 1994. A useful role for theory in conservation. *Ecology* 75: 615–626.

Doak, D. F., Marino, P. C. and Kareiva, P. M. 1992. Spatial scale mediates the influence of habitat fragmentation on dispersal success: implications for conservation. *Theor. Pop. Biol.* 41: 315–336.

Dobson, A. P. and Lyles, A. M. 1989. The population dynamics and conservation of primate populations. *Cons. Biol.* 3: 362–380.

Dobson, F. S. 1982. Competition for mates and predominant juvenile male dispersal in mammals. *Anim. Beh.* 30: 1183–1192.

Dobzhansky, T. and Wright, S. 1943. Genetics of natural populations. X. Dispersion rates in *Drosophila pseudoobscura. Genetics* 28: 304–340.

Dodd, A. P. 1959. The biological control of prickly pear in Australia. *Monographiae Biologicae* 8: 565–577.

Doebeli, M. 1995. Dispersal and dynamics. *Theor. Pop. Biol.* 47: 82–106.

Doncaster, C. P., Micol, T. and Jensen, S. P. 1996. Determining minimum habitat requirements in theory and practice. *Oikos* 75: 335–339.

van Dorp, D. and Opdam, P. F. M. 1987. Effects of patch size, isolation and regional abundance on forest bird communities. *Landscape Ecology* 1: 59–73.

Draffan, R. D. W., Garnett, S. T. and Malone, G. J. 1983. Birds of the Torres Strait: an annotated list and biogeographical analysis. *Emu* 83: 207–234.

Drake, J. A. 1988. Models of community assembly and the structure of ecological landscapes. In *Mathematical ecology* (ed. T. Hallam, L. Gross and S. Levin), pp. 584–604. World Press, Singapore.

Drake, J. A., Mooney, H. A. and di Castro, F. 1989. *Biological invasions: a global perspective.* Wiley, Chichester.

Drechsler, M. and Wissel, C. 1997. Separability of local and regional dynamics in metapopulations. *Theor. Pop. Biol.* 51: 9–21.

Drechsler, M. and Wissel, C. 1998. Trade-offs between local and regional scale management of metapopulations. *Biol. Cons.* 83: 31–41.

Dunning Jr, J. B., Borgella Jr, R., Clements, K. and Meffe, G. K. 1995. Patch isolation, corridor effects, and colonization by a resident sparrow in a managed pine woodland. *Cons. Biol.* 9: 542–550.

Durrett, R. and Levin, S. 1994. The importance of being discrete (and spatial). *Theor. Pop. Biol.* 46: 363–394.

Dybdal, M. F. 1994. Extinction, recolonisation, and the genetic structure of tidepool copepod populations. *Evol. Ecology* 8: 113–124.

Dytham, C. 1995. The effect of habitat destruction pattern on species persistence: a cellular model. *Oikos* 74: 340–344.

Ebenhard, T. 1987. An experimental test of the island colonization survival model: bank vole (*Clethrionomys glareolus*) populations with different demographic parameter values. *Journal of Biogeography* 14: 213–223.

Ebenhard, T. 1991. Colonization in metapopulations: a review of theory and observations. *Biol. J. Linn. Soc.* 42: 105–121.

Ebenhard, T. 1995. Wetland butterflies in a fragmented landscape: the survival of small populations. *Entom. Tidskr.* 116: 73–82.

Ebenman, B. and Persson, L. 1988. *Size-structured populations.* Wadsworth, Belmont, California.

Eber, S. and Brandl, R. 1994. Ecological and genetic spatial patterns of *Urophora cardui* (Diptera: Tephritidae) as evidence for population structure and biogeographical processes. *J. Anim. Ecol.* 63: 187–190.

Eber, S. and Brandl, R. 1996. Metapopulation dynamics of the tephritid fly *Urophora cardui*: an evaluation of incidence function model assumptions with field data. *J. Anim. Ecol.* 65: 621–630.

Edmunds, G. F. J. and Aalstad, D. N. 1978. Coevolution in insect herbivores and conifers. *Science* 199: 941–945.

Ehrlich, P. R. 1961. Intrinsic barriers to dispersal in checkerspot butterfly. *Science* 134: 108–109.

Ehrlich, P. R. 1965. The population biology of the butterfly, *Euphydryas editha*. II. The structure of the Jasper Ridge colony. *Evolution* 19: 327–336.

Ehrlich, P. R. 1983. Genetics and extinction of butterfly populations. In *Genetics and conservation—a reference for managing wild animal populations* (ed. C. M. Schonewald-Cox, S. M. Chambers, B. MacBryde and L. Thomas), pp. 152–163. Benjamin Cummings, Menlo Park, CA.

Ehrlich, P. R. 1984. The structure and dynamics of butterfly populations. In *The biology of butterflies* (ed. R. I. Vane-Wright and P. R. Ackery), pp. 25–40. Academic Press, London.

Ehrlich, P. R. 1986. Which animal will invade? In *Ecology of biological invasions of North America and Hawaii* (ed. H. A. Mooney and J. A. Drake), pp. 79–95. Springer, New York.

Ehrlich, P. R. 1992. Population biology of checkerspot butterflies and the preservation of global biodiversity. *Oikos* 63: 6–12.

Ehrlich, P. R. and Birch, L. C. 1967. The 'balance of nature' and 'population control'. *Am. Nat.* 101: 97–107.

Ehrlich, P. R. and Murphy, D. D. 1981. The population biology of butterflies (*Euphydryas*). *Biologisches Zentralblatt* 100: 613–629.

Ehrlich, P. R. and Murphy, D. D. 1987. Conservation lessons from long-term studies of checkerspot butterflies. *Cons. Biol.* 1: 122–131.

Ehrlich, P. R., White, R. R., Singer, M. C., McKechnie, S. W. and Gilbert, L. E. 1975. Checkerspot butterflies: a historical perspective. *Science* 188: 221–228.

Ehrlich, P. R., Murphy, D. D., Singer, M. C., Sherwood, C. B., White, R. R. and Brown, I. L. 1980. Extinction, reduction, stability and increase: the responses of checkerspot butterfly (*Euphydryas*) populations to the California drought. *Oecologia* 46: 101–105.

Eliasson, C. 1991. Occurrence and biology of *Euphydryas maturna* (L.) (Lepidoptera, Nymphalidae) in Central Sweden (in Swedish with English summary). *Entom. Tidskr.* 112: 113–120.

Eliasson, C. 1995. *Projekt Natfjärilar*. Unpublished report.

Ellner, S. and Turchin, P. 1995. Chaos in a noisy world: new methods and evidence from time-series analysis. *Am. Nat.* 145: 343–375.

Ellstrand, L. C. and Elam, D. R. 1993. Population genetic consequences of small population size: implications for plant conservation. *Ann. Rev. Ecol. Syst.* 24: 217–242.

Elton, C. 1958. The ecology of invasions by animals and plants. Chapman and Hall, London.

Enoksson, B., Angelstam, P. and Larsson, K. 1995. Deciduous trees and resident species—the problem of fragmentation within a coniferous forest landscape. *Landscape Ecology* 10: 267–275.

Eriksson, O. 1996. Regional dynamics of plants: a review of evidence for remnant, source–sink and metapopulations. *Oikos* 77: 248–258.

Errington, P. L. 1943. An analysis of mink predation upon muskrats in north-central United States. *Research Bull. Iowa Agr. Exper. Station* 320: 797–924.

Esch, G., Bush, A. and Aho, J. 1990. *Parasite communities: patterns and processes*. Chapman and Hall, London.

Fahrig, L. 1997. Relative effects of habitat loss and fragmentation on population extinction. *J. Wildlife Manage.* 61: 603–610.

Fahrig, L. and Merriam, G. 1994. Conservation of fragmented populations. *Cons. Biol.* 8: 50–59.

Falk, D. A. and Holsinger, K. E. 1991. *Genetics and conservation of rare plants.* Oxford University Press.

Feldman, M. W. 1989. *Mathematical evolutionary theory.* Princeton University Press.

Felsenstein, J. 1976. The theoretical population genetics of variable selection and migration. *Annual Reviews of Genetics* 10: 253–288.

Fiedler, P. L. and Jain, S. K. 1992. *Conservation biology: theory and practise of nature conservation, preservation and management.* Chapman and Hall, London.

Fischer, M. and Stöcklin, J. 1997. Local extinctions of plants in remnants of extensively used calcareous grasslands 1950–85. *Cons. Biol.* 11: 727–737.

Fisher, R. A. 1930. *The genetical theory of natural selection.* Oxford University Press.

Fisher, R. A. 1937. The wave of advance of advantageous genes. *Ann. Eugen. (London)* 7: 355–369.

Foley, P. 1994. Predicting extinction times from environmental stochasticity and carrying capacity. *Cons. Biol.* 8: 124–137.

Foley, P. 1997. Extinction models for local populations. In *Metapopulation biology* (ed. I. A. Hanski and M. E. Gilpin), pp. 215–246. Academic Press, San Diego.

Ford, E. B. 1945. *Butterflies.* Collins, London.

Ford, H. D. and Ford, E. B. 1930. Fluctuations in numbers and its influence on variation in *Melitaea aurinia. Trans. Entomol. Soc. London* 78: 345–351.

Forman, R. T. T. 1995. Some general principles of landscape and regional ecology. *Landscape Ecology* 10: 133–142.

Forman, R. T. T. and Godron, M. 1986. *Landscape ecology.* John Wiley and Sons, New York.

Forney, K. A. and Gilpin, M. E. 1990. Spatial structure and population extinction: a study with *Drosophila* flies. *Cons. Biol.* 3: 45–51.

Forsman, E. D., Meslow, E. C. and Wight, H. M. 1984a. Distribution and biology of the spotted owl in Oregon. *Wildlife Monographs* 87: 1–64.

Frank, K. and Wissel, C. 1998. Spatial aspects of metapopulation survival—from model results to rules of thumb for landscape management. *Landscape Ecology,* in press.

Frank, S. A. 1997. Spatial processes in host–parasite genetics. In *Metapopulation biology* (ed. I. A. Hanski and M. E. Gilpin), pp. 325–358. Academic Press, San Diego.

Frankel, O. H. and Soulé, M. E. 1981. *Conservation and evolution.* Cambridge University Press.

Frankham, R. 1995. Inbreeding and extinction: a threshold effect. *Cons. Biol.* 9: 792–799.

Frankham, R. 1996. Relationship of genetic variation to population size in wildlife. *Cons. Biol.* 10: 1500–1508.

Freedman, H. I. and Waltman, D. 1977. Mathematical models of population interactions with dispersal. I. Stability of two habitats with and without predator. *SIAM J. Appl. Math.* 32: 631–648.

Fretwell, S. D. 1975. The impact of Robert MacArthur on ecology. *Ann. Rev. Ecol. Syst.* 6: 1–13.

Fritz, R. S. 1979. Consequences of insular population structure: distribution and extinction of spruce grouse populations. *Oecologia* 42: 57–65.

Fulto, B. B. 1940. The hornworm parasite, *Apanteles congregatus* Say and the hyperparasite, *Hypoteromalus tabacum* (Fitch). *Ann. Entomol. Soc. Amer.* 33: 231–244.

Gabriel, W., Burger, R. and Lynch, M. 1991. Population extinction by mutational load and demographic stochasticity. In *Species conservation: a population biological approach* (ed. A. Seitz and V. Loeschcke), pp. 49–59. Birkhäuser, Basel.

Gadgil, M. 1971. Dispersal: population consequences and evolution. *Ecology* 52: 253–261.

Gaston, K. J. 1994. *Rarity*. Chapman and Hall, London.

Gaston, K. J. and Blackburn, T. M. 1995. Mapping biodiversity using surrogates for species richness: macro-scales and New World birds. *Proceedings of the Royal Society of London, Series B* 262: 335–341.

Gaston, K. J. and Lawton, J. H. 1987. A test of statistical techniques for detecting density dependence in sequential censuses of animal populations. *Oecologia* 74: 404–410.

Gaston, K. J. and Lawton, J. H. 1989. Insect herbivores on bracken do not support the core–satellite hypothesis. *Am. Nat.* 134: 761–777.

Gaston, K. J. and Lawton, J. H. 1990. Effects of scale and habitat on the relationship between regional distribution and local abundance. *Oikos* 58: 329–335.

Gaston, K. J. and McArdle, B. H. 1993. Measurement of variation in the size of populations in space and time: Some points of clarification. *Oikos* 68: 357–360.

Gaston, K. J. and McArdle, B. H. 1994. The temporal variability in animal abundances: Measures, methods and patterns. *Philosophical Transactions of the Royal Society, London B* 345: 335–358.

Gaston, K. J., Quinn, R. M., Wood, S. and Arnold, H. R. 1996. Measures of geographic range size: the effects of sample size. *Ecography* 19: 259–263.

Gaston, K. J., Blackburn, T. M. and Lawton, J. H. 1997. Interspecific abundance-range size relationships: an appraisal of mechanisms. *J. Anim. Ecol.* 66: 579–601.

Gerber, A. S. and Templeton, A. R. 1996. Population sizes and within-deme movement of *Trimerotropis saxatilis* (Acrididae), a grasshopper with a fragmented distribution. *Oecologia* 105: 343–350.

Gilbert, F. S. 1980. The equilibrium theory of island biogeography: fact or fiction? *Journal of Biogeography* 7: 209–235.

Gilbert, L. E. 1991. Biodiversity of a Central American *Heliconius* community: pattern, process, and problems. In *Plant–animal interactions: evolutionary ecology in tropical and temperate regions* (ed. P. W. Price, T. M. Lewinsohn, G. W. Fernandes and W. W. Benson), pp. 403–427. John Wiley and Sons, New York.

Gilbert, L. E. and Singer, M. C. 1973. Dispersal and gene flow in a butterfly species. *Am. Nat.* 107: 58–72.

Giles, B. E. and Goudet, J. 1997. A case study of genetic structure in a plant metapopulation. In *Metapopulation biology* (ed. I. A. Hanski and M. E. Gilpin), pp. 429–454. Academic Press, San Diego.

Gilligan, D. M., Woodworth, L. M., Montgomery, M. E., Briscoe, D. A. and Frankham, R. 1997. Is mutation accumulation a threat to the survival of endangered populations? *Cons. Biol.* 11: 1235–1241.

Gilpin, M. E. 1988. A comment on Quinn and Hastings: extinction in subdivided habitats. *Cons. Biol.* 2: 290–292.

Gilpin, M. E. 1990. Extinction of finite metapopulations in correlated environments. In *Living in a patchy environment* (ed. B. Shorrocks and I. R. Swingland), pp. 177–186. Oxford Science Publications.

Gilpin, M. E. and Diamond, J. M. 1976. Calculation of immigration and extinction curves from the species–area–distance relation. *Proc. Nat. Acad. Sci. USA* 73: 4130–4134.

Gilpin, M. E. and Diamond, J. M. 1980. Subdivision of nature reserves and the maintenance of species diversity. *Nature* 285: 567–568.

Gilpin, M. E. and Diamond, J. M. 1981. Immigration and extinction probabilities for individual species: relation to incidence functions and species colonization curves. *Proc. Nat. Acad. Sci. USA* 78: 392–396.

Gilpin, M. E. and Soulé, M. E. 1986. Minimum viable populations: process of species extinction. In *Conservation biology: the science of scarcity and diversity* (ed. M. E. Soulé), pp. 19–34. Sinauer, Sunderland, Massachusetts.

Gleason, H. A. 1922. On the relation of species and area. *Ecology* 3: 158–162.

Gleick, J. 1987. *Chaos*. Sphere Books Ltd, London.

Godfray, H. C. J. and Hassell, M. P. 1992. Long time series reveal density dependence. *Nature* 359: 673–674.

Gonzalez-Andujar, J. L. and Perry, J. N. 1993. Chaos, metapopulations and dispersal. *Ecol. Model.* 65: 255–263.

Goodall, D. W. 1952. Quantitative aspects of plant distribution. *Biological Reviews* 27: 194–245.

Goodman, D. 1987a. The demography of change extinction. In *Viable populations for conservation* (ed. M. E. Soulé), pp. 11–34. Cambridge University Press.

Goodman, D. 1987b. Consideration of stochastic demography in the design and management of biological reserves. *Natur. Res. Mod.* 1: 205–234.

Gotelli, N. J. and Simberloff, D. 1987. The distribution and abundance of tallgrass prairie plants: a test of the core–satellite hypothesis. *Am. Nat.* 130: 18–35.

Grant, P. R. 1970. Colonization of islands by ecologically dissimilar species of mammals. *Can. J. Zool.* 48: 545–553.

Greenwood, P. J. 1980. Mating systems, philopatry and dispersal in birds and mammals. *Anim. Beh.* 28: 1140–1162.

Greig-Smith, P. 1957. *Quantitative plant ecology*. Butterworth Scientific, London.

Grenfell, B. and Harwood, J. 1997. (Meta)population dynamics of infectious diseases. *Trends Ecol. Evol.* 12: 395–404.

Gressit, J. L. 1964. Insects on Campbell Island. *Pacific Insect Monographs* 7: 551–600.

Griffith, B., Scott, J. M., Carpenter, J. W. and Reed, C. 1989. Translocation as a species conservation tool: status and strategy. *Science* 245: 477–480.

Grubb, P. J. 1977. The maintenance of species richness in plant communities: the importance of the regeneration niche. *Biological Reviews* 52: 107–145.

Grubb, P. J. 1986. Problems posed by sparse and patchily distributed species in species-rich plant communities. In *Community ecology* (ed. J. Diamond and T. J. Case), pp. 207–125. Harper and Row, New York.

Grubb, P. J. 1988. The uncoupling of disturbance and recruitment, two kinds of seed bank, and persistence of plant populations at the regional and local scales. *Ann. Zool. Fenn.* 25: 23–36.

Gurney, W. S. C. and Nisbet, R. M. 1978. Single species population fluctuations in patchy environments. *Am. Nat.* 112: 1075–1090.

Gustafson, E. J. and Parker, G. R. 1992. Relationship between landcover proportion and indices of landscape spatial pattern. *Landscape Ecology* 7: 101–110.

Gutierrez, R. J. and Harrison, S. 1996. Applications of metapopulation theory to spotted owl management: a history and critique. In *Metapopulations and wildlife conservation management* (ed. C. McCullough), pp. 167–185. Island Press, Covelo.

Gutierrez, R. J., Franklin, A. B. and Lahaye, W. S. 1995. Spotted owl (*Strix occidentalis*). *Birds of North America* 179: 1–28.

Gyllenberg, M. and Hanski, I. 1992. Single-species metapopulation dynamics: a structured model. *Theor. Pop. Biol.* 72: 35–61.

Gyllenberg, M. and Hanski, I. 1997. Habitat deterioration, habitat destruction and metapopulation persistence in a heterogeneous landscape. *Theor. Pop. Biol.* 52: 198–215.

Gyllenberg, M. and Silvestrov, D. S. 1994. Quasi-stationary distribution of a stochastic metapopulation model. *J. Math. Biol.* 33: 35–70.

Gyllenberg, M., Söderbacka, G. and Ericsson, S. 1993. Does migration stabilize local population dynamics? Analysis of a discrete metapopulation model. *Math. Biosci.* 118: 25–49.

Gyllenberg, M., Osipov, A. V. and Söderbacka, G. 1996. Bifurcation analysis of metapopulation model with sources and sinks. *Journal of Nonlinear Science* 6: 329–366.

Gyllenberg, M., Hanski, I. and Hastings, A. 1997. Structured metapopulation models. In *Metapopulation biology* (ed. I. A. Hanski and M. E. Gilpin), pp. 93–122. Academic Press, San Diego.

Haas, C. A. 1995. Dispersal and use of corridors by birds in wooded patches of agricultural landscape. *Cons. Biol.* 9: 845–854.

Haccou, P. and Iwasa, Y. 1996. Establishment probability in fluctuating environments: a branching process model. *Theor. Pop. Biol.* 50: 254–280.

Haldane, J. B. S. 1939. The equilibrium between mutation and random extinction. *Annals of Eugenics* 9: 400–405.

Hamilton, W. D. 1963. The evolution of altruistic behaviour. *Am. Nat.* 97: 354–356.

Hamilton, W. D. 1964. The genetical evolution of social behaviour. I. *J. Theor. Biol.* 7: 1–16.

Hamilton, W. D. 1967. Extraordinary sex ratios. *Science* 156: 477–488.

Hamilton, W. D. and May, R. M. 1977. Dispersal in stable habitats. *Nature* 269: 578–581.

Hanski, I. 1981. Coexistence of competitors in patchy environment with and without predation. *Oikos* 37: 306–312.

Hanski, I. 1982a. Dynamics of regional distribution: the core and satellite species hypothesis. *Oikos* 38: 210–221.

Hanski, I. 1982b. Distributional ecology of anthropochorous plants in villages surrounded by forest. *Ann. Bot. Fenn.* 19: 1–15.

Hanski, I. 1982c. Structure in bumblebee communities. *Ann. Zool. Fenn.* 19: 319–326.

Hanski, I. 1983a. Coexistence of competitors in patchy environment. *Ecology* 64: 493–500.

Hanski, I. 1983b. Distributional ecology and abundance of dung and carrion-feeding beetles (Scarabaeidae) in tropical rain forests in Sarawak, Borneo. *Acta Zool. Fennica* 167: 1–45.

Hanski, I. 1985. Single-species spatial dynamics may contribute to long-term rarity and commonness. *Ecology* 66: 335–343.

Hanski, I. 1986. Population dynamics of shrews on small islands accord with the equilibrium theory. *Biol. J. Linn. Soc.* 28: 23–36.

Hanski, I. 1987. Carrion fly community dynamics: patchiness, seasonality and coexistence. *Ecol. Entomol.* 12: 257–266.

Hanski, I. 1989. Does it help to have more of the same? *Trends Ecol. Evol.* 4: 113–114.

Hanski, I. 1990. Density dependence, regulation and variability in animal populations. *Philosophical Transactions of the Royal Society, London B* 330: 141–150.

Hanski, I. 1991a. Single-species metapopulation dynamics: Concepts, models and observations. *Biol. J. Linn. Soc.* 42: 17–38.

Hanski, I. 1991b. Reply to Nee, Gregory and May. *Oikos* 62: 88–89.

Hanski, I. 1992. Inferences from ecological incidence functions. *Am. Nat.* 139: 657–662.

Hanski, I. 1993. Dynamics of small mammals on islands. *Ecography* 16: 372–375.

Hanski, I. 1994a. A practical model of metapopulation dynamics. *J. Anim. Ecol.* 63: 151–162.

Hanski, I. 1994b. Spatial scale, patchiness and population dynamics on land. *Philosophical Transactions of the Royal Society, London B* 343: 19–25.

Hanski, I. 1994c. Patch-occupancy dynamics in fragmented landscapes. *Trends Ecol. Evol.* 9: 131–135.

Hanski, I. 1996. Metapopulation ecology. In *Population dynamics in ecological space and time* (ed. O. E. Rhodes, R. K. Chesser and M. H. Smith), pp. 13–43. Chicago University Press.

Hanski, I. 1998. Connecting the parameters of local extinction and metapopulation dynamics. *Oikos* 83: 390–396.

Hanski, I. and Cambefort, Y. 1991. *Ecology of dung beetles*. Princeton University Press.

Hanski, I. and Gilpin, M. E. 1997. *Metapopulation biology: ecology, genetics, and evolution*. Academic Press, San Diego.

Hanski, I. and Gyllenberg, M. 1993. Two general metapopulation models and the core–satellite species hypothesis. *Am. Nat.* 142: 17–41.

Hanski, I. and Gyllenberg, M. 1997. Uniting two general patterns in the distribution of species. *Science* 275: 397–400.

Hanski, I. and Hammond, P. 1995. Biodiversity in boreal forests. *Trends Ecol. Evol.* 10: 5–6.

Hanski, I. and Kuitunen, J. 1986. Shrews on small islands: epigenetic variation elucidates population stability. *Hol. Ecol.* 9: 193–204.

Hanski, I. and Kuussaari, M. 1995. Butterfly metapopulation dynamics. In *Population dynamics: New approaches and synthesis* (ed. N. Cappuccino and P. W. Price), pp. 149–171. Academic Press, London.

Hanski, I. and Ranta, E. 1983. Coexistence in a patchy environment: three species of *Daphnia* in rock pools. *J. Anim. Ecol.* 52: 263–279.

Hanski, I. and Simberloff, D. 1997. The metapopulation approach, its history, conceptual domain and application to conservation. In *Metapopulation biology* (ed. I. A. Hanski and M. E. Gilpin), pp. 5–26. Academic Press, San Diego.

Hanski, I. and Thomas, C. D. 1994. Metapopulation dynamics and conservation: a spatially explicit model applied to butterflies. *Biol. Conserv.* 68: 167–180.

Hanski, I. and Woiwod, I. P. 1993a. Mean-related stochasticity and population variability. *Oikos* 67: 29–39.

Hanski, I. and Woiwod, I. P. 1993b. Spatial synchrony in the dynamics of moth and aphid populations. *J. Anim. Ecol.* 62: 656–668.

Hanski, I. and Zhang, D.-Y. 1993. Migration, metapopulation dynamics and fugitive coexistence. *J. Theor. Biol.* 163: 491–504.

Hanski, I., Turchin, P., Korpimäki, E. and Henttonen, H. 1993a. Population oscillations of boreal rodents: regulation by mustelid predators leads to chaos. *Nature* 364: 232–235.

Hanski, I., Woiwod, I. P. and Perry, J. 1993b. Density dependence, population persistence and largely futile arguments. *Oecologia* 95: 595–598.

Hanski, I., Kouki, J. and Halkka, A. 1993c. Three explanations of the positive relationship between distribution and abundance of species. In *Community diversity: historical and geographical perspectives* (ed. R. E. Ricklefs and D. Schluter), pp. 108–116. Chicago University Press.

Hanski, I., Kuussaari, M. and Nieminen, M. 1994. Metapopulation structure and migration in the butterfly *Melitaea cinxia*. *Ecology* 75: 747–762.

Hanski, I., Pakkala, T., Kuussaari, M. and Lei, G. 1995a. Metapopulation persistence of an endangered butterfly in a fragmented landscape. *Oikos* 72: 21–28.

Hanski, I., Pöyry, J., Kuussaari, M. and Pakkala, T. 1995b. Multiple equilibria in metapopulation dynamics. *Nature* 377: 618–621.

Hanski, I., Foley, P. and Hassell, M. P. 1996a. Random walks in a metapopulation: how much density dependence is necessary for long-term persistence? *J. Anim. Ecol.* 65: 274–282.

Hanski, I., Moilanen, A. and Gyllenberg, M. 1996b. Minimum viable metapopulation size. *Am. Nat.* 147: 527–541.

Hanski, I., Moilanen, A., Pakkala, T. and Kuussaari, M. 1996c. The quantitative incidence function model and persistence of an endangered butterfly metapopulation. *Cons. Biol.* 10: 578–590.

Hanski, I., Alho, J. and Moilanen, A. 1999. Estimating the parameters of survival and migration of individuals in metapopulations. *Ecology*, in press.

Hansson, L. 1991. Dispersal and connectivity in metapopulations. *Biol. J. Linn. Soc.* 42: 89–103.

Hansson, L. and Henttonen, H. 1985. Gradients in density variations of small rodents: the importance of latitude and snow cover. *Oecologia* 67: 394–402.

Hardin, G. 1960. The competitive exclusion principle. *Science* 131: 1292–1297.

Harms, B. and Opdam. P, 1990. Woods as habitat patches for birds: Application in landscape planning in The Netherlands. In *Changing landscapes: An ecological perspective* (ed. I. S. Zonneveld and R. T. T. Forman), pp. 73–97. Springer, Berlin.

Harper, J. L. 1969. The role of predation in vegetational diversity. In *Diversity and stability in ecological systems. Brookhaven symposium in biology 22*, pp. 48–62.

Harris, L. D. 1988. Edge effect and conservation of biotic diversity. *Cons. Biol.* 2: 330–332.

Harris, R. B., Metzger, L. H. and Bevin, C. D. 1986. *GAPPS. Version 3.0 Montana co-operative research unit*. University of Montana, Missoula.

Harrison, R. G. 1980. Dispersal polymorphism in insects. *Ann. Rev. Ecol. Syst.* 11: 95–118.

Harrison, S. 1989. Long-distance dispersal and colonization in the Bay checkerspot butterfly, *Euphydryas editha bayensis*. *Ecology* 70: 1236–1243.

Harrison, S. 1991. Local extinction in a metapopulation context: an empirical evaluation. *Biol. J. Linn. Soc.* 42: 73–88.

Harrison, S. 1994. Metapopulations and conservation. In *Large-scale ecology and conservation biology* (ed. P. J. Edwards, R. M. May and N. R. Webb), pp. 111–128. Blackwell Scientific Press, Oxford.

Harrison, S. 1997. Persistent, local outbreaks in the western tussock moth (*Orgyia vetusta*): roles of resource quality, predation and poor dispersal. *Ecol. Entomol.* 22: 158–166.

Harrison, S. and Hastings, A. 1996. Genetic and evolutionary consequences of metapopulation structure. *Trends Ecol. Evol.* 11: 180–183.

Harrison, S. and Quinn, J. F. 1990. Correlated environments and the persistence of metapopulations. *Oikos* 56: 293–298.

Harrison, S. and Taylor, A. D. 1997. Empirical evidence for metapopulation dynamics. In *Metapopulation biology* (ed. I. A. Hanski and M. E. Gilpin), pp. 27–42. Academic Press, San Diego.

Harrison, S. and Wilcox, C. 1995. Evidence that predator satiation may restrict the spatial spread of a tussock moth (*Orgyia vetusta*), a flightless defoliator. *Oecologia* 10199: 309–316.

Harrison, S., Murphy, D. D. and Ehrlich, P. R. 1988. Distribution of the Bay checkerspot butterfly, *Euphydryas editha bayensis*: evidence for a metapopulation model. *Am. Nat.* 132: 360–382.

Harrison, S., Quinn, J. F., Baughman, J. F., Murphy, D. D. and Ehrlich, P. R. 1991. Estimating the effects of scientific study on two butterfly populations. *Am. Nat.* 137: 227–243.

Harrison, S., Doak, D. F. and Stahl, A. 1993. Spatial models and spotted owls: exploring some biological issues behind recent events. *Cons. Biol.* 7: 950–953.

Harrison, S., Thomas, C. D. and Lewinsohn, T. M. 1995. Testing a metapopulation model of coexistence in the insect community on ragwort (*Senecio jacobaea*). *Am. Nat.* 145: 545–561.

Hassell, M. P. 1978. *The dynamics of arthropod predator–prey systems*. Princeton University Press.

Hassell, M. P. and May, R. M. 1985. From individual behaviour to population dynamics. In *Behavioural ecology* (ed. R. M. Sibly and R. H. Smith), pp. 3–32. Blackwell, Oxford.

Hassell, M. P. and Pacala, S. W. 1990. Heterogeneity and the dynamics of host–parasitoid interactions. In *Population regulation and dynamics* (ed. M. P. Hassell and R. M. May), pp. 81–98. Royal Society, London.

Hassell, M. P., Latto, J. and May, R. M. 1989. Seeing the wood for the trees: detecting density dependence from existing life-table studies. *J. Anim. Ecol.* 58: 883–892.

Hassell, M. P., Comins, H. N. and May, R. M. 1991. Spatial structure and chaos in insect population dynamics. *Nature* 353: 255–258.

Hassell, M. P., Miramontes, O., Rohani, P. and May, R. M. 1995. Appropriate formulations for dispersal in spatially structured models. *J. Anim. Ecol.* 64: 662–664.

Hastings, A. 1983. Can spatial variation alone lead to selection for dispersal? *Theor. Pop. Biol.* 24: 244–251.

Hastings, A. 1991. Structured models of metapopulation dynamics. In *Metapopulation dynamics* (ed. M. Gilpin and I. Hanski), pp. 57–71. Academic Press, London.

Hastings, A. 1992. Age-dependent dispersal is not a simple process: density dependence, stability, and chaos. *Theor. Pop. Biol.* 41: 388–400.

Hastings, A. 1993. Complex interactions between dispersal and dynamics—lessons from coupled logistic equations. *Ecology* 74: 1362–1372.

Hastings, A. and Harrison, S. 1994. Metapopulation dynamics and genetics. *Ann. Rev. Ecol. Syst.* 25: 167–188.

Hastings, A. and Higgins, K. 1994. Persistence of transients in spatially structured models. *Science* 263: 1133–1136.

Haukisalmi, V. and Henttonen, H. 1994. Distribution patterns and microhabitat segregation in gastrointestinal helminths of *Sorex* shrews. *Oecologia* 97: 236–242.

Hedrick, P. W. 1983. *Genetics of populations*. Science Book International, Boston.

Hedrick, P. W. 1986. Genetic polymorphism in heterogenous environments: a decade later. *Ann. Rev. Ecol. Syst.* 17: 535–566.

Hedrick, P. W. and Gilpin, M. E. 1997. Genetic effective size of a metapopulation. In *Metapopulation biology* (ed. I. A. Hanski and M. E. Gilpin), pp. 166–182. Academic Press, San Diego.

Hedrick, P. W., Lacy, R. C., Allendorf, F. W. and Soulé, M. E. 1996. Directions in conservation biology: comments on Caughley. *Cons. Biol.* 10: 1312–1320.

Heikkilä, J., Below, A. and Hanski, I. 1994. Synchronous dynamics of microtine rodent populations on islands in Lake Inari in northern Fennoscandia: evidence for regulation by mustelid predators. *Oikos* 70: 245–252.

Hendry, R. J. and McGlade, J. M. 1995. The role of memory in ecological systems. *Proceedings of the Royal Society of London B* 259: 153–159.

Hengeveld, R. 1990. *Dynamic biogeography*. Cambridge University Press.

Hengeveld, R. and Haeck, J. 1982. The distribution of abundance. I. Measurements. *Journal of Biogeography* 9: 303–316.

Hering, F. 1995. Habitat patches of the threatened butterfly species *Melitaea cinxia* (L.) on the Åland islands, Finland: vegetation characteristics and caterpillar–host plant interactions. M.Sc. Dissertation, Münster, Germany.

Herzig, A. L. 1995. Effects of population density on long-distance dispersal in the goldenrod beetle *Trirhabda virgata*. *Ecology* 76: 2044–2054.

Heschel, M. S. and Paige, K. N. 1995. Inbreeding depression, environmental stress, and population size variation in Scarlet Gilia (*Ipomopsis aggregata*). *Cons. Biol.* 9: 126–133.

Hess, G. 1996. Disease in metapopulation models: implications for conservation. *Ecology* 77: 1617–1632.

Hesse, R., Allee, W. C. and Schmidt, K. P. 1951. *Ecological animal geography*. Chapman and Hall, New York.

Hestbeck, J. B., Nichols, J. D. and Malecki, R. A. 1991. Estimates of movement and site fidelity using mark–resight data of wintering Canada geese. *Ecology* 72: 523–533.

Higgins, L. G. 1981. A revision of *Phycoides* Hübner and related genera, with a review of the classification of the Melitaeinae (Lepidoptera: Nymphalidae). *Bulletin of the British Museum of Natural History* 43: 77–243.

Higgs, A. J. and Usher, M. B. 1980. Should nature reserves be large or small? *Nature* 285: 568–569.

Hilborn, R. 1975. Similarities in dispersal tendency among siblings in four species of voles (*Microtus*). *Ecology* 56: 1221–1225.

Hill, C. J. 1995. Linear strips of rainforest vegetation as potential dispersal corridors for rainforest insects. *Cons. Biol.* 9: 1559–1566.

Hill, J. K., Thomas, C. D. and Lewis, O. T. 1996. Effects of habitat patch size and isolation on dispersal by *Hesperia comma* butterflies: implications for metapopulation structure. *J. Anim. Ecol.* 65: 725–735.

Hjermann, D. O. and Ims, R. A. 1996. Landscape ecology of the wart-biter *Decticus verrucivorus* in a patchy landscape. *J. Anim. Ecol.* 65: 768–780.

Holmes, E. E., Lewis, M. A., Banks, J. E. and Veit, R. R. 1994. Partial differential equations in ecology: spatial interactions and population dynamics. *Ecology* 75: 17–29.

Holt, R. D. 1977. Predation, apparent competition, and the structure of prey communities. *Theor. Pop. Biol.* 12: 197–229.

Holt, R. D. 1984. Spatial heterogeneity, indirect interactions, and the coexistence of prey species. *Am. Nat.* 124: 377–406.

Holt, R. D. 1985. Population dynamics in two-patch environments: some anomalous consequences of optimal habitat distribution. *Theor. Pop. Biol.* 28: 181–208.

Holt, R. D. 1992. A neglected facet of island biogeography: The role of internal spatial dynamics in area effects. *Theor. Popul. Biol.* 41: 354–371.

Holt, R. D. 1993. Ecology at the mesoscale: The influence of regional processes on local communities. In *Species diversity in ecological communities* (ed. R. E. Ricklefs and D. Schluter), pp. 77–88. University of Chicago Press.

Holt, R. D. 1995. Demographic constraints in evolution: towards unifying the evolutionary theories of senescence and the niche conservatism. *Evol. Ecology* 10: 1–11.

Holt, R. D. 1996. Adaptive environments in source–sink environments: direct and indirect effects of density dependence on niche evolution. *Oikos* 75: 182–192.

Holt, R. D. 1997. From metapopulation dynamics to community structure: some consequences of spatial heterogeneity. In *Metapopulation biology* (ed. I. A. Hanski and M. E. Gilpin), pp. 149–165. Academic Press, San Diego.

Holt, R. D. and Gaines, M. S. 1992. Analysis of adaptation in heterogeneous landscapes: implications for the evolution of fundamental niches. *Evol. Ecology* 6: 433–447.

Holt, R. D. and Hassell, M. P. 1993. Environmental heterogeneity and the stability of host–parasitoid interactions. *J. Anim. Ecol.* 62: 89–100.

Holt, R. D. and McPeek, M. A. 1996. Chaotic population dynamics favors the evolution of dispersal. *Am. Nat.* 148: 709–718.

Holyoak, M. and Lawler, S. P. 1996. Persistence of an extinction-prone predator–prey interaction through metapopulation dynamics. *Ecology* 77: 1867–1879.

Horn, H. S. and MacArthur, R. H. 1972. Competition among fugitive species in a harlequin environment. *Ecology* 53: 749–752.

Houston, A. I. and McNamara, J. M. 1992. Phenotypic plasticity as a state-dependent life-history decision. *Evol. Ecology* 6: 243–253.

Howe, H. F. and Westley, L. C. 1986. Ecology of pollination and seed dispersal. In *Plant ecology* (ed. M. J. Crawley), pp. 185–216. Blackwell, Oxford.

Hubbell, S. P. and Foster, R. B. 1986. Biology, chance, and history and the structure of tropical rain forest tree communities. In *Community ecology* (ed. J. Diamond and T. J. Case), pp. 314–329. Harper and Row, New York.

Huffaker, C. B. 1958. Experimental studies on predation: dispersion factors and predator–prey oscillations. *Hilgardia* 27: 343–383.

Husband, B. C. and Barrett, S. C. 1996. A metapopulation perspective in plant population biology. *J. Ecol.* 84: 461–469.

Huston, M. 1979. A general hypothesis of species diversity. *Am. Nat.* 113: 81–101.

Hutchinson, G. E. 1951. Copepodology for the ornithologist. *Ecology* 32: 571–577.

Hutchinson, G. E. 1957. Concluding remarks. *Cold Spring Harbour Symposium on Quantitative Biology* 22: 415–427.

Hutchinson, G. E. 1978. *An introduction to population ecology.* Yale University Press, New Haven.

Ims, R. A. and Steen, H. 1990. Geographical synchrony in microtine population cycles: a theoretical evaluation of the role of nomadic avian predators. *Oikos* 57: 381–387.

Ims, R. A. and Yoccoz, N. G. 1997. Studying transfer processes in metapopulations: emigration, migration, and colonization. In *Metapopulation biology* (ed. I. A. Hanski and M. E. Gilpin), pp. 247–266. Academic Press, San Diego.

Ims, R. A., Rolstad, J. and Wegge, P. 1993. Predicting space use responses to habitat fragmentation: can voles *Microtus oeconomus* serve as an experimental model system (EMS) for capercaillie grouse *Tetrao urogallus* in boreal forest. *Biol. Cons.* 63: 261–268.

Inghe, O. 1989. Genet and ramet survivorship under different mortality regimes—a cellular automata model. *J. Theor. Biol.* 138: 257–270.

Ingvarsson, P. K. 1997. The effect of delayed population growth on the genetic differentiation of local populations subject to frequent extinctions and recolonisations. *Evolution* 51: 29–35.

Ingvarsson, P. K., Olsson, K. and Ericsson, L. 1997. Extinction–recolonization dynamics in the mycophagous beetle *Phalacrus substriatus*. *Evolution* 51: 187–195.

Ives, A. R. 1988. Aggregation and the coexistence of competitors. *Ann. Zool. Fennici* 25: 75–88.

Ives, A. R. 1991. Aggregation and coexistence in a carrion fly community. *Ecol. Monogr.* 61: 75–94.

Ives, A. R. and May, R. M. 1985. Competition within and between species in a patchy environment: relations between microscopic and macroscopic models. *J. Theor. Biol.* 115: 65–92.

Ives, A. R. and Settle, W. H. 1997. Metapopulation dynamics and pest control in agricultural systems. *Am. Nat.* 149: 220–246.

Jain, S. K. 1976. The evolution of inbreeding in plants. *Ann. Rev. Ecol. Syst.* 7: 469–495.

Järvinen, A. and Väisänen, R. A. 1984. Reproduction of pied flycatchers (*Ficedula hypoleuca*) in good and bad breeding seasons in a northern marginal area. *Auk* 101: 439–450.

Järvinen, O. and Ranta, E. 1987. Patterns and processes in species assemblages on northern Baltic islands. *Ann. Zool. Fenn.* 24: 249–266.

Järvinen, O. and Vepsäläinen, K. 1976. Wing dimorphism as an adaptive strategy in water striders (*Gerris*). *Hereditas* 84: 61–68.

Jiménez, J. A., Hughes, K. A., Alaks, G., Graham, L. and Lacy, R. C. 1994. An experimental study of inbreeding depression in a natural habitat. *Science* 266: 271–273.

Johnson, C. G. 1969. *Insect migration and dispersal by flight.* Methuen, London.

Johnson, M. L. and Gaines, M. S. 1990. Evolution of dispersal: theoretical models and empirical tests using birds and mammals. *Ann. Rev. Ecol. Syst.* 21: 449–480.

Johst, K. and Wissel, C. 1997. Extinction risk in a temporally correlated fluctuating environment. *Theor. Pop. Biol.* 52: 91–100.

Jones, H. L. and Diamond, J. M. 1976. Short-time-base studies of turnover in breeding bird populations on the California Channel Islands. *Condor* 78: 526–549.

de Jong, G. 1979. The influence of the distribution of juveniles over patches of food on the dynamics of a population. *Neth. J. Zool.* 29: 33–51.

Jordan III, W. R., Gilpin, M. E. and Aber, J. D. 1987. *Restoration ecology: a synthetic approach to ecological research.* Cambridge University Press.

Kaitala, A. 1988. Wing muscle dimorphism: two reproductive pathways of water strider *Gerris thoracicus* in relation to habitat instability. *Oikos* 53: 222–228.

Kaneko, K. 1992. Overview of coupled map lattices. *Chaos* 2: 279–282.

Kaneko, K. 1993. The coupled map lattice: introduction, phenomenology, Lyapunov analysis, thermodynamics and applications. In *Theory and application of coupled map lattices* (ed. K. Kaneko), pp. 1–49. John Wiley, Chichester.

Karban, R. 1989. Fine-scale adaptation of herbivorous thrips in individual host plants. *Nature* 340: 60–61.

Kareiva, P. 1983. Influence of vegetation texture on herbivore populations: resource concentration and herbivore movements. In *Variable plants and herbivores in natural and managed systems* (ed. R. F. Denno and M. S. McClure), pp. 259–290. Academic Press, New York.

Kareiva, P. 1985. Finding and loosing plants by *Phyllotreta*: patch size and surrounding habitat. *Ecology* 66: 1810–1816.

Kareiva, P. 1990. Population dynamics in spatially complex environments: theory and data. In *Population regulation and dynamics* (ed. M. P. Hassell and R. M. May), pp. 53–68. Royal Society, London.

Kareiva, P., Skelly, D. and Ruckelshaus, M. 1997. Reevaluating the use of models to predict the consequences of habitat loss and fragmentation. In *The ecological basis of conservation* (ed. A. T. Pickett, R. S. Ostfeld, M. Schachak and G. E. Likens), pp. 156–166. Chapman and Hall, New York.

Kawecki, T. J. 1993. Age and size at maturity in a patchy environment: fitness maximization versus evolutionary stability. *Oikos* 66: 309–317.

Kawecki, T. J. 1995. Demography of source–sink populations and the evolution of ecological niches. *Evol. Ecol.* 9: 38–44.

Kawecki, T. J. and Stearns, S. C. 1993. The evolution of life histories in spatially heterogeneous environments: optimal reaction norms revisited. *Evol. Ecology* 7: 155–174.

Keddy, P. A. 1981. Experimental demography of the sand dune annual *Cakile edentula*, growing along an environmental gradient in Nova Scotia. *J. Ecol.* 69: 615–630.

Keeling, M. J. and Grenfell, B. T. 1997. Disease extinction and community size: modeling the persistence of measles. *Science* 275: 65–67.

Keller, L. F., Arcese, P., Smith, J. N. M., Hochachka, W. M. and Stearns, S. C. 1994. Selection against inbred song sparrows during a natural population bottleneck. *Nature* 372: 356–357.

Kermack, W. O. and McKendrick, A. G. 1927. A contribution to the mathematical theory of epidemics. *Proceedings of the Royal Society of London A* 155: 700–721.

Kimura, M. and Ohta, T. 1971. *Theoretical aspects of population genetics.* Princeton University Press.

Kindvall, O. 1995. The impact of extreme weather on habitat preference and survival in a metapopulation of the bush cricket *Metrioptera bicolor* in Sweden. *Biol. Cons.* 73: 51–58.

Kindvall, O. 1996a. Ecology of the bush cricket *Metrioptera bicolor* with implications for metapopulation theory and conservation. Ph.D. Dissertation, Uppsala University, Sweden.

Kindvall, O. 1996b. Habitat heterogeneity and survival in a bush cricket metapopulation. *Ecology* 77: 207–214.

Kindvall, O. and Ahlén, I. 1992. Geometrical factors and metapopulation dynamics of the bush cricket, *Metrioptera bicolor* Philippi (Orthoptera: Tettigoniidae). *Cons. Biol.* 6: 520–529.

Kingland, S. 1985. *Modeling nature.* University of Chicago Press.

Kirkpatrick, M. and Barton, N. H. 1997. Evolution of a species' range. *Am. Nat.* 150: 1–23.

Kleiman, D. G. 1989. Reintroduction of captive mammals for conservation. *BioScience* 39: 152–161.

Klemetti, T. 1998. Punakeltaverkkoperhosen (*Euphydryas aurinia*) metapopulaatiorakenne. M.Sc. Thesis, University of Turku, Finland.

Kolasa, J. and Pickett, S. T. A. 1990. *Ecological heterogeneity.* Springer, New York.

Komonen, A. 1997. Uhanalaisten kirjo- ja punakeltaverkkoperhosen loiskillan rakenne. M.Sc. Thesis, University of Helsinki, Finland.

Kot, M. 1989. Diffusion-driven period-doubling bifurcations. *BioSystems* 22: 279–287.

Kot, M. and Schaffer, W. M. 1986. Discrete-time growth–dispersal models. *Math. Biosc.* 80: 109–136.

Kot, M., Lewis, M. A. and van den Drieschke, P. 1996. Dispersal data and the spread of invading organisms. *Ecology* 77: 2027–2042.

Krebs, C. J. 1994. Ecology. *The experimental analysis of distribution and abundance.* Harper & Row, New York.

Krebs, C. J., Keller, B. L. and Tamarin, R. H. 1969. *Microtus* population biology: demographic changes in fluctuating populations of *M. ochrogaster* and *M. pennsylvanicus* in southern Indiana. *Ecology* 50: 557–607.

Krebs, J. R. and Davies, N. B. 1984. *Behavioral ecology: an evolutionary approach.* Blackwell, Oxford.

Kruess, A. and Tscharntke, T. 1994. Habitat fragmentation, species loss, and biological control. *Science* 264: 1581–1584.

Kuhn, T. S. 1970. *The structure of scientific revolutions*. University of Chicago Press.

Kuno, E. 1981. Dispersal and the persistence of populations in unstable habitats: a theoretical note. *Oecologia* 49: 123–126.

Kuussaari, M. 1998. Biology of the Glanville fritillary butterfly (*Melitaea cinxia*). Ph.D. Dissertation, University of Helsinki, Finland.

Kuussaari, M., Nieminen, M., Pöyry, J. and Hanski, I. 1995. Life history and distribution of the Glanville fritillary *Melitaea cinxia* (Nymphalidae) in Finland. *Baptria* 20: 167–180.

Kuussaari, M., Nieminen, M. and Hanski, I. 1996. An experimental study of migration in the butterfly *Melitaea cinxia*. *J. Anim. Ecol.* 65: 791–801.

Kuussaari, M., Saccheri, I., Camara, M. and Hanski, I. 1998. Allee effect and population dynamics in the Glanville fritillary butterfly. *Oikos* 82: 384–392.

Kwan, W. and Whitmore, T. C. 1971. On the influence of soil properties on species distribution in a Malayan lowland Dipterocarp rain forest. *Malayan Forest* 33: 42–54.

Lack, D. 1954. *The natural regulation of animal numbers*. Clarendon Press, Oxford.

Lacy, R. C. 1993. VORTEX: A computer population simulation model for population viability analysis. *Wildlife Research* 20: 45–65.

Lahaye, W. S., Gutierrez, R. J. and Akçakaya, H. R. 1994. Spotted owl metapopulation dynamics in Southern California. *J. Anim. Ecol.* 63: 775–785.

Lamberson, R. H., McKelvey, R., Noon, B. R. and Voss, C. 1992. A dynamic analysis of northern spotted owl viability in a fragmented forest landscape. *Cons. Biol.* 6: 505–512.

Lamberson, R. H., McKelvey, R., Noon, B. R. and Voss, C. 1994. The effects of varying dispersal capabilities on the population dynamics of the northern spotted owl. *Cons. Biol.* 8: 185–195.

Lambrechts, M. M. and Dias, P. C. 1993. Differences in the onset of laying between island and mainland Mediterranean blue tits *Parus caeruleus*: phenotypic plasticity or genetic differences. *Ibis* 135: 451–455.

Lamont, B. R., Klinkhamer, P. G. L. and Witkowski, E. T. F. 1993. Population fragmentation may reduce fertility to zero in *Banksia goodii*: a demonstration of the Allee effect. *Oecologia* 94: 446–450.

Lande, R. 1979. Effective deme sizes during long-term evolution estimated from rates of chromosomal rearrangement. *Evolution* 33: 234–251.

Lande, R. 1987. Extinction thresholds in demographic models of territorial populations. *Am. Nat.* 130: 624–635.

Lande, R. 1988a. Demographic models of the northern spotted owl (*Strix occidentallis caurina*). *Oecologia* 75: 601–607.

Lande, R. 1988b. Genetics and demography in biological conservation. *Science* 241: 1455–1460.

Lande, R. 1993. Risks of population extinction from demographic and environmental stochasticity and random catastrophes. *Am. Nat.* 142: 911–927.

Lande, R. 1995. Risk of population extinction from fixation of new deleterious mutations. *Evolution* 48: 1460–1469.

Lande, R. and Barrowclough, G. F. 1987. Effective population size, genetic variation, and their use in population management. In *Viable populations for conservation* (ed. M. E. Soulé), pp. 87–123. Chicago University Press.

Lande, R. and Orzack, S. H. 1988. Extinction dynamics of age-structured populations in a fluctuating environment. *Proc. Nat. Acad. Sci. USA* 85: 7418–7421.

Lande, R., Engen, S. and Saether, B. E. 1998. Extinction times in finite metapopulation models with stochastic local dynamics. *Oikos* in press.

Larsen, K. W. and Boutin, S. 1994. Movements, survival, and settlement of red squirrel (*Tamiasciurus hudsonicus*) offspring. *Ecology* 75: 214–223.

Laurance, W. F. 1991. Ecological correlates of extinction proneness in Australian tropical rain forest mammals. *Cons. Biol.* 5: 79–89.

Lavorel, S., Gardner, R. H. and O'Neill, R. V. 1993. Analysis of patterns in hierarchically structured landscapes. *Oikos* 67: 521–528.

Lawton, J. H. 1993. Range, population abundance and conservation. *Trends Ecol. Evol.* 8: 409–413.

Lawton, J. H. and May, R. M. 1983. The birds of Selborne. *Nature* 306: 732–733.

Lawton, J. H. and May, R. M. 1994. *Extinction rates.* Oxford University Press.

Lawton, J. H. and Woodroffe, G. L. 1991. Habitat and the distribution of water voles: why are there gaps in a species' range? *J. Anim. Ecol.* 60: 79–91.

Lawton, J. H., Nee, S., Letcher, A. J. and Harvey, P. H. 1994. Animal distributions: pattern and process. In *Large-scale ecology and conservation biology* (ed. P. J. Edwards, R. M. May and N. R. Webb), pp. 41–58. Blackwell Scientific Press, Oxford.

Lei, G. C. 1997. Metapopulation dynamics of host–parasitoid interactions. Ph.D. Dissertation, University of Helsinki, Finland.

Lei, G. and Hanski, I. 1997. Metapopulation structure of *Cotesia melitaearum*, a specialist parasitoid of the butterfly *Melitaea cinxia*. *Oikos* 78: 91–100.

Lei, G. and Hanski, I. 1998. Spatial dynamics of two competing specialist parasitoids in a host metapopulation. *J. Anim. Ecol.* 67: 422–433.

Lei, G. C., Vikberg, V., Nieminen, M. and Kuussaari, M. 1997. The parasitoid complex attacking Finnish populations of the Glanville fritillary *Melitaea cinxia* (Lep: Nymphalidae), an endangered butterfly. *J. Nat. Hist.* 31: 635–648.

Leigh, E. G. 1981. The average lifetime of a population in a varying environment. *J. Theor. Biol.* 90: 231–239.

Leimar, O. and Norberg, U. 1997. Metapopulation extinction and genetic variation in dispersal-related traits. *Oikos* 80: 448–458.

Lennon, J. J., Turner, J. R. G. and Connell, D. 1997. A metapopulation model of species boundaries. *Oikos* 78: 486–502.

Levene, H. 1953. Genetic equilibrium when more than one ecological niche is available. *Am. Nat.* 87: 331–333.

Levin, S. A. 1974. Dispersion and population interactions. *Am. Nat.* 108: 207–228.

Levin, S. A. 1979. Nonuniform stable solutions to reaction–diffusion equations: applications to ecological pattern formation. In *Pattern formation by dynamic systems and pattern recognition* (ed. H. Haken), pp. 210–222. Springer, Berlin.

Levin, S. A. 1992. The problem of pattern and scale in ecology. *Ecology* 73: 1943–1967.

Levin, S. A. 1994. Patchiness in marine and terrestrial systems: from individuals to populations. *Philosophical Transactions of the Royal Society, London B* 343: 99–103.

Levin, S. A. and Pacala, S. W. 1997. Theories of simplification on scaling of spatially distributed processes. In *Spatial ecology* (ed. D. Tilman and P. Kareiva), pp. 271–295. Princeton University Press.

Levin, S. A. and Segel, L. 1976. Hypothesis for the origin of plankton patchiness. *Nature* 259: 659.

Levin, S. A., Cohen, D. and Hastings, A. 1984. Dispersal strategies in patchy environments. *Theor. Pop. Biol.* 26: 165–191.

Levin, S. A., Powell, T. M. and Steele, J. H. 1993. *Patch dynamics.* Springer, Berlin.

Levins, R. 1968. *Evolution in changing environments*. Princeton University Press.

Levins, R. 1969. Some demographic and genetic consequences of environmental heterogeneity for biological control. *Bull. Entomol. Soc. Amer.* 15: 237–240.

Levins, R. 1970. Extinction. *Lect. Notes Math.* 2: 75–107.

Levins, R. and Culver, D. 1971. Regional coexistence of species and competition between rare species. *Proc. Nat. Acad. Sci. USA* 68: 1246–1248.

Lewis, M. A. 1997. Variability, patchiness, and jump dispersal in the spread of an invading population. In *Spatial ecology* (ed. D. Tilman and P. Kareiva), pp. 46–74. Princeton University Press.

Lewis, O. T., Thomas, C. D., Hill, J. K., Brookes, M. I., Robin Crane, T. P., Graneau, Y. A., *et al.* 1997. Three ways of assessing the metapopulation structure in the butterfly *Plebejus argus*. *Ecol. Entomol.* 22: 283–293.

Lindenmayer, D. B. and Possingham, H. P. 1994. *The risk of extinction*. The Australian National University, Canberra.

Lindenmayer, D. B. and Possingham, H. P. 1995. Modelling the viability of metapopulations of the endangered Leadbeater's possum in southeastern Australia. *Biodiv. Conserv.* 4: 984–1018.

Lindenmayer, D. B., Clark, T. W., Lacy, R. C. and Thomas, V. C. 1993. Population viability analysis as a tool in wildlife conservation policy: with reference to Australia. *Environ. Manag.* 17: 745–758.

Lindenmayer, D. B., Burgman, M. A., Akçakaya, H. R., Lacy, R. C. and Possingham, H. P. 1995. A review of the generic computer programs ALEX, RAMAS/Space and Vortex for modelling the viability of wildlife populations. *Ecol. Mod.* 82: 161–174.

Linkola, K. 1916. Studien Über den Einfluss der Kultur auf die Flora in den Gegenden nordlich von Ladogasee. I. Allgemeiner Teil. *Acta Soc. Fauna Flora Fenn.* 45: 1–432.

Liu, J., Dunning Jr, J. B. and Pulliam, H. R. 1995. Potential effects of a forest management plan on Bachmann's sparrow (*Aimophila aestivalis*): linking a spatially explicit model with GIS. *Cons. Biol.* 9: 62–75.

Lloyd, A. L. and May, R. M. 1996. Spatial heterogeneity in epidemic models. *J. Theor. Biol.* 179: 1–11.

Lomolino, M. V. 1993. Winter filtering, immigrant selection and species composition of insular mammals of Lake Huron. *Ecography* 16: 24–30.

Lotka, A. J. 1925. *Elements of physical biology*. Williams and Wilkins, Baltimore.

Lubina, J. A. and Levin, S. A. 1988. The spread of a reinvading species: range expansion in the California sea otter. *Am. Nat.* 131: 526–543.

Luckens, C. J. 1985. *Hypodryas intermedia* Menetries in Europe: an account of the life history. *Entomol. Rec. J. Var.* 97: 37–45.

Ludwig, D. 1996. The distribution of population survival times. *Am. Nat.* 147: 506–526.

Lynch, J. F. and Johnson, N. K. 1974. Turnover and equilibria in insular avifaunas, with special reference to the California Channel Islands. *Condor* 76: 370–384.

Lynch, M., Conery, J. and Buerger, R. 1995. Mutation accumulation and the extinction of small populations. *Am. Nat.* 146: 489–514.

van der Maarel, E. and Sykes, M. T. 1993. Small-scale plant species turnover in limestone grassland: the carousel model and some comments on the niche concept. *J. Veget. Sc.* 4: 179–188.

MacArthur, R. H. 1957. On the relative abundance of bird species. *Proc. Nat. Acad. Sci. USA* 43: 293–295.

MacArthur, R. H. and Wilson, E. O. 1963. An equilibrium theory of insular zoogeography. *Evolution* 17: 373–387.

MacArthur, R. H. and Wilson, E. O. 1967. *The theory of island biogeography*. Princeton University Press.

Machtans, C. S., Villard, M. A. and Hannon, S. J. 1996. Use of riparian buffer strips as movement corridors by forest birds. *Cons. Biol.* 10: 1366–1379.

Mangel, M. and Tier, C. 1993a. Dynamics of metapopulations with demographic stochasticity and environmental catastrophes. *Theor. Popul. Biol.* 44: 1–31.

Mangel, M. and Tier, C. 1993b. A simple direct method for finding persistence times of populations and application to conservation problems. *Proc. Nat. Acad. Sci. USA* 90: 1083–1086.

Mangel, M. and Tier, C. 1994. Four facts every conservation biologist should know about persistence. *Ecology* 75: 607–614.

Marcot, B. G. and Holthausen, R. 1987. Analyzing population viability of the spotted owl in the Pacific Northwest. *Transaction of North American Wildlife Nature Research Conference* 52: 333–347.

Margules, C. R., Nicholls, A. O. and Usher, M. B. 1994. Apparent species turnover, probability of extinction and the selection of nature reserves: a case study of the Ingleborough limestone pavements. *Cons. Biol.* 8: 398–409.

Marino, P. C. 1991. Dispersal and coexistence of mosses (Splachnaceae) in patchy habitats. *J. Ecol.* 79: 1047–1060.

Marttila, O., Haahtela, T., Aarnio, H. and Ojalainen, P. 1990. *Suomen päiväperhoset*. Kirjayhtymä, Helsinki.

Maruyama, T. 1970. On the fixation probability of mutant genes in a subdivided population. *Genet. Res.* 15: 221–225.

Maruyama, T. 1971. Analysis of population structure. II. Two dimensional stepping stone models of finite length and other geographically structured populations. *Ann. Hum. Genet.* 35: 179–196.

Maurer, B. A. and Villard, M. A. 1994. Geographic variation in abundance of North American birds. *Research Exploration* 10: 306–317.

May, R. M. 1973. *Stability and complexity in model ecosystems*. Princeton University Press.

May, R. M. 1974. Biological populations with non-overlapping generations: stable points, stable cycles and chaos. *Science* 186: 645–647.

May, R. M. 1975. Patterns of species abundance and diversity. In *Ecology and evolution of communities* (ed. M. L. Cody and J. M. Diamond), pp. 81–120. Belknap, Cambridge, Mass.

May, R. M. 1976. Models for single populations. In *Theoretical ecology* (ed. R. M. May), pp. 5–29. Blackwell, Oxford.

May, R. M. 1991. The role of ecological theory in planning reintroduction of endangered species. *Symp. Zool. Soc. London* 62: 145–163.

May, R. M. 1994. The effects of spatial scale on ecological questions and answers. In *Large-scale ecology and conservation biology* (ed. P. J. Edwards, R. M. May and N. R. Webb), pp. 1–18. Blackwell Scientific Press, Oxford.

May, R. M. and Nowak, M. A. 1994. Superinfection, metapopulation dynamics, and the evolution of diversity. *J. Theor. Biol.* 170: 95–114.

May, R. M. and Oster, G. F. 1976. Bifurcations and dynamic complexity in simple ecological models. *Am. Nat.* 110: 573–599.

May, R. M., Lawton, J. H. and Stork, N. E. 1995. Assessing extinction rates. In *Extinction rates* (ed. J. H. Lawton and R. M. May), pp. 1–24. Oxford University Press.

Maynard Smith, J. 1964. Group selection and kin selection. *Nature* 201: 1145–1147.

Maynard Smith, J. 1974. *Models in ecology*. Cambridge University Press.

McArdle, B. H. and Gaston, K. J. 1992. Comparing population viabilities. *Oikos* 64: 610–612.

McCallum, H. I. 1992. Effects of immigration on chaotic population dynamics. *J. Theor. Biol.* 154: 277–284.

McCarthy, M. A., Lindenmayer, D. B. and Dreschler, M. 1997. Extinction debts and risks faced by abundant species. *Cons. Biol.* 11: 221–226.

McCauley, D. E. 1993. Evolution in metapopulations with frequent local extinction and recolonization. *Oxford Surv. Evol. Biol.* 10: 109–134.

McCauley, D. E., Raveill, J. and Antonovics, J. 1995. Local founding events as determinants of genetic structure in a plant metapopulation. *Heredity* 75: 630–636.

McCullough, D. R. 1996. Metapopulations and Wildlife Conservation. *Island Press, Washington D.C.*

McGarrahan, E. 1997. Much-studied butterfly winks out on Stanford preserve. *Science* 275: 479–480.

McIntosh, R. P. 1985. *The background to ecology: concept and theory*. Cambridge University Press.

McIntosh, R. P. 1991. Concept and terminology of homogeneity and heterogeneity in ecology. In *Ecological heterogeneity. Ecological Studies 86* (ed. J. Kolasa and S. T. A. Pickett), pp. 24–46. Springer, Berlin.

McPeek, M. A. and Holt, R. D. 1992. The evolution of dispersal in spatially and temporally varying environments. *Am. Nat.* 140: 1010–1027.

Mehlman, D. W. 1997. Change of avian abundance across the geographic range in response to environmental change. *Ecol. Appl.* 7: 614–624.

Menendez, R. 1993. Patterns of distribution and abundance in dung-beetle species. *Stud. Oecol.* 10–11: 395–400.

Menges, E. S. 1988. Conservation biology of Furbish's lousewort. In *Final report to Region 5*, U.S. Fish and wildlife service, Holcomb Research Institute Report No. 126. Butler University, Indianapolis, Indiana.

Menges, E. S. 1990. Population viability analysis for an endangered plant. *Cons. Biol.* 4: 52–62.

Merriam, G. 1991. Corridors and connectivity: animal populations in heterogenous environments. In *Nature conservation 2: The role of corridors* (ed. D. A. Saunders and R. J. Hobbs), pp. 133–142. Surrey Beatty and Sons, Chipping Norton, N.S.W.

Merriam, G. and Lanoue, A. 1990. Corridor use by small mammals: field measurement for three experimental types of *Peromyscus leucopus*. *Landscape Ecology* 4: 123–131.

Metz, J. A. J. and Diekmann, O. 1986. *The dynamics of physiologically structured populations*. Springer, Berlin.

Metz, J. A. J., de Jong, T. J. and Klinkhamer, P. G. L. 1983. What are the advantages of dispersing; a paper by Kuno explained and extended. *Oecologia* 57: 166–169.

Middleton, D. A. and Nisbet, R. M. 1997. Population persistence time: estimates, models, and mechanisms. *Ecol. Appl.* 7: 107–118.

Middleton, D. A., Veitch, A. R. and Nisbet, R. M. 1995. The effect of an upper limit to population size on persistence time. *Theor. Pop. Biol.* 48: 277–305.

Miller, G. L. and Carroll, B. W. 1989. Modeling vertebrate dispersal distances: alternatives to the geometric distribution. *Ecology* 70: 977–986.

Mills, M. L. 1995. Edge effects and isolation: red-backed voles on forest remnants. *Cons. Biol.* 9: 395–403.

Milne, A. 1957. The natural control of insect populations. *Can. Entomol.* 89: 193–213.

Milne, A. 1962. On the theory of natural control of insect populations. *J. Theor. Biol.* 3: 19–50.

Mimura, M. and Murray, J. D. 1978. On a diffusive prey–predator model which exhibits patchiness. *J. Theor. Biol.* 75: 249–262.

Moilanen, A. 1995. Parameterisation of a metapopulation model: an empirical comparison of several different genetic algorithms, simulated annealing and tabu search. In *Proceedings of the 2nd IEEE conference on evolutionary computing*, pp. 551–556. IEEE Press, New Jersey.

Moilanen, A. 1998. Patch occupancy models of metapopulation dynamics: efficient parameter estimation with implicit statistical inference. *Ecology* in press.

Moilanen, A. and Hanski, I. 1995. Habitat destruction and coexistence of competitors in a spatially realistic metapopulation model. *J. Anim. Ecol.* 64: 141–144.

Moilanen, A. and Hanski, I. 1998. Metapopulation dynamics: effects of habitat quality and landscape structure. *Ecology* 79: 2503–2515.

Moilanen, A., Smith, A. T. and Hanski, I. 1998. Long-term dynamics in a metapopulation of the American pika. *Am. Nat.* 152: 530–542.

Monro, J. 1967. The exploitation and conservation of resources by populations of insects. *J. Anim. Ecol.* 36: 531–547.

Moore, S. D. 1989. Patterns of juvenile mortality within an oligophagous butterfly population. *Ecology* 70: 1726–1737.

Mopper, S. and Strauss, S. Y. 1998. *Genetic structure and local adaptation in natural insect populations.* Chapman and Hall, New York.

Morton, R. D., Law, R., Pimm, S. L. and Drake, J. A. 1996. On models for assembling ecological communities. *Oikos* 75: 493–499.

Moulton, M. P. and Pimm, S. L. 1983. The introduced Hawaiian avifauna: biogeographic evidence for competition. *Am. Nat.* 121: 669–690.

Moulton, M. P. and Pimm, S. L. 1986. Species introduction to Hawaii. *Ecology of the biological invasions of North America and Hawaii* 58: 231–249.

Murdoch, W. W. and Walde, S. J. 1989. Analysis of insect population dynamics. In *Towards a more exact ecology* (ed. P. J. Grubb and J. B. Whittaker), pp. 113–140. Blackwell, Oxford.

Murphy, D. D., Freas, K. E. and Weiss, S. B. 1990. An environment–metapopulation approach to population viability analysis for a threatened invertebrate. *Cons. Biol.* 4: 41–51.

Murray, J. D. 1988. Spatial dispersal of species. *Trends Ecol. Evol.* 3: 307–309.

Murray Jr, B. G. 1979. *Population dynamics: alternative models.* Academic Press, New York.

Nachman, G. 1987. Systems analysis of acarine predator–prey interactions. I. A stochastic simulation model of spatial processes. *J. Anim. Ecol.* 56: 247–265.

Nachman, G. 1988. Regional persistence of locally unstable predator–prey interactions. *Experimental and Applied Acarology* 5: 293–318.

Nachman, G. 1991. An acarine predator–prey metapopulation system inhabiting greenhouse cucumbers. *Biol. J. Linn. Soc.* 42: 285–303.

Nagylaki, T. 1992. Rate of evolution of a quantitative character. *Proc. Nat. Acad. Sci. USA* 89: 8121–8124.

Nash, D. R., Agassiz, D. L. J., Godfray, H. C. J. and Lawton, J. H. 1995. The pattern of spread of invading species: two leaf-mining moths colonizing Great Britain. *J. Anim. Ecol.* 64: 225–233.

Nee, S. 1994. How populations persist. *Nature* 367: 123–124.

Nee, S. and May, R. M. 1992. Dynamics of metapopulations: habitat destruction and competitive coexistence. *J. Anim. Ecol.* 61: 37–40.

Nee, S., Gregory, R. D. and May, R. M. 1991. Core and satellite species: theory and artefacts. *Oikos* 62: 83–87.

Nee, S., May, R. M. and Hassell, M. P. 1997. Two-species metapopulation models. In *Metapopulation biology* (ed. I. A. Hanski and M. E. Gilpin), pp. 123–148. Academic Press, San Diego.

Neubert, M. G., Kot, M. and Lewis, M. A. 1995. Dispersal and pattern formation in a discrete-time predator–prey model. *Theor. Pop. Biol.* 48: 7–43.

Nève, G., Barascud, B., Hughes, R., Aubert, J., Descimon, H., Lebrun, P. *et al.* 1996. Dispersal, colonisation power and metapopulation structure in the vulnerable butterfly *Proclossiana eunomia* (Lepidoptera: Nymphalidae). *J. Appl. Ecol.* 33: 14–22.

New, T. R. 1991. *Butterfly conservation*. Oxford University Press, Melbourne, Australia.

New, T. R., Pyle, R. M., Thomas, J. A., Thomas, C. D. and Hammond, P. C. 1995. Butterfly conservation management. *Ann. Rev. Entomol.* 40: 57–83.

Newman, D. and Pilson, D. 1997. Increased probability of extinction due to decreased effective population size: experimental populations of *Clarkia pulchella*. *Evolution* 51: 354–362.

Newmark, W. D. 1995. Extinction of mammal populations in western North American national parks. *Cons. Biol.* 9: 512–526.

Newmark, W. D. 1996. Insularization of Tanzanian parks and the local extinction of large mammals. *Cons. Biol.* 10: 1549–1556.

Nichols, J. D. 1992. Capture–recapture models. *BioScience* 42: 94–102.

Nichols, J. D., Brownie, C., Hines, J. E., Pollock, K. H. and Hestbeck, J. B. 1993. The estimation of exchanges among populations or subpopulations. In *Marked individuals in the study of bird populations* (ed. J. D. Lebreton and P. M. North), pp. 265–179. Birkhäuser, Basel.

Nicholson, A. J. 1933. The balance of animal populations. *J. Anim. Ecol.* 2: 132–178.

Nicholson, A. J. 1947. Fluctuation of animal populations. Rep. 26th Meet. A. N. Z. A. A. S. Perth, 1947.

Nicholson, A. J. 1954. An outline of the dynamics of animal populations. *Austral. J. Zool.* 2: 9–65.

Nicholson, A. J. 1957. The self-adjustment of populations to change. *Cold Spring Harbor Symp. Quant. Biol.* 22: 153–173.

Nicholson, A. J. and Bailey, V. A. 1935. The balance of animal populations. *Proc. Zool. Soc. Lond.* 3: 551–598.

Nieminen, M. 1996a. Metapopulation dynamics of moths. Ph.D. Dissertation, University of Helsinki, Finland.

Nieminen, M. 1996b. Migration of moth species in a network of small islands. *Oecologia* 108: 643–651.

Nieminen, M. 1996c. Risk of population extinction in moths: effect of host plant characteristics. *Oikos* 76: 475–484.

Nieminen, M. and Hanski, I. 1998. Metapopulations of moths on islands: a test of two contrasting models. *Journal of Animal Ecology* 67: 149–160.

Nilsson, T. 1997. Spatial population dynamics of the black tinder fungus beetle, *Bolitophagus reticulatus* (Coleoptera: Tenebriniodae). Ph.D. Dissertation, Uppsala University, Sweden.

Nisbet, R. M. and Gurney, W. S. C. 1982. *Modelling fluctuating populations*. John Wiley and Sons, New York.

Noordwijk van, A. J. 1994. The interaction of inbreeding depression and environmental stochasticity in the risk of extinction of small populations. In *Conservation genetics* (ed. V. Loeschcke, J. Tomiuk and S. K. Jain), pp. 131–146. Birkhäuser, Basel.

Noss, R. F. 1987. Corridors in real landscapes: a reply to Simberloff and Cox. *Cons. Biol.* 1: 159–164.

Nunney, L. and Campbell, K. A. 1993. Assessing minimum viable population size: demography meets population genetics. *Trends Ecol. Evol.* 8: 234–239.

Nürnberger, B. and Harrison, R. G. 1995. Spatial population structure in the whirling beetle *Dineutus assimilis*: evolutionary inferences based on mitochondrial DNA and field data. *Evolution* 49: 266–275.

Obeso, J. R. 1992. Geographic distribution and community structure of bumblebees in the northern Iberian Peninsula. *Oecologia* 89: 244–252.

O'Connor, R. J. 1987. Organization of avian assemblages—the influence of intraspecific habitat dynamics. In *Organization of communities: past and present* (ed. J. H. R. Gee and P. S. Giller), pp. 163–183. Blackwell Scientific, Oxford.

Okubo, A. 1980. *Diffusion and ecological problems: mathematical models*. Springer, Berlin.

Okubo, A., Murray, J. and Williamson, M. 1989. On the spatial spread of grey squirrels in Britain. *Proceedings of the Royal Society of London B* 238: 113–125.

Olivieri, I. and Gouyon, P.-H. 1985. Seed dimorphism for dispersal: theory and implications. In *Structure and functioning of plant populations* (ed. J. Haeck and J. W. Woldendorp), pp. 77–90. North Holland, Amsterdam.

Olivieri, I. and Gouyon, P.-H. 1997. Evolution of migration rate and other traits: the metapopulation effect. In *Metapopulation biology* (ed. I. A. Hanski and M. E. Gilpin), pp. 293–324. Academic Press, San Diego.

Olivieri, I., Swan, M. and Gouyon, P.-H. 1983. Reproductive system and colonizing strategy of two species of *Carduus* (Compositae). *Oecologia* 60: 114–117.

Olivieri, I., Couvet, D. and Gouyon, P.-H. 1990. The genetics of transient populations: research at the metapopulation level. *Trends Ecol. Evol.* 5: 207–210.

Olivieri, I., Michalakis, Y. and Gouyon, P.-H. 1995. Metapopulation genetics and the evolution of dispersal. *Am. Nat.* 146: 202–228.

Oostermeijer, J. G. 1996a. Population size, genetic variation, and related parameters in small, isolated plant populations: a case study. In *Species survival in plant populations* (ed. J. Settele, C. Margules, P. Poshlod and K. Henle), pp. 61–68. Kluwer Academic Publishers, Dordrecht.

Oostermeijer, J. G. 1996b. Population viability of *Gentiana pneumonathe*: the relative importance of demography, genetics, and reproductive biology. Ph.D. Dissertation, University of Amsterdam, The Netherlands.

Oostermeijer, J. G., Eijck van, M. W. and den Nijs, J. C. 1994. Offspring fitness in relation to population size and genetic variation in the rare perennial plant species *Gentiana pneumonathe*. *Oecologia* 97: 289–296.

Oostermeijer, J. G., van Eijck, M. W., van Leeuwen, N. C. and den Nijs, J. C. 1995. Analysis of the relationship between allozyme heterozygosity and fitness in the rare *Gentiana pheumonathe* L. *Evol. Biol.* 8: 739–759.

Ouborg, N. J. 1993. Isolation, population size and extinction: the classical and meta-population approaches applied to vascular plants along the Dutch Rhine-system. *Oikos* 66: 298–308.

Ouborg, N. J. and Treuren van, R. 1994. The significance of genetic erosion in the process of extinction. 4. Inbreeding load and heterosis in relation to population size in the mint *Salvia pratensis*. *Evolution* 48: 996–1008.

Pacala, S. W. 1997. Dynamics of plant communities. In *Plant ecology* (ed. M. E. Crawley), pp. 532–555. Blackwell, Oxford.

Pacala, S. W. and Deutschman, D. H. 1995. Details that matter: the spatial distribution of individual trees maintains forest ecosystem function. *Oikos* 74: 357–365.

Pacala, S. W. and Levin, S. A. 1997. Biologically generated spatial pattern and the coexistence of competing species. In *Spatial ecology* (ed. D. Tilman and P. Kareiva), pp. 204–232. Princeton University Press.

Pacala, S. W., Canham, C. D., Saponara, J., Silander Jr, J. A., Kobe, R. K. and Ribbens, E. 1996. Forest models defined by field measurements: estimation, error analysis and dynamics. *Ecol. Monogr.* 66: 1–43.

Pagel, M. and Payne, J. H. 1996. High migration affects estimation of the extinction threshold. *Oikos* 76: 323–329.

Paine, R. T. 1966. Food web complexity and species diversity. *Am. Nat.* 100: 65–75.

Pajunen, V. I. 1979. Competition between rockpool corixids. *Ann. Zool. Fenn.* 16: 138–143.

Pajunen, V. I. 1982. Replacement analysis of non-equilibrium competition between rock pool Corixids (Hemiptera; Corixidae). *Oecologia* 52: 153–155.

Pajunen, V. I. 1986. Distributional dynamics of *Daphnia* species in a rock-pool environment. *Ann. Zool. Fenn.* 23: 131–140.

Palo, J., Varvio, S.-L., Hanski, I. and Väinölä, R. 1995. Developing microsatellite markers for insect population structure: complex variation in a checkerspot butterfly. *Hereditas* 123: 295–300.

Paradis, E. 1995. Survival, immigration and habitat quality in the Mediterranean pine voles. *J. Anim. Ecol.* 64: 579–591.

Parmesan, C. 1991. Evidence against plant 'apparency' as a constraint on evolution of insect search efficiency (Lepidoptera: Nymphalidae). *Journal of Insect Behavior* 4: 417–430.

Parmesan, C., Singer, M. and Harris, I. 1995. Absence of adaptive learning from the oviposition foraging behaviour of a checkerspot butterfly. *Anim. Beh.* 50: 161–175.

Patterson, B. D. 1987. The principle of nested subsets and its implications for biological conservation. *Cons. Biol.* 1: 323–334.

Peltonen, A. and Hanski, I. 1991. Patterns of island occupancy explained by colonization and extinction rates in shrews. *Ecology* 72: 1698–1708.

Peroni, P. A. 1994. Invasion of red maple (*Acer rubrum* L.) during old field succession in the North Carolina Piedmont: age structure of red maple in young pine stands. *Bull. Torrey Bot. Club* 121: 357–359.

Pettersson, B. 1985. Extinction of an isolated population of the middle-spotted woodpecker *Dendrocopus medius* (L.) in Sweden and its relation to general theories on extinction. *Biol. Cons.* 32: 335–353.

Phillips, P. C. 1993. Peak shifts and polymorphism during phase three of Wright's shifting balance process. *Evolution* 47: 1733–1743.

Pimm, S. L. 1992. *The balance of nature*. The University of Chicago Press.

Pokki, J. 1981. Distribution, demography and dispersal of the field vole, *Microtus agrestis* (L.), in the Tvärminne archipelago, Finland. *Acta Zool. Fennica* 164: 1–48.

Pollard, E. and Yates, T. J. 1993. *Monitoring butterflies for ecology and conservation*. Chapman and Hall, London.

Pollard, E., Hall, M. L. and Bibby, T. J. 1986. *Monitoring the abundance of butterflies, 1976–1985*. Nature Conservation Council, Great Britain.

Pollard, E., Lakhani, K. H. and Rothery, P. 1987. The detection of density dependence from series of annual censuses. *Ecology* 68: 2046–2055.

Porter, J. H. and Dooley, J. L. 1993. Animal dispersal patterns: a reassessment of simple mathematical models. *Ecology* 74: 2436–2443.

Porter, K. 1981. The population dynamics of small colonies of the butterfly *Euphydryas aurinia*. Ph.D. Dissertation, Oxford, UK.

Porter, K. 1983. Multivoltinism in *Apanteles bignellii* and the influence of weather on synchronisation with its host *Euphydryas aurinia*. *Entomol. Exper. Appl.* 34: 155–162.

Possingham, H. P. and Noble, I. R. 1991. An evaluation of population viability analysis for assessing the risk of extinction. In *Research consultancy for the research assessment commission, Forest and timber inquiry*. Canberra, Australia.

Possingham, H. P., Davies, I., Noble, I. R. and Norton, T. W. 1992. A metapopulation simulation model for assessing the likelihood of plant and animal extinctions. *Math. Comput. Simul.* 33: 367–372.

Post, W. M. and Pimm, S. L. 1983. Community assembly and food web stability. *Math. Biosci.* 64: 169–182.

Pöyry, J. 1996. Täpläverkkoperhosen (*Melitaea cinxia* L.) populaatiorakenteen ja liikkeiden tutkimus merkintä-jälleenpyynti-menetelmän avulla. M.Sc. Thesis, University of Helsinki, Finland.

Pressey, R. L., Humphries, C. J., Margules, C. R., Vane-Wright, R. F. and Williams, P. H. 1996. Beyond opportunism: key principles for systematic reserve collection. *Trends Ecol. Evol.* 8: 124–128.

Preston, F. W. 1948. The commonness, and rarity, of species. *Ecology* 29: 254–283.

Preston, F. W. 1960. Time and space and the variation of species. *Ecology* 41: 611–627.

Preston, F. W. 1962. The canonical distribution of commonness and rarity. *J. Ecol.* 43: 185–215.

Primack, R. B. and Miao, S. L. 1992. Dispersal can limit local plant distribution. *Cons. Biol.* 6: 513–519.

Prince, S. D. and Carter, R. N. 1985. The geographical distribution of prickly lettuce (*Lactuca serriola*). 3. Its performance in transplant sites beyond its distribution limit in Britain. *Journal of Ecology* 73: 49–64.

Prober, S. M. and Brown, A. H. D. 1994. Conservation of the grassy white box woodlands: Population genetics and fragmentation of *Eucalyptus albens*. *Cons. Biol.* 8: 1003–1013.

Pulliam, H. R. 1988. Sources, sinks, and population regulation. *Am. Nat.* 132: 652–661.

Pulliam, H. R. 1996. Sources and sinks: empirical evidence and population consequences. In *Population dynamics in ecological space and time* (ed. O. E. Rhodes Jr, R. K. Chester and M. H. Smith), pp. 45–70. University of Chicago Press.

Pulliam, H. R. and Danielson, B. J. 1991. Sources, sinks, and habitat selection: a landscape perspective on population dynamics. *Am. Nat.* 137: 50–66.

Pulliam, H. R., Dunning Jr, J. B. and Liu, J. 1992. Population dynamics in complex landscapes: a case study. *Ecol. Appl.* 2: 165–177.

Pullin, A. S. 1995. *Ecology and conservation of butterflies*. Chapman and Hall, London.

Pusey, A. E. 1987. Sex-biased dispersal and inbreeding avoidance in birds and mammals. *Trends Ecol. Evol.* 2: 295–299.

Quinn, J. F. and Hastings, A. 1987. Extinction in subdivided habitats. *Cons. Biol.* 1: 198–208.

Quintana-Ascension, P. F. and Menges, E. S. 1996. Inferring metapopulation dynamics from patch-level incidence of Florida scrub plants. *Cons. Biol.* 10: 1210–1219.

Rabinowitz, D., Cairns, S. and Dillon, T. 1986. Seven forms of rarity and their frequency in the flora of the British isles. In *Conservation biology. The science of scarcity and diversity* (ed. M. E. Soulé), pp. 182–204. Sinauer, Sunderland, Mass.

Ralls, K., Ballou, J. D. and Templeton, A. 1988. Estimates of lethal equivalents and the cost of inbreeding in mammals. *Cons. Biol.* 2: 185–193.

Ranta, E. 1979. Niche of *Daphnia* species in rock pools. *Archiv für Hydrobiologie* 87: 205–223.

Ranta, E., Kaitala, V., Lindström, J. and Lindén, H. 1995. Synchrony in population dynamics. *Proceedings of the Royal Society of London B* 262: 113–118.

Rapoport, E. H. 1982. *Areography: Geographical strategies of species*. Pergamon Press, Oxford.

Raunkiaer, C. 1910. Investigations and statistics of plant formations. *Botanisk Tidsskrift* 30.

Raunkiaer, C. 1934. *The life forms of plants and statistical plant geography*. Clarendon Press, Oxford.

Rausher, M. D., Mackay, D. A. and Singer, M. C. 1981. Pre- and post-alighting host discrimination by *Euphydryas editha* butterflies: the behavioral mechanisms causing clumped distribution of egg clusters. *Anim. Beh.* 29: 1220–1228.

Ray, C., Gilpin, M. E. and Smith, A. T. 1991. The effect of conspecific attraction on metapopulation dynamics. *Biol. J. Linn. Soc.* 42: 123–134.

Reading, R. P., Grensten, J. J., Beissinger, S. R. and Clark, T. W. 1993. Attributes of black-tailed prairie dog colonies in north-central Montana, with management recommendations for the conservation of biodiversity. In *Proceedings of the symposium on the management of prairie dog complexes for the reintroduction of the black-footed ferret* (ed. J. L. Oldemayer, D. E. Biggins, B. J. Miller and R. Crete), pp. 9–10.

Reed, T. M. 1980. Turnover frequencies in island birds. *Journal of Biogeography* 7: 329–335.

Reeve, J. D. 1988. Environmental variability, migration, and persistence in host–parasitoid systems. *Am. Nat.* 132: 810–836.

Rhodes Jr, O. E., Chester, R. K. and Smith, M. H. 1996. *Population dynamics in ecological space and time*. Chicago University Press.

Rich, A. C., Dobkin, D. S. and Niles, L. J. 1994. Defining forest fragmentation by corridor width: the influence of narrow forest dividing corridors on forest-nesting birds in southern New Jersey. *Cons. Biol.* 8: 1109–1121.

Ricklefs, R. E. 1990. *Ecology*. Freeman, New York.

Rieman, B. E. and McIntyre, J. D. 1995. Occurrence of bull trout in naturally fragmented habitat patches of varied size. *Transactions of the American Fisheries Society* 124: 285–296.

Roderick, G. K. and Caldwell, R. S. 1992. An entomological perspective on animal dispersal. In *Animal dispersal—small mammals as a model* (ed. N. C. Stenseth and W. Z. Lidicker), pp. 274–290. Chapman and Hall, London.

Roff, D. A. 1975. Population stability and the evolution of dispersal in a heterogeneous environment. *Oecologia* 19: 217–237.

Roff, D. A. 1986. The evolution of wing dimorphism in insects. *Evolution* 40: 1009–1020.

Roff, D. A. 1990. The evolution of flightlessness in insects. *Ecol. Monogr.* 60: 389–421.

Roff, D. A. 1994. Habitat persistence and the evolution of wing dimorphism in insects. *Am. Nat.* 144: 772–798.

Roff, D. A. 1996. The evolution of threshold traits in animals. *Quarterly Reviews in Biology* 71: 3–35.

Rohani, P., May, R. M. and Hassell, M. P. 1996. Metapopulations and local stability: the effects of spatial structures. *J. Theor. Biol.* 181: 97–109.

Rolstad, J. and Wegge, P. 1987. Distribution and size of capercaillie leks in relation to old forest fragmentation. *Oecologia* 72: 389–394.

Rolstad, J. and Wegge, P. 1989. Capercaillie, *Tetrao urogallus*, populations and modern forestry—a case study for landscape ecological studies. *Finnish Game Research* 46: 43–52.

Ronce, O. and Olivieri, I. 1997. Evolution of reproductive effort in a metapopulation with local extinctions and ecological succession. *Am. Nat.* 150: 220–249.

Rosenzweig, M. L. 1995. *Species diversity in space and time*. Cambridge University Press.

Rosenzweig, M. L. and Clark, C. W. 1994. Island extinction rates from regular censuses. *Cons. Biol.* 8: 491–494.

Ross, R. 1909. *The prevention of malaria*. Murray, London.

Royama, T. 1992. *Analytical population dynamics*. Chapman and Hall, London.

Russell, G. J., Diamond, J. M., Pimm, S. L. and Reed, T. M. 1995. A century of turnover: community dynamics at three timescales. *J. Anim. Ecol.* 64: 628–641.

Ruxton, G. D. 1993. Linked populations can still be chaotic. *Oikos* 68: 347–348.

Ruxton, G. D. 1996. Dispersal and chaos in spatially structured models: an individual-level approach. *J. Anim. Ecol.* 65: 161–169.

Ruxton, G. D. and Doebeli, M. 1996. Spatial self-organization and persistence of transients in a metapopulation model. *Proc. Royal Soc. London B* 263: 1153–1158.

Ruxton, G. D., Gonzales-Andujar, J. L. and Perry, J. N. 1997. Mortality during dispersal stabilizes local population fluctuations. *Oikos* 66: 289–292.

Saarinen, P. 1993. Kalliosinisiiven (*Scolitantides orion*) ekologia ja esiintyminen Lohjalla. M.Sc. Thesis, University of Helsinki, Finland.

Saccheri, I. J. 1995. An experimental study of the effects of population bottlenecks on genetic variation and fitness in the butterfly *Bicyclus anynana* (Satyridae). Ph.D. Dissertation, University of London, UK.

Saccheri, I. J., Brakefield, P. M. and Nichols, R. A. 1996. Severe inbreeding depression and rapid fitness rebound in the butterfly *Bicyclus anynana* (Satyridae). *Evolution* 50: 2000–2013.

Saccheri, I. J., Kuussaari, M., Kankare, M., Vikman, P., Fortelius, W. and Hanski, I. 1998. Inbreeding and extinction in a butterfly metapopulation. *Nature* 392: 491–494.

Sale, P. F. 1977. Maintenance of high diversity in coral reef fish communities. *Am. Nat.* 111: 337–359.

Saunders, D. A. and Hobbs, R. J. 1991. *The role of corridors*. Surrey Beatty and Sons, NSW, Australia.

Saunders, D. A., Hobbs, R. J. and Margules, C. R. 1991. Biological consequences of ecosystem fragmentation: a review. *Cons. Biol.* 5: 18–32.

Saurola, P. 1995. *Suomen pöllöt*. Kirjayhtymä, Helsinki.

Scheuring, I. and Jánosi, I. M. 1996. When two and two make four: a structured population without chaos. *J. Theor. Biol.* 178: 89–97.

Schoener, T. W. 1976. The species–area relation within archipelagos: models and evidence from island land birds. 16th international ornithological congress, pp. 629–641.

Schoener, T. W. 1983. Rate of species turnover declines from lower to higher organisms: a review of the data. *Oikos* 41: 372–377.

Schoener, T. W. 1991. Extinction and the nature of the metapopulation: a case system. *Acta Oecol.* 12: 53–75.

Schoener, T. W. and Schoener, A. 1983. The time to extinction of a colonizing propagule of lizards increases with island area. *Nature* 302: 332–334.

Schoener, T. W. and Spiller, D. A. 1987. Effect of lizards on spider populations: manipulative reconstruction of a natural experiment. *Science* 236: 949–952.

Schöps, K., Emberson, R. M. and Wratten, S. D. 1998. Does host-plant exploitation influence the population dynamics of a rare weevil? In *Proceedings of the ecology and population dynamics section of the 20th congress of entomology, Florence. 1996* (ed. B. J. F. Manly, J. Baumgärtner and F. Brandlmayer), AA Balkema, Rotterdam.

Schroeder, M. 1991. *Fractals, chaos and power laws.* W. H. Freedman and Co., New York.

Schwerdtfeger, F. 1958. Is the density of animal populations regulated by mechanisms or by chance? *Proceedings of the 10th international congress of entomology* 4: 115–122.

Scott, J. A. 1986. *The butterflies of North America.* Stanford University Press.

Serena, M. 1995. *Reintroduction biology of Australian and New Zealand fauna.* Surrey Beatty and Sons, Chipping Norton, Australia.

Shaffer, M. L. 1981. Minimum population size for species conservation. *BioScience* 31: 131–134.

Shaffer, M. L. 1985. The metapopulation and species conservation: the special case of the northern spotted owl. In *Ecology and management of the spotted owl in the Pacific Northwest. General technical report PNW-185* (ed. R. J. Gutierrez and A. B. Carey), pp. 86–99. USDA Forest Service, Portland, Oregon.

Shaffer, M. L. 1990. Population viability analysis. *Cons. Biol.* 4: 39–40.

Sheppe, W. 1965. Island populations and gene flow in the deer mouse, *Peromyscus leucopus*. *Evolution* 19: 480–495.

Shields, G. F. 1982. Comparative avian cytogenetics: a review. *Condor* 84: 45–58.

Shigesada, N., Kawasaki, K. and Teramoto, E. 1979. Spatial segregation of interacting species. *J. Theor. Biol.* 79: 83–99.

Shigesada, N., Kawasaki, K. and Teramoto, E. 1987. The speeds of travelling frontal waves in heterogeneous environments. In *Mathematical topics in population biology, morphogenesis and neuroscience* (ed. E. Teramoto and M. Yamaguti), pp. 88–97. Springer, Berlin.

Shorrocks, B. 1990. Coexistence in a patchy environment. In *Living in a patchy environment* (ed. B. Shorrocks and I. R. Swingland), pp. 91–106. Oxford University Press.

Shorrocks, B. and Rosewell, J. 1987. Spatial patchiness and community structure: coexistence and guild size of drosophilids on ephemeral resources. In *Organization of communities: past and present* (ed. J. H. R. Gee and P. S. Giller), pp. 29–51. Blackwell Scientific Publications, Oxford.

Shorrocks, B. and Swingland, I. R. 1990. *Living in a patchy environment.* Oxford University Press.

Short, J. and Turner, B. 1994. A test of the vegetation mosaic hypothesis: a hypothesis to explain the decline and extinction of Australian mammals. *Cons. Biol.* 8: 439–449.

Siitonen, J. and Martikainen, P. 1994. Coleoptera and *Aradus* (Hemiptera) collected from aspen in Finnish and Russian Karelia: notes on rare and threatened species. *Scand. J. For. Res.* 9: 185–191.

Simberloff, D. 1969. Experimental zoogeography of islands. A model for insular colonization. *Ecology* 50: 296–314.

Simberloff, D. 1974. Equilibrium theory of island biogeography and ecology. *Ann. Rev. Ecol. Syst.* 5: 161–182.

Simberloff, D. 1976. Species turnover and equilibrium biogeography. *Science* 193: 572–578.

Simberloff, D. 1983. When is an island community in equilibrium? *Science* 220: 1275–1277.

Simberloff, D. 1986. Design of nature reserves. In *Wildlife conservation evaluation* (ed. M. B. Usher), pp. 315–237. Chapman and Hall, Cambridge.

Simberloff, D. 1988. The contribution of population and community biology to conservation science. *Ann. Rev. Ecol. Syst.* 19: 473–512.

Simberloff, D. 1994. The ecology of extinction. *Acta Palaeontologica Polonica* 38: 159–174.

Simberloff, D. 1995. Introduced species. In *Encyclopedia of environmental biology* (ed. W. A. Nierenberg), pp. 323–336. Academic Press, San Diego.

Simberloff, D. S. and Abele, L. G. 1976. Island biogeography theory and conservation practice. *Science* 191: 285–286.

Simberloff, D. and Cox, J. 1987. Consequences and costs of conservation corridors. *Cons. Biol.* 1: 63–71.

Simberloff, D., Farr, J. A., Cox, J. and Mehlman, D. W. 1992. Movement corridors: conservation bargains or poor investments? *Cons. Biol.* 6: 493–504.

Sinclair, A. R. E. 1989. Population regulation in animals. In *Ecological concepts* (ed. J. M. Cherrett), pp. 197–242. Blackwell, Oxford.

Singer, M. C. 1971. Evolution of food-plant preference in the butterfly *Euphydryas editha*. *Evolution* 25: 383–389.

Singer, M. C. 1972. Complex components of habitat suitability within a butterfly colony. *Science* 176: 75–77.

Singer, M. C. 1983. Determinants of multiple host use by a phytophagous insect population. *Evolution* 37: 389–403.

Singer, M. C. 1994. Behavioural constraints on the evolutionary expansion of insect diet: a case history from checkerspot butterflies. In *Behavioural mechanisms in evolutionary ecology* (ed. L. Real), pp. 279–296. University of Chicago Press.

Singer, M. C. and Ehrlich, P. R. 1979. Population dynamics of the checkerspot butterfly *Euphydryas editha*. *Fortschr. Zool.* 25: 53–60.

Singer, M. C. and Thomas, C. D. 1996. Evolutionary responses of a butterfly metapopulation to human and climate-caused environmental variation. *Am. Nat.* 148: 9–39.

Singer, M. C., Ng, D., Vasco, D. and Thomas, C. D. 1992. Rapidly evolving associations among oviposition preferences fail to constrain evolution of insect diet. *Am. Nat.* 139: 9–20.

Singer, M. C., Thomas, C. D. and Parmesan, C. 1993. Rapid human-induced evolution of insect diet. *Nature* 366: 681–683.

Singer, M. C., Thomas, C. D., Billington, H. L. and Parmesan, C. 1994. Correlates of speed of evolution of host preference in a set of twelve populations of the butterfly *Euphydryas editha*. *Ecoscience* 1: 107–114.

Sjögren, P. 1991. Extinction and isolation gradients in metapopulations: the case of the pool frog (*Rana lessonae*). *Biol. J. Linn. Soc.* 42: 135–147.

Sjögren Gulve, P. 1994. Distribution and extinction patterns within a northern metapopulation of the pool frog, *Rana lessonae*. *Ecology* 75: 1357–1367.

Sjögren Gulve, P. and Ray, C. 1996. Using logistic regression to model metapopulation dynamics: large-scale forestry extirpates the pool frog. In *Metapopulations and wildlife conservation* (ed. D. R. McCullough), pp. 111–138. Island Press, Washington, D. C.

Skellam, J. G. 1951. Random dispersal in theoretical populations. *Biometrika* 38: 196–218.

Slatkin, M. 1974. Competition and regional coexistence. *Ecology* 55: 128–134.

Slatkin, M. 1977. Gene flow and genetic drift in a species subject to frequent local extinctions. *Theor. Pop. Biol.* 12: 253–262.

Slatkin, M. 1985. Gene flow in natural populations. *Ann. Rev. Ecol. Syst.* 16: 393–430.

Slatkin, M. 1994. Gene flow and population structure. In *Ecological genetics* (ed. L. A. Real), pp. 3–17. Princeton University Press.

Slatkin, M. and Barton, N. H. 1989. A comparison of three indirect methods for estimating average levels of gene flow. *Evolution* 43: 1349–1368.

Small, R. J., Holzwart, J. C. and Rusch, D. H. 1993. Are ruffed grouse more vulnerable to mortality during dispersal? *Ecology* 74: 2020–2026.

Smith, A. T. 1974. The distribution and dispersal of pikas: consequences of insular population structure. *Ecology* 55: 1112–1119.

Smith, A. T. 1980. Temporal changes in insular populations of the pika *Ochotona princeps*. *Ecology* 61: 8–13.

Smith, A. T. and Gilpin, M. E. 1997. Spatially correlated dynamics in a Pika metapopulation. In *Metapopulation biology* (ed. I. A. Hanski and M. E. Gilpin), pp. 407–428. Academic Press, San Diego.

Smith, A. T. and Peacock, M. M. 1990. Conspecific attraction and the determination of metapopulation colonization rates. *Cons. Biol.* 4: 320–323.

Smith, F. E. 1975. Ecosystems and evolution. *Bull. Ecol. Soc. Am.* 56: 2.

Söderström, L. 1989. Regional distribution patterns of bryophyte species on spruce logs in northern Sweden. *Bryologist* 92: 349–355.

Solbreck, C. 1991. Unusual weather and insect population dynamics: *Lygaeus equestris* during an extinction and recovery period. *Oikos* 60: 343–350.

Soulé, M. E. 1976. Allozyme variation, its determinants in space and time. In *Molecular evolution* (ed. F. J. Ayala), pp. 60–77. Sinauer Associates, Sunderland, Mass.

Soulé, M. E. 1980. Thresholds for survival: maintaining fitness and evolutionary potential. In *Conservation biology: an evolutionary–ecological perspective* (ed. M. E. Soulé and B. A. Wilcox), pp. 111–124. Sinauer, Sunderland, Mass.

Soulé, M. E. 1987. *Viable populations for conservation.* Cambridge University Press, New York.

Soulé, M. E. and Kohm, K. A. 1989. *Research priorities for conservation biology.* Island Press, Covelo.

Soulé, M. E. and Simberloff, D. 1986. What do genetics and ecology tell us about the design of nature reserves? *Biol. Cons.* 35: 19–40.

Soulé, M. E., Bolger, D. T., Alberts, A. C., Wright, J., Sorice, M. and Hill, S. 1988. Reconstructed dynamics of rapid extinctions of chaparral-requiring birds in urban habitat islands. *Cons. Biol.* 2: 75–92.

Soulé, M. E., Alberts, A. C. and Bolger, D. T. 1992. The effects of habitat fragmentation on chaparral plants and vertebrates. *Oikos* 63: 39–47.

Southwood, T. R. E. 1962. Migration of terrestrial arthropods in relation to habitat. *Biol. Rev.* 37: 171–214.

Stacey, P. B. 1979. Habitat saturation and communal breeding in the acorn woodpecker. *Animal Behavior* 27: 1153–1167.

Stacey, P. B. and Taper, M. 1992. Environmental variation and the persistence of small populations. *Ecol. Appl.* 2: 18–29.

Stacey, P. B., Johnson, V. A. and Taper, M. L. 1997. Migration within metapopulations: the impact upon local population dynamics. In *Metapopulation biology* (ed. I. A. Hanski and M. E. Gilpin), pp. 267–192. Academic Press, San Diego.

Stamp, N. E. 1981. Effect of group size on parasitism in a natural population of the Baltimore checkerspot (*Euphydryas phaeton*). *Oecologia* 49: 201–206.

Stamp, N. E. 1982a. Behavioural interactions of parasitoids and Baltimore checkerspot caterpillars (*Euphydryas phaeton*). *Env. Entomol.* 11: 100–104.

Stamp, N. E. 1982b. Searching behaviour of parasitoids for web-making caterpillars: a test of optimal searching theory. *J. Anim. Ecol.* 51: 387–396.

Stamps, J. A. 1991. The effects of conspecific on habitat selection in territorial species. *Behavioural Ecology and Sociobiology* 28: 29–36.

Stamps, J. A., Buechner, M. and Krishnan, V. V. 1987. The effects of edge permeability and habitat geometry on emigration from patches of habitat. *Am. Nat.* 129: 533–552.

Steen, H. 1994. Low survival of long distance dispersers of the root vole (*Microtus oeconomus*). *Ann. Zool. Fenn.* 31: 271–274.

Steinberg, E. K. and Kareiva, P. 1997. Challenges and opportunities for empirical evaluation of 'Spatial Theory'. In *Spatial ecology* (ed. D. Tilman and P. Kareiva), pp. 318–332. Princeton University Press.

Stelter, C., Reich, M., Grimm, V. and Wissel, C. 1997. Modelling persistence in dynamic landscapes: lessons from a metapopulation of the grasshopper *Bryodema tuberculata*. *J. Anim. Ecol.* 66: 508–518.

Stenberg, I. and Hochstad, O. 1992. Habitat use and density of breeding woodpeckers in the 1990s in More og Romsdal county, western Norway. *Cinclus* 15: 49–61.

Stephens, D. W. and Krebs, J. R. 1986. *Foraging theory*. Princeton University Press.

Stiling, P. D. 1987. The frequency of density dependence in insect host–parasitoid systems. *Ecology* 68: 844–856.

Stock, T. M. and Holmes, J. C. 1988. Functional relationships and microhabitat distributions of enteric helminths of grebes (Podicipedidae): the evidence for interactive communities. *J. Parasitol.* 74: 214–227.

Stone, L. 1995. Biodiversity and habitat destruction: a comparative study of model forest and coral reef ecosystems. *Proceedings of the Royal Society of London B* 261: 381–388.

Strong, D. R. 1984. Density-vague ecology and liberal population regulation in insects. In *A New ecology: novel approaches to interactive systems* (ed. P. W. Price, C. N. Slobodchikoff and W. S. Gaud), pp. 313–327. Wiley, New York.

Strong, D. R. 1986. Density vagueness: adding the variance in the demography of real populations. In *Community ecology* (ed. J. M. Diamond and T. J. Case), pp. 257–268. Harper and Row, New York.

Sutcliffe, O. L. and Thomas, C. D. 1996. Open corridors appear to facilitate dispersal by ringlet butterflies (*Aphantopus hyperantus*) between woodland clearings. *Cons. Biol.* 10: 1359–1365.

Sutcliffe, O. L., Thomas, C. D. and Moss, D. 1996. Spatial synchrony and asynchrony in butterfly population dynamics. *J. Anim. Ecol.* 65: 85–95.

Sutcliffe, O. L., Thomas, C. D., Yates, T. J. and Greatorex-Davis, J. N. 1997. Correlated extinctions, colonisations and population fluctuations in a highly connected ringlet butterfly metapopulation. *Oecologia* 109: 235–241.

Svensson, B. W. 1992. Changes in occupancy, niche breadth and abundance of three *Gyrinus* species as their respective range limits are approached. *Oikos* 63: 147–156.

van Swaay, C. A. M. 1990. An assessment of the changes in the butterfly abundance in The Netherlands during the 20th century. *Biol. Cons.* 52: 287–302.

van Swaay, C. A. M. 1995. Measuring changes in butterfly abundance in The Netherlands. In *Ecology and conservation of butterflies* (ed. A. S. Pullin), pp. 230–247. Chapman and Hall, London.

Swingland, I. R. and Greenwood, P. J. 1983. *The ecology of animal movements*. Oxford University Press.

Taylor, A. D. 1988. Large-scale spatial structure and population dynamics in arthropod predator–prey systems. *Ann. Zool. Fenn.* 25: 63–74.

Taylor, L. R. 1961. Aggregation, variance and the mean. *Nature* 189: 732–735.

Taylor, P. D. 1988. An inclusive fitness model for dispersal of offspring. *J. Theor. Biol.* 130: 363–378.

Taylor, R. A. J. 1978. The relationship between density and distance of dispersing insects. *Ecol. Entomol.* 3: 63–70.

Taylor, R. A. J. and Taylor, L. R. 1979. A behavioural model for the evolution of spatial dynamics. In *Population dynamics* (ed. R. M. Anderson, B. D. Turner and L. R. Taylor), pp. 1–28. Blackwell, Oxford.

Temple, S. A. and Cary, J. R. 1988. Modelling dynamics of habitat interior bird populations in fragmented landscapes. *Cons. Biol.* 2: 340–347.

Terborgh, J. 1974. Preservation of natural diversity: the problem of species extinction. *BioScience* 24: 715–722.

Terborgh, J. 1975. Faunal equilibria and the design of wildlife preserves. In *Tropical ecological systems: trends in terrestrial and aquatic research* (ed. F. Golley and E. Medina), pp. 369–380. Springer, New York.

Terborgh, J. and Faaborg, J. 1973. Turnover and ecological release in the avifauna of Mona Island, Puerto Rico. *Auk* 90: 759–779.

Thomas, C. D. 1994a. Extinction, colonization, and metapopulations: environmental tracking by rare species. *Cons. Biol.* 8: 373–378.

Thomas, C. D. 1994b. Local extinctions, colonizations and distributions: habitat tracking by British butterflies. In *Individuals, populations and patterns in ecology* (ed. S. R. Leather, A. D. Watt, N. J. Mills and K. F. A. Walters), pp. 319–336. Intercept, Andover.

Thomas, C. D. 1994c. The ecology and conservation of butterfly metapopulations in the fragmented British landscape. In *Ecology and conservation of butterflies* (ed. A. S. Pullin), pp. 46–63. Chapman and Hall, London.

Thomas, C. D. and Hanski, I. 1997. Butterfly metapopulations. In *Metapopulation biology* (ed. I. A. Hanski and M. E. Gilpin), pp. 359–386. Academic Press, San Diego.

Thomas, C. D. and Harrison, S. 1992. Spatial dynamics of a patchily-distributed butterfly species. *J. Anim. Ecol.* 61: 437–446.

Thomas, C. D. and Jones, T. M. 1993. Partial recovery of a skipper butterfly (*Hesperia comma*) from population refuges: lessons for conservation in a fragmented landscape. *J. Anim. Ecol.* 62: 472–481.

Thomas, C. D. and Singer, M. C. 1987. Variation in host preference affects movement patterns within a butterfly population. *Ecology* 68: 1262–1267.

Thomas, C. D. and Singer, M. C. 1998. Scale-dependant evolution of specialization in a checkerspot butterfly: from individuals to metapopulations and ecotypes. In *Genetic structure and local adaptation in natural insect populations* (ed. S. Mopper and S. Y. Strauss), pp. 343–374. Chapman and Hall, New York.

Thomas, C. D., Thomas, J. A. and Warren, M. S. 1992. Distributions of occupied and vacant butterfly habitats in fragmented landscapes. *Oecologia* 92: 563–567.

Thomas, C. D., Singer, M. C. and Boughton, D. A. 1996. Catastrophic extinction of population sources in a butterfly metapopulation. *Am. Nat.* 148: 957–975.

Thomas, C. D., Hill, J. K. and Lewis, O. T. 1998. Evolutionary consequences in a localised butterfly. *J. Anim. Ecol.* 67: 485–497.

Thomas, J. A. 1984. The conservation of butterflies in temperate countries: past efforts and lessons for the future. *11th Symp. Royal Entom. Soc. London*, pp. 333–353.

Thomas, J. A. 1991. Rare species conservation: case studies of European butterflies. In *The scientific management of temperate communities for conservation* (ed. I. Spellenberg, B. Goldsmith and M. G. Morris), pp. 149–197. Blackwell, Oxford.

Thomas, J. A., Moss, D. and Pollard, E. 1994. Increased fluctuations by butterfly populations towards the northern edges of species' ranges. *Ecography* 17: 215–220.

Thomas, J. W., Forsman, E. D., Lint, J. B., Meslow, E. C., Noon, B. R. and Verner, J. 1990. A conservation strategy for the northern spotted owl. 1990–791–171/2002b. U.S. Government Printing Office, Washington, D.C.

Thompson, W. R. 1956. The fundamental theory of natural and biological control. *Ann. Rev. Entomol.* 1: 379–402.

Thornhill, N. W. 1993. *The natural history of inbreeding and outbreeding.* The University of Chicago Press.

Thrall, P. H. and Antonovics, J. 1995. Theoretical and empirical studies of metapopulations: population and genetic dynamics of the *Silene–Ustilago* system. *Can. J. Bot.* 73: 1249–1258.

Tilman, D. 1982. *Resource competition and community structure.* Princeton University Press.

Tilman, D. and Kareiva, P. 1997. *Spatial ecology.* Princeton University Press.

Tilman, D. and Lehman, C. L. 1997. Habitat destruction and species extinction. In *Spatial ecology* (ed. D. Tilman and P. Kareiva), pp. 233–249. Princeton University Press.

Tilman, D., May, R. M., Lehman, C. L. and Nowak, M. A. 1994. Habitat destruction and the extinction debt. *Nature* 371: 65–66.

Tilman, D., Lehman, C. L. and Yin, C. 1997. Habitat destruction, dispersal, and deterministic extinction in competitive communities. *Am. Nat.* 149: 407–435.

Tolman, T. 1997. *Butterflies of Britain and Europe.* Collins, London.

Tscharntke, T. 1992. Fragmentation of *Phragmites* habitats, minimum viable population size, habitat suitability, and local extinction of moths, midges, flies, aphids, and birds. *Cons. Biol.* 6: 530–536.

Turchin, P. 1986. Modelling the effect of patch host size on Mexican bean beetle emigration. *Ecology* 67: 124–132.

Turchin, P. 1990. Rarity of density dependence or population regulation with lags? *Nature* 344: 660–663.

Turchin, P. 1995. Population regulation: old arguments and a new synthesis. In *Population dynamics: new approaches and synthesis* (ed. N. Cappuccino and P. W. Price), pp. 19–40. Academic Press, London.

Turchin, P., Reeve, J. D., Cronin, J. T. and Wilkens, R. T. 1998. Spatial pattern formation in ecological systems: Bridging theoretical and empirical approaches. In *Modelling spatiotemporal dynamics in ecology* (ed. J. Bascompte and R. V. Solé), pp. 199–213. Springer, New York.

Turchin, P. and Taylor, A. D. 1992. Complex dynamics in ecological time series. *Ecology* 73: 289–305.

Turchin, P. and Thoeny, W. T. 1993. Quantifying dispersal of southern pine beetles with mark–recapture experiments and a diffusion model. *Ecol. Appl.* 3: 187–198.

Turelli, M. 1977. Random environments and stochastic calculus. *Theor. Pop. Biol.* 12: 140–178.

Turelli, M. and Hoffmann, A. A. 1991. Rapid spread of an inherited incompatibility factor in California *Drosophila. Nature* 353: 440–442.

Turing, A. M. 1952. The chemical basis of morphogenesis. *Philosophical Transactions of the Royal Society, London B* 237: 37–72.

Turner, M. G. 1989. Landscape ecology: the effect of pattern on process. *Ann. Rev. Ecol. Syst.* 20: 171–197.

Urban, D. L., O'Neill, R. V. and Shugart, H. H. 1987. Landscape ecology. *BioScience* 37: 119–127.

U.S. Department of Agriculture, Forest Service. 1988. Final supplement to the environmental impact statement for an amendment to the Pacific Northwest Regional Guide. Portland, Oregon.

Val, J., Verboom, J. and Metz, J. A. J. 1995. A deterministic size-structured metapopulation model. Preprint.

van Valen, L. 1971. Group selection and the evolution of dispersal. *Evolution* 25: 591–598.

Valverde, T. and Silvertown, J. 1997. A metapopulation model for *Primula vulgaris*, a temperate forest understorey herb. *Journal of Ecology* 85: 193–210.

Varley, G. C., Gradwell, G. R. and Hassell, M. P. 1973. *Insect population ecology*. Blackwell, Oxford.

Veltman, C. J., Nee, S. and Crawley, M. J. 1996. Correlates of introduction success in exotic New Zealand birds. *Am. Nat.* 147: 542–557.

Venable, D. L. 1979. The demographic consequences of achene polymorphism in *Heterotheca latifolia* Buckl. (Compositae): germination, survivorship, fecundity and dispersal. Ph.D. Dissertation, University of Texas, Austin.

Venable, D. L. and Lawlor, L. 1980. Delayed germination and dispersal in desert annuals: escape in space and time. *Oecologia* 46: 272–282.

Venier, L. A. and Fahrig, L. 1996. Habitat availability causes the species abundance–distribution relationship. *Oikos* 76: 564–570.

Vepsäläinen, K. 1974. Determination of wing lengths and diapause in water striders (*Gerris* Fabr., Heteroptera). *Hereditas* 77: 163–176.

Verboom, J. and van Apeldoorn, R. 1990. Effects of habitat fragmentation on the red squirrel, *Sciurus vulgaris* L. *Landscape Ecology* 4: 171–176.

Verboom, J., Schotman, A., Opdam, P. and Metz, J. A. J. 1991. European nuthatch metapopulations in a fragmented agricultural landscape. *Oikos* 61: 149–156.

Volterra, V. 1926. Variations and fluctuations of the numbers of individuals in animal species living together. *J. Cons. Perm. Int. Ent. Mer.* 3: 3–51.

Wade, M. J. 1990. Genotype–environment interaction for climate and competition in a natural population of flour beetles, *Tribolium castaneum*. *Evolution* 44: 2004–2011.

Wade, M. J. and McCauley, D. E. 1988. Extinction and recolonization: their effects on the differentiation of local populations. *Evolution* 42: 995–1005.

Wahlberg, N. 1995. One day in the life of a butterfly; a study of the biology of the Glanville fritillary *Melitaea cinxia*. M.Sc. Thesis, University of Helsinki, Finland.

Wahlberg, N. 1997. The life history and ecology of *Melitaea diamina* (Nymphalidae) in Finland. *Nota Lepid.* 20: 70–81.

Wahlberg, N. 1998. The life history and ecology of *Euphydryas maturna* (Nymphalidae: Melitaeini) in Finland. *Nota Lepid.* 21: 154–169.

Wahlberg, N., Moilanen, A. and Hanski, I. 1996. Predicting the occurrence of endangered species in fragmented landscapes. *Science* 273: 1536–1538.

Walde, S. J. 1994. Immigration and the dynamics of a predator–prey interaction in biological control. *J. Anim. Ecol.* 63: 337–346.

Waller, D. M., O'Malley, D. M. and Gawler, S. C. 1987. Genetic variation in the extreme endemic *Pedicularis furbisiae* (Schrophulariaceae). *Cons. Biol.* 1: 335–340.

Warren, M. S. 1987. The ecology and conservation of the heath fritillary butterfly, *Mellicta athalia*. I. Host selection and phenology. *J. Appl. Ecol.* 24: 467–482.

Warren, M. S. 1992. The conservation of British butterflies. In *The ecology of butterflies in Britain* (ed. R. L. H. Dennis), pp. 246–274. Oxford University Press.

Warren, M. S. 1993. A review of butterfly conservation in central southern Britain: I. protection, evaluation and extinction on prime sites. *Biol. Cons.* 64: 25–35.

Warren, M. S. 1994. The UK status and suspected metapopulation structure of a threatened European butterfly, the marsh fritillary *Eurodryas aurinia*. *Biol. Cons.* 67: 239–249.

Warren, M. S., Thomas, C. D. and Thomas, J. A. 1984. The status of the Heath Fritillary Butterfly *Mellicta athalia* Rott. in Britain. *Biol. Cons.* 29: 287–305.

Waser, P. M. 1985. Does competition drive dispersal? *Ecology* 66: 1170–1175.

Watkinson, A. R. 1985. On the abundance of plants along an environmental gradient. *J. Ecol.* 73: 569–578.

Watkinson, A. R. and Sutherland, W. J. 1995. Sources, sinks and pseudo-sinks. *J. Anim. Ecol.* 64: 126–130.

Watkinson, A. R., Lonsdale, W. M. and Andrew, M. H. 1989. Modelling the population dynamics of an annual plant: *Sorghum intrans* in the wet–dry tropics. *J. Ecol.* 77: 162–181.

Wauters, L., Casale, P. and Dhondt, A. A. 1994. Space use and dispersal of red squirrels in fragmented habitats. *Oikos* 69: 140–146.

Weddell, B. J. 1991. Distribution and movements of Columbian ground squirrels (*Spermophilus columbianus* (Ord.)): are habitat patches like islands? *Journal of Biogeography* 18: 385–394.

Weiss, S. B., Murphy, D. D. and White, R. R. 1988. Sun, slope and butterflies: topographic determinants of habitat quality in *Euphydryas editha*. *Ecology* 69: 1486–1496.

Weiss, S. B., Murphy, D. D., Ehrlich, P. R. and Metzler, C. F. 1993. Adult emergence phenology in checkerspot butterflies: the effects of macroclimate, topography, and population history. *Oecologia* 96: 261–270.

Wennergren, U., Ruckelshaus, M. and Kareiva, P. 1995. The promise and limitations of spatial models in conservation biology. *Oikos* 74: 349–356.

Western, D. and Pearl, M. 1989. *Conservation biology in the 21st Century.* Oxford University Press.

White, A., Begon, M. and Bowers, R. G. 1996. Host–pathogen systems in a spatially patchy environment. *Proceedings of the Royal Society of London B* 263: 325–332.

Whitlock, M. C. 1992. Nonequilibrium population structure in forked fungus beetles: extinction, colonization, and genetic variation among populations. *Am. Nat.* 139: 952–970.

Whitlock, M. C. 1995. Variance-induced peak shifts. *Evolution* 49: 252–259.

Whitlock, M. C. and Barton, N. H. 1997. The effective size of a subdivided population. *Genetics* 146: 427–441.

Whitlock, M. C. and Fowler, K. 1996. The distribution among populations of phenotypic variance with inbreeding. *Evolution* 50: 1919–1926.

Whitlock, M. C. and McCauley, D. E. 1990. Some population genetic consequences of colony formation and extinction: genetic correlations within founding groups. *Evolution* 44: 1717–1724.

Whitlock, M. C., Phillips, P. C. and Wade, M. J. 1993. Gene interaction affects the additive genetic variance in subdivided populations with migration and extinction. *Evolution* 47: 1758–1769.

Whitmore, T. C. and Sayer, J. A. 1992. *Tropical deforestation and species extinction.* Chapman and Hall, London.

Wiens, J. A. 1977. On competition and variable environments. *Am. Sci.* 65: 590–597.

Wiens, J. A. 1984. Resource systems, populations, and communities. In *A new ecology: novel approaches to interactive systems* (ed. P. W. Price, C. N. Slobodchikoff and W. S. Gand), pp. 397–436. Wiley, New York.

Wiens, J. A. 1989. Spatial scaling in ecology. *Functional Ecology* 3: 385–397.

Wiens, J. A. 1997. Metapopulation dynamics and landscape ecology. In *Metapopulation biology* (ed. I. A. Hanski and M. E. Gilpin), pp. 43–68. Academic Press, San Diego.

Wiklund, C. 1984. Egg-laying patterns in butterflies in relation to their phenology and the visual apparency and abundance of their host plants. *Oecologia* 63: 23–29.

Williams, C. B. 1943. Area and number of species. *Nature* 152: 264–267.

Williams, C. B. 1950. The application of the logarithmic series to the frequency of occurrence of plant species in quadrants. *Journal of Ecology* 38: 107–138.

Williams, E. H. 1988. Habitat and range of *E. gillettii* (Nymphalidae). *Journal of the Lepidopterists' Society* 42: 37–45.

Williams, M. R. 1995. An extreme-value function model of the species incidence and species–area relations. *Ecology* 76: 2607–2616.

Williamson, J. A. and Charlesworth, B. 1976. The effect of age of founder on the probability of survival of a colony. *J. Theor. Biol.* 56: 175–190.

Williamson, M. 1989. The MacArthur–Wilson theory today: true but trivial? *Journal of Biogeography* 16: 3–4.

Wilson, D. S. 1980. *The natural selection of populations and communities*. The Benjamin/Cummings Publishing Company, Mento Park, California, USA.

Wilson, E. O. 1994. *Naturalist*. Allen Lane, London.

Wilson, E. O. and Willis, E. O. 1975. Applied biogeography. In *Ecology and evolution of communities* (ed. M. L. Cody and J. M. Diamond), pp. 523–534. Harvard University Press, Cambridge.

Wilson, H. B., Godfray, C., Hassell, M. P. and Pacala, S. 1998. Deterministic and stochastic host parasitoid dynamics in spatially extended systems. In *Modeling spatiotemporal dynamics in ecology* (ed. J. Bascompte and R. V. Solé), pp. 63–82. Springer, New York.

Wilson, K. 1995. Insect migration in heterogeneous environments. In *Insect migration: tracking resources through space and time* (ed. V. A. Drake and A. G. Gatehouse), pp. 243–263. Cambridge University Press.

Wilson, W. G., de Roos, A. M. and McCauley, E. 1993. Spatial instabilities within the diffusive Lotka–Volterra system: individual-based simulation results. *Theor. Pop. Biol.* 43: 91–127.

Wissel, C. and Maier, B. 1992. A stochastic model for the species–area relationship. *Journal of Biogeography* 19: 355–362.

Wissel, C. and Stöcker, S. 1991. Extinction of populations by random influences. *Theor. Pop. Biol.* 39: 315–328.

With, K. A. and King, A. W. 1997. The use and misuse of neutral landscape models in ecology. *Oikos* 79: 219–229.

With, K. A., Gardner, R. H. and Turner, M. G. 1997. Landscape connectivity and population distributions in heterogeneous landscapes. *Oikos* 78: 151–169.

Witkowski, Z. and Płonka, P. 1984. The species/area relationship in plants and birds on protected and unprotected isolates in Poland. *Bull. Pol. Acad. Sci. Biol. Sci.* 32: 241–249.

Woiwod, I. P. and Hanski, I. 1992. Patterns of density dependence in moths and aphids. *J. Anim. Ecol.* 61: 619–629.

Wolda, H. 1989. The equilibrium concept and density dependence tests: what does it all mean? *Oecologia* 81: 430–432.

Wolda, H. and Dennis, B. 1993. Density dependence tests, are they? *Oecologia* 95: 581–591.

Wolda, H., Dennis, B. and Taper, M. L. 1994. Density dependence tests, and largely futile comments: Answers to Holyoak and Lawton (1993) and Hanski, Woiwod and Perry (1993). *Oecologia* 98: 229–234.

Wright, D. H. 1991. Correlations between incidence and abundance are expected by chance. *Journal of Biogeography* 18: 463–466.

Wright, D. H., Patterson, B. D., Mikkelson, G. M., Cutler, A. and Atmar, W. 1997. A comparative analysis of nested subset patterns of species composition. *Oecologia* 113: 1–20.

Wright, S. 1931. Evolution in Mendelian populations. *Genetics* 16: 97–169.

Wright, S. 1932. The roles of mutation, inbreeding, crossbreeding and selection in evolution. *Proc. Sixth Int. Cong. Genet.* 1: 356–366.

Wright, S. 1940. Breeding structure of populations in relation to speciation. *Am. Nat.* 74: 232–248.

Wu, J. and Levin, S. A. 1994. A spatial patch dynamic modeling approach to pattern and process in an annual grassland. *Ecol. Monogr.* 64: 447–464.

Wuorenrinne, H. 1978. Metsä urbaanipaineen puristuksessa. Perusselvitysraportti. Espoon kunta, Espoo, Finland.

Yahner, R. H. 1988. Changes in wildlife communities near edges. *Cons. Biol.* 2: 333–339.

Zwölfer, H. 1982. Das Verbreitungsareal der Bohrfliege *Urophora cardui* L. (Diptera: Tephritidae) als hinweis auf die ursprungliche Habitate der Ackerdistel (*Cirsium arvense* L.) (Scop.). *Verhandlungen der Deutschen Zoologischen Gesellschaft* 298.

Index

Note: page numbers in *italics* refer to figures and tables.